## Note from th

In June 1993, A. Wiles announced the proof of Fermat's last theorem. However, this proof contained a flaw. In October 1994, A. Wiles made public two new preprints. One contained a modified version of the proof of Fermat's theorem. The second paper, in collaboration with R. Taylor, was intended to establish an important step in the proof of the theorem. These manuscripts are presently under scrutiny and, if approved by the specialists, they provide the long searched first proof of Fermat's theorem.

This reprint of the 1979 edition offers a historical view of the attempts to prove the theorem.

In the new final section, the most significant recent developments are reported and ample references are listed, where the reader may learn about the theories and methods that led Wiles to his proof.

# Springer

*New York*
*Berlin*
*Heidelberg*
*Barcelona*
*Budapest*
*Hong Kong*
*London*
*Milan*
*Paris*
*Santa Clara*
*Singapore*
*Tokyo*

**Pierre de Fermat**

1601–1665

Paulo Ribenboim

# 13 Lectures on Fermat's Last Theorem

Springer

Paulo Ribenboim
Department of Mathematics and Statistics
Jeffery Hall
Queen's University
Kingston
Canada K7L 3N6

AMS Subject Classifications (1980): 10-03, 12-03, 12Axx

**Library of Congress Cataloguing in Publication Data**

Ribenboim, Paulo.
   13 lectures on Fermat's last theorem.

   Includes bibliographies and indexes.
   1.  Fermat's theorem.  I. Title.
QA244.R5    512′.74    79-14874

ISBN 978-1-4419-2809-2

Printed in the United States of America.

9 8 7 6 5 4 3

Hommage à André Weil
pour sa Leçon: goût,
rigueur et pénétration.

# Preface

Fermat's problem, also called Fermat's last theorem, has attracted the attention of mathematicians for more than three centuries. Many clever methods have been devised to attack the problem, and many beautiful theories have been created with the aim of proving the theorem. Yet, despite all the attempts, the question remains unanswered.

The topic is presented in the form of lectures, where I survey the main lines of work on the problem. In the first two lectures, there is a very brief description of the early history, as well as a selection of a few of the more representative recent results. In the lectures which follow, I examine in succession the main theories connected with the problem. The last two lectures are about analogues to Fermat's theorem.

Some of these lectures were actually given, in a shorter version, at the Institut Henri Poincaré, in Paris, as well as at Queen's University, in 1977.

I endeavoured to produce a text, readable by mathematicians in general, and not only by specialists in number theory. However, due to a limitation in size, I am aware that certain points will appear sketchy.

It is for me gratifying to acknowledge the help and encouragement I received in the preparation of this book: A. J. Coleman and the Mathematics Department at Queen's University—for providing excellent working conditions; E. M. Wight—for her diligent and skillful typing of the manuscript; G. Cornell—who read the book and helped very much in improving the style; The Canada Council—for partial support; C. Pisot and J. Oesterlé— who arranged for my lectures at the Institut Henri Poincaré.

It is also my pleasure to report here various suggestions, criticisms and comments from several specialists, whom I consulted on specific points or to whom I have sent an earlier typescript version of this book. In alphabetical order: A. Baker, D. Bertrand, K. Inkeri, G. Kreisel, H. W. Lenstra Jr., J. M.

Masley, M. Mendès-France, B. Mazur, T. Metsänkylä, A. Odlyzko, K. Ribet, A. Robert, P. Samuel, A. Schinzel, E. Snapper, C. L. Stewart, G. Terjanian, A. J. van der Poorten, S. S. Wagstaff Jr., M. Waldschmidt, L. C. Washington.

*Kingston, March, 1979*                                    Paulo Ribenboim

# General Bibliography

There have been several editions of Fermat's works. The first printing was supervised by his son Samuel de Fermat.

1670
> Diophanti Alexandrini Arithmeticorum libri sex, et de Numeris Multangulis liber unus. Cum commentariis C.G. Bacheti V.C. et observationibus D. P. de Fermat senatoris Tolosani. Accessit Doctrinae Analyticae inventum novum, collectum ex variis ejusdem D. de Fermat, epistolis. B. Bosc, in-folio, Tolosae.

1679
> Varia Opera Mathematica D. Petri de Fermat, Senatoris Tolosani. J. Pech, in-folio, Tolosae. Reprinted in 1861, in Berlin, by Friedländer & Sohn, and in 1969, in Brussels, by Culture et Civilisation.

1891/1894/1896/1912/1922
> Oeuvres de Fermat, en 4 volumes et un supplément. Publiées par les soins de MM. Paul Tannery et Charles Henry. Gauthier-Villars, Paris.

In 1957 the old boys high school of Toulouse was renamed "Lycée Pierre de Fermat". For the occasion the Toulouse Municipal Library and the Archives of Haute-Garonne organized an exhibit in honor of Fermat. A brochure was published, describing considerable "Fermatiana":

1957
> Un Mathématicien de Génie: Pierre de Fermat (1601–1665). Lycée Pierre de Fermat, Toulouse, 1957.

Many books, surveys and articles have been devoted totally or in part to a historical or mathematical study of Fermat's work, and more specially, to the last theorem. The following selection is based on their interest and availability to the modern reader:

1883    Tannery, P.
> Sur la date des principales découvertes de Fermat. *Bull. Sci. Math.*, sér. 2, **7**, 1883, 116–128. Reprinted in *Sphinx-Oedipe*, **3**, 1908, 169–182.

1860   Smith, H. J. S.
       Report on the Theory of Numbers, part II, art. 61, Report of the British Asso-
       ciation, 1860, *Collected Mathematical Works*, Clarendon Press, Oxford, 1894,
       131–137. Reprinted by Chelsea Publ. Co., New York, 1965.

1910   Bachmann, P.
       *Niedere Zahlentheorie*. Teubner, Leipzig, 1910. Reprinted by Chelsea Publ. Co.,
       New York, 1966.

1910   Lind, B.
       *Uber das letzte Fermatsche Theorem*, Teubner, Leipzig, 1910.

1917   Dickson, L. E.
       Fermat's Last Theorem and the Origin and Nature of the Theory of Algebraic
       Numbers. *Annals of Math.*, **18**, 1917, 161–187.

1919   Bachmann, P.
       *Das Fermatproblem in seiner bisherigen Entwicklung*, Walter de Gruyter, Berlin,
       1919. Reprinted by Springer-Verlag, Berlin, 1976.

1920   Dickson, L. E.
       *History of the Theory of Numbers*, II, Carnegie Institution, Washington, 1920.
       Reprinted by Chelsea Publ. Co., New York, 1971.

1921   Mordell, L. J.
       *Three Lectures on Fermat's Last Theorem*, Cambridge University Press, Cam-
       bridge, 1921. Reprinted by Chelsea Publ. Co., New York, 1962, and by VEB
       Deutscher Verlag d. Wiss. Berlin, 1972.

1925   Ore, Ø.
       Fermats Teorem. *Norske Mat. Tidtskrift* 7, 1925, 1–10.

1927   Khinchine, A. I.
       *Velikai Teorema Ferma* (*The Great Theorem of Fermat*). State Editor, Moskow–
       Leningrad, 1927.

1928   Vandiver, H. S. and Wahlin, G. E.
       Algebraic Numbers, II. Bull. Nat. Research Council, 62, 1928. Reprinted by
       Chelsea Publ. Co., New York, 1967.

†1934   Morishima, T.
       *Fermat's Problem* (in Japanese), Iwanami Shoten, Tokyo, 1934, 54 pages.

1948   Got, T.
       Une énigme mathématique. Le dernier théorème de Fermat. (A çhapter in *Les
       Grands Courants de la Pensée Mathématique*, edited by F. Le Lionnais). Cahiers
       du Sud., Marseille, 1948. Reprinted by A. Blanchard, Paris, 1962.

1961   Bell, E. T.
       *The Last Problem*, Simon and Schuster, New York, 1961.

1966   Noguès, R.
       *Théorème de Fermat, son Histoire*, A. Blanchard, Paris, 1966.

1970   Smadja, R.
       Le Théorème de Fermat (Thèse de 3ᵉ cycle), Université de Paris VI, 1970.

1973   Besenfelder, H. J.
       Das Fermat-Problem. Diplomarbeit, Universität Karlsruhe, 61 pages, 1973.

1973   Fournier, J. C.
       Sur le Dernier Théorème de Fermat (Thèse de 3ᵉ cycle), Université de Paris VI, 1973.

1973   Mahoney, M. S.
       *The Mathematical Career of Pierre de Fermat*, Princeton University Press, Prince-
       ton, 1973.

1974/5   Ferguson, R. P.
On Fermat's Last Theorem, I, II, III. *J. Undergrad. Math.*, **6**, 1974, 1–14, 85–98
and **7**, 1975, 35–45.

1977   Edwards, H. M.
*Fermat's Last Theorem. A Genetic Introduction to Algebraic Number Theory.*
Springer-Verlag, New York, 1977.

For the basic facts about algebraic number theory, the reader may consult:

1966   Borevich, Z. I. and Shafarevich, I. R.
*Number Theory*, Academic Press, New York, 1966.

1972   Ribenboim, P.
*Algebraic Numbers*, Wiley-Interscience, New York, 1972.

This last book will be quoted as [Ri].

The sign † in front of a bibliography entry indicates that I was unable to examine the
item in question. All the information gathered in this book stems directly from the
original sources.

# Contents

# The Early History of Fermat's Last Theorem

## 1. The Problem

Pierre de Fermat (1601–1665) was a French judge who lived in Toulouse. He was a universal spirit, cultivating poetry, Greek philology, law but mainly mathematics. His special interest concerned the solutions of equations in integers.

For example, Fermat studied equations of the type

$$X^2 - dY^2 = \pm 1,$$

where $d$ is a positive square-free integer (that is, without square factors different from 1) and he discovered the existence of infinitely many solutions. He has also discovered which natural numbers $n$ may be written as the sum of two squares, namely those with the following property: every prime factor $p$ of $n$ which is congruent to 3 modulo 4 must divide $n$ to an even power.

In the margin of his copy of Bachet's edition of the complete works of Diophantus, Fermat wrote:

> It is impossible to separate a cube into two cubes, or a biquadrate into two biquadrates, or in general any power higher than the second into powers of like degree; I have discovered a truly remarkable proof which this margin is too small to contain.

This copy is now lost, but the remark appears in the 1670 edition of the works of Fermat, edited in Toulouse by his son Samuel de Fermat. It is stated in Dickson's *History of the Theory of Numbers*, volume II, that Fermat's assertion was made about 1637. Tannery (1883) mentions a letter from Fermat to Mersenne (for Sainte-Croix) in which he wishes to find two cubes whose sum is a cube, and two biquadrates whose sum is a biquadrate. This letter appears, with the date June 1638, in volume 7 of *Correspondance du Père Marin Mersenne* (1962); see also Itard (1948). The same problem was proposed to Frénicle de Bessy (1640) in a letter to Mersenne, and to Wallis and Brouncker in a letter to Digby, written in 1657, but there is no mention of the remarkable proof he had supposedly found.

In modern language, Fermat's statement means:

*The equation $X^n + Y^n = Z^n$, where n is a natural number larger than 2, has no solution in integers all different from 0.*

No proof of this statement was ever found among Fermat's papers. He did, however, write a proof that the equations $X^4 - Y^4 = Z^2$ and $X^4 + Y^4 = Z^4$ have no solutions in integers all different from 0. In fact, this is one of two proofs by Fermat in number theory which have been preserved[1]. With very few exceptions, all Fermat's other assertions have now been confirmed. So this problem is usually called Fermat's last theorem, despite the fact that it has never been proved.

Fermat's most notable erroneous belief concerns the numbers $F_n = 2^{2^n} + 1$, which he thought were always prime. But Euler showed that $F_5$ is not a prime. Sierpiński and Schinzel pointed out some other false assertions made by Fermat.

Mathematicians have debated whether Fermat indeed possessed the proof of the theorem. Perhaps, at one point, he mistakenly believed he had found such a proof. Despite Fermat's honesty and frankness in acknowledging imperfect conclusions, it is very difficult to understand today, how the most distinguished mathematicians could have failed to rediscover a proof, if one had existed.

To illustrate Fermat's candor, we quote from his letter of October 18, 1640 to Frénicle de Bessy:

> Mais je vous advoue tout net (car par advance je vous advertis que comme je suis pas capable de m'attribuer plus que je ne sçay, je dis avec même franchise ce que je ne sçay pas) que je n'ay peu encore démonstrer l'exclusion de tous diviseurs en cette belle proposition que je vous avois envoyée, et que vous m'avez confirmée touchant les nombres 3, 5, 17, 257, 65537 & c. Car bien que je réduise l'exclusion à la plupart des nombres, et que j'aye même des raisons probables pur le reste, je n'ay peu encore démonstrer nécessairement la vérité de cette proposition, de laquelle pourtant je ne doute non plus à cette heure que je faisois auparavant. Si vous en avez la preuve assurée, vous m'obligerez de me la communiquer: car après cela rien ne m'arrestera en ces matières.

---

[1] The other proof, partial but very interesting, was brought to light and reproduced by Hofmann (1943, pages 41–44). Fermat showed that the only solutions in integers of the system $x = 2y^2 - 1$, $x^2 = 2z^2 - 1$ are $x = 1$ and $x = 7$.

Again, in a letter to Pascal from August 29, 1654, Fermat proposes the same problem:

> Au reste, il n'est rien à l'avenir que je ne vous communique avec toute franchise. Songez cependant, si vous le trouvez à propos, à cette proposition: les puissances carrées de 2, augmentées de l'unité, sont toujours des nombres premiers: $2^2 + 1 = 5$, $2^{2^2} + 1 = 17$, $2^{2^3} + 1 = 257$, $2^{2^4} + 1 = 65537$, sont premiers, et ainsi à l'infini. C'est une proposition de la verité de laquelle je vous répond. La démonstration en est trés malaisée, et je vous avoue que je n'ai pu encore la trouver pleinement; je ne vous la proposerois pas pour la chercher si j'en étois venu à bout.

Incidentally Pascal has written to Fermat stating:

> Je vous tiens pour le plus grand géomètre de toute l'Europe.

It is also highly improbable that Fermat would have claimed to have proved his last theorem, just because he succeeded in proving it for a few small exponents.

In contrast, Gauss believed that Fermat's assertions were mostly extrapolations from paticular cases. In 1807, Gauss wrote:

> Higher arithmetic has this special feature that many of its most beautiful theorems may be easily discovered by induction, while any proof can be only obtained with the utmost difficulty. Thus, it was one of the great merits of Euler to have proved several of Fermat's theorems which he obtained, it appears, by induction.

Even though he himself gave a proof for the case of cubes, Gauss did not hold the problem in such high esteem. On March 21, 1816, he wrote to Olbers about the recent mathematical contest of the Paris Academy on Fermat's last theorem:

> I am very much obliged for your news concerning the Paris prize. But I confess that Fermat's theorem as an isolated proposition has very little interest for me, because I could easily lay down a multitude of such propositions, which one could neither prove nor dispose of.

In trying to prove Fermat's theorem for every positive integer $n \geq 3$, I make the following easy observation. If the theorem holds for an integer $m$ and $n = lm$ is a multiple of $m$, then it holds also for $n$. For, if $x, y, z$ are non-zero integers and $x^n + y^n = z^n$ then $(x^l)^m + (y^l)^m = (z^l)^m$, contradicting the hypothesis. Since every integer $n \geq 3$ is a multiple of 4 or of a prime $p \neq 2$, it suffices to prove Fermat's conjecture for $n = 4$ and for every prime $p \neq 2$. However, I shall occasionally also mention some proofs for exponents of the form $2p$, or $p^n$ where $p$ is an odd prime.

The statement of Fermat's last theorem is often subdivided further into two cases:

The *first case* holds for the exponent $p$ when there do not exist integers $x, y, z$ such that $p \nmid xyz$ and $x^p + y^p = z^p$.

The *second case* holds for the exponent $p$ when there do not exist integers $x$, $y$, $z$, all different from 0, such that $p|xyz$, $\gcd(x,y,z) = 1$ and $x^p + y^p = z^p$.

## 2. Early Attempts

It was already known in antiquity that a sum of two squares of integers may well be the square of another integer. Pythagoras was supposed to have proven that the lengths $a$, $b$, $c$ of the sides of a right-angle triangle satisfy the relation

$$a^2 + b^2 = c^2;$$

so the above fact just means the existence of such triangles with sides measured by integers.

But the situation is already very different for cubes, biquadrates and so on. Fermat's proof for the case of biquadrates is very ingenious and proceeds by the method which he called *infinite descent*. Roughly, it goes as follows: Suppose a certain equation $f(X,Y,Z) = 0$ has integral solutions $a$, $b$, $c$, with $c > 0$, the method just consists in finding another solution in integers $a'$, $b'$, $c'$ with $0 < c' < c$. Repeating this procedure a number of times, one would reach a solution $a''$, $b''$, $c''$, with $0 < c'' < 1$, which is absurd. This method of infinite descent is nothing but the well-ordering principle of the natural numbers.

Little by little Fermat's problem aroused the interest of mathematicians and a dazzling array of the best minds turned to it.

Euler considered the case of cubes. Without loss of generality, one may assume $x^3 + y^3 = z^3$ where $x$, $y$, $z$ are pairwise relatively prime integers, $x$, $y$ are odd, so $x = a - b$, $y = a + b$. Then $x + y = 2a$, $x^2 - xy + y^2 = a^2 + 3b^2$ and $z^3 = x^3 + y^3 = 2a(a^2 + 3b^2)$, where the integers $2a$, $a^2 + 3b^2$ are either relatively prime or have their greatest common divisor equal to 3. Euler was led to studying odd cubes $a^2 + 3b^2$ (with $a$, $b$ relatively prime), and forms of their divisors; he concluded the proof by the method of infinite descent. The properties of the numbers $a^2 + 3b^2$ which were required had to be derived from a detailed study of divisibility, and therefore were omitted from the proof published in Euler's book on algebra (1822). This proof, with the same gap, was reproduced by Legendre. Later, mathematicians intrigued by the missing steps were able without much difficulty, to reconstruct the proof on a sound basis. In today's language, numbers of the form $a^2 + 3b^2$ are norms of algebraic integers of the quadratic extension $\mathbb{Q}(\sqrt{-3})$ of the rational field $\mathbb{Q}$ and the required properties can be deduced from the unique factorization theorem, which is valid in that field.

Gauss gave another proof for the case of cubes. His proof was not "rational" since it involved complex numbers, namely those generated by the cube root of unity $\zeta = (-1 + \sqrt{-3})/2$, i.e., numbers from the quadratic field $\mathbb{Q}(\sqrt{-3})$. He consciously used the arithmetic properties of this field. The underlying idea was to call "integers" all numbers of the form $(a + b\sqrt{-3})/2$ where $a$, $b$ are integers of the same parity; then to define divisibility and the prime integers, and to use the fact that every integer is, in a unique way, the product of powers of primes. Of course some new facts appeared. First, the integers $\pm\zeta$, $\pm\zeta^2$ that divide 1 are "units" since $\zeta\zeta^2 = 1$ and therefore should not be taken into account so to speak, in questions of divisibility. Thus, all the properties have to be stated "up to units". Secondly, the unique factorization, which was taken for granted, was by no means immediate—in fact it turned out to be false in general. I shall return to this later.

Gauss's proof was an early incursion into the realm of number fields, i.e., those sets of complex numbers obtained from the roots of polynomials by the operations of addition, subtraction, multiplication, and division.

In the 1820s a number of distinguished French and German mathematicians were trying intensively to prove Fermat's theorem.

In 1825, G. Lejeune Dirichlet read at the Académie des Sciences de Paris a paper where he attempted to prove the theorem for the exponent 5. In fact his proof was incomplete, as pointed out by Legendre, who provided an independent and complete proof. Dirichlet then completed his own proof, which was published in Crelle Journal, in 1828.

Dirichlet's proof is "rational", and involves numbers of the form $a^2 - 5b^2$. He carefully analyzed the nature of such numbers which are 5th powers when either $a$, $b$ are odd, or $a$, $b$ have different parity, and 5 does not divide $a$, 5 divides $b$, and $a$, $b$ are relatively prime. Nowadays the properties he derived can be obtained from the arithmetic of the field $\mathbb{Q}(\sqrt{5})$. In this field too, every integer has a unique factorization. Moreover every unit is a power of $(1 + \sqrt{5})/2$, which is of crucial importance in the proof. Of course, for Dirichlet this knowledge took the form of numerical manipulations which lead to the same result.

In 1832 Dirichlet settled the theorem for the exponent 14.

The next important advance was due to Lamé, who, in 1839 proved the theorem for $n = 7$. Soon after, Lebesgue simplified Lamé's proof considerably by a clever use of the identity,

$$(X + Y + Z)^7 - (X^7 + Y^7 + Z^7)$$
$$= 7(X + Y)(X + Z)(Y + Z)$$
$$\times [(X^2 + Y^2 + Z^2 + XY + XZ + YZ)^2 + XYZ(X + Y + Z)]$$

already considered by Lamé.

While these special cases of small exponents were being studied, a very remarkable theorem was proved by Sophie Germain, a French mathematician.

Previously Barlow, and then Abel, had indicated interesting relations that $x$, $y$, $z$ must satisfy if $x^p + y^p = z^p$ (and $x$, $y$, $z$ are not zero). Through clever manipulations, Sophie Germain proved:

*If $p$ is an odd prime such that $2p + 1$ is also a prime then the first case of Fermat's theorem holds for $p$.*

These results were communicated by letter to Legendre and Cauchy since the regulations of the Academy prevented women from presenting the discoveries in person.

There are many primes $p$ for which $2p + 1$ is also prime, but it is still not known whether there are infinitely many such primes.

Following Sophie Germain's ideas, Legendre proved the following theorem.

*Let $p$, $q$ be distinct odd primes, and assume the following two conditions*:

1. *$p$ is never congruent modulo $q$ to a $p$th power.*
2. *the congruence $X^p + Y^p + Z^p \equiv 0 \pmod{q}$ has no solution $x$, $y$, $z$, unless $q$ divides $xyz$.*

Then the first case of Fermat's theorem holds for $p$.

With this result, Legendre extended Sophie Germain's theorem as follows:

*If $p$ is a prime such that $4p + 1, 8p + 1, 10p + 1, 14p + 1$, or $16p + 1$ is also a prime then the first case of Fermat's theorem holds for the exponent $p$.*

This was sufficient to establish the first case for all prime exponents $p < 100$.

# 3. Kummer's Monumental Theorem

By 1840, Cauchy and Lamé were working with values of polynomials at roots of unity, trying to prove Fermat's theorem for arbitrary exponents. Already in 1840 Cauchy published a long memoir on the theory of numbers, which however was not directly connected with Fermat's problem. In 1847, Lamé presented to the Academy a "proof" of the theorem and his paper was printed in full in Liouville's journal. However, Liouville noticed that the proof was not valid, since Lamé had tacitly assumed that the decomposition of certain polynomial expressions in the $n$th root of unity into irreducible factors was unique.

Lamé attributed his use of complex numbers to a suggestion from Liouville, while Cauchy claimed that he was about to achieve the same results, given more time. Indeed, during that same year, Cauchy had 18 communications printed by the Academy on complex numbers, or more specifically, on radical polynomials. He tried to prove what amounted to the euclidean algorithm, and hence unique factorization for cyclotomic integers. Then, assuming unique factorization, he drew wrong conclusions. Eventually Cauchy recognized his mistake. In fact, his approach led to results which were later rediscovered by Kummer with more suitable terminology. A noteworthy proposition of Cauchy was the following one, (*C. R. Acad. Sci. Paris*, **25**, 1847, page 181) later also found by Genocchi and by Kummer:

*If the first case of Fermat's theorem fails for the exponent p, then the sum*

$$1^{p-4} + 2^{p-4} + \cdots + \left(\frac{p-1}{2}\right)^{p-4}$$

*is a multiple of p.*

By the year 1847, mathematicians were aware of both the subtlety and importance of the unique decomposition of cyclotomic integers into irreducible factors.

In Germany, Kummer devoted himself to the study of the arithmetic of cyclotomic fields. Already, in 1844, he recognized that the unique factorization theorem need not hold for the cyclotomic field $\mathbb{Q}(\zeta_p)$. The first such case occurs for $p = 23$. However, while trying to rescue the unique factorization he was led to the introduction of new "ideal numbers". Here is an excerpt of a letter from Kummer to Liouville (1847):

> ... Encouraged by my friend Mr. Lejeune Dirichlet, I take the liberty of sending you a few copies of a dissertation which I have written three years ago, at the occasion of the century jubileum of the University of Königsberg, as well as of another dissertation of my friend and student Mr. Kronecker, a young and distinguished geometer. In these memoirs, which I beg you to accept as a sign of my deep esteem, you will find developments concerning certain points in the theory of complex numbers composed of roots of unity, i.e., roots of the equation $r^n = 1$, which have been recently the subject of some discussions at your illustrious Academy, at the occasion of an attempt by Mr. Lamé to prove the last theorem of Fermat.
>
> Concerning the elementary proposition for these complex numbers, that *a composite complex number may be decomposed into prime factors in only one way,* which you regret so justly in this proof, which is also lacking in some other points, I may assure you that *it does not hold in general* for complex numbers of the form
>
> $$a_0 + a_1 r + a_2 r^2 + \cdots + a_{n-1} r^{n-1},$$
>
> but it is possible to rescue it, by introducing a new kind of complex numbers, which I have called *ideal complex number.* The results of my research on this matter have been communicated to the Academy of Berlin and printed in the *Sitzungsberichte* (March 1846); a memoir on the same subject will appear soon in the Crelle Journal. I have considered already long ago the applications of this theory to the proof of Fermat's theorem and I succeeded in deriving the impossibility of the equation $x^n + y^n = z^n$ from two properties of the prime number $n$, so that it remains only to find out whether these properties are shared by all prime numbers. In case these results seem worth some of your attention, you may find them published in the *Sitzungsberichte* of the Berlin Academy, this month.

The theorem which Kummer mentioned in this letter represented a notable advance over all his predecessors.

The ideal numbers correspond to today's *divisors.* Dedekind rephrased this concept, introducing the *ideals,* which are sets $I$ of algebraic integers of the cyclotomic field such that $0 \in I$; if $\alpha$, $\beta \in I$ then $\alpha + \beta$, $\alpha - \beta \in I$; if $\alpha \in I$ and $\beta$ is any cyclotomic integer then $\alpha\beta \in I$. Ideals may be multiplied in a very natural way.

Each cyclotomic integer $\alpha$ determines a *principal ideal* consisting of all elements $\beta\alpha$, where $\beta \in A$, the set of cyclotomic integers.

If all ideals are principal there is unique factorization in the cyclotomic field, and conversely. For the cases when not all ideals are principal, Kummer wanted to "measure" to what extent some of the ideals were not principal. So he considered two nonzero ideals $I$, $I'$ equivalent when $I'$ consists of all multiples of the elements of $I$ by some nonzero element $\alpha$ in the cyclotomic field. Thus, there is exactly one equivalence class when all ideals are principal. Kummer proved that there are only finitely many equivalence classes of ideals in each cyclotomic field $\mathbb{Q}(\zeta_p)$.

Let $h_p$ denote the number of such classes. If $p$ does not divide $h_p$ then $p$ is said to be a *regular* prime. In this case, if the ideal $I^p$ is a principal ideal then $I$ is itself a principal ideal. But the main property used by Kummer is the following lemma:

*If $p$ is a regular prime, $p \neq 2$, if $\omega$ is a unit in the ring $A$ of cyclotomic integers of $\mathbb{Q}(\zeta_p)$, and if there exists an ordinary integer $m$ such that $\omega - m \in A(1 - \zeta)^{p-1}$, then $\omega$ is the $p$th power of another unit.*

The proof of this lemma requires deep analytical methods.

Armed with this formidable weapon, Kummer proved:

*Fermat's last theorem holds for every exponent $p$ which is a regular prime.*

This is the theorem which Kummer mentioned in his letter to Liouville. At first Kummer believed that there exist infinitely many regular primes. But, he later realized that this is far from evident—and in fact, it has, as yet, not been proved.

A well-known story concerning a wrong proof of Fermat's theorem submitted by Kummer, originates with Hensel. Specifically, in his address to commemorate the first centennial of Kummer's birth, Hensel (1910) stated:

> Although it is not well known, Kummer at one time believed he had found a complete proof of Fermat's theorem. (This is attested to by reliable witnesses including Mr. Gundelfinger who heard the story from the mathematician Grassmann.) Seeking the best critic for his proof, Kummer sent his manuscript to Dirichlet, author of the insuperably beautiful proof for the case $\lambda = 5$. After a few days, Dirichlet replied with the opinion that the proof was excellent and certainly correct, provided the numbers in $\alpha$ could not only be decomposed into indecomposable factors, as Kummer proved, but that this could be done in only one way. If however, the second hypothesis couldn't be satisfied, most of the theorems for the arithmetic of numbers in $\alpha$ would be unproven and the proof of Kummer's theorem would fall apart. Unfortunately, it appeared to him that the numbers in $\alpha$ didn't actually possess this property in general.

This is confirmed in a letter, which is not dated (but likely from the summer of 1844), written by Eisenstein to Stern, a mathematician from Göttingen.

In a recent paper, Edwards (1975) analyzes this information, in the light of a letter from Liouville to Dirichlet and expresses doubts about the existence of such a "false proof" by Kummer.

# 4. Regular Primes

To decide whether a prime is regular it is necessary to compute the number of equivalence classes of ideals of the cyclotomic field. Kummer succeeded in deriving formulas for the class number $h_p$ which were good enough to allow an explicit computation for fairly high exponents $p$. In this way, he discovered that 37, 59, 67 were irregular primes—actually these are the only ones less than 100.

One of the most interesting features in this study was the appearance of the Bernoulli numbers. In the derivation of the class number formula, there was an expression of the type

$$1^k + 2^k + \cdots + (n-1)^k$$

which had to be computed for large values of $k$ and $n$. First it is easy to show that there is a unique polynomial $S_k(X)$ with rational coefficients of degree $k + 1$, having leading coefficient $1/(k + 1)$ and such that for every $n \geq 1$ its value is $S_k(n) = 1^k + 2^k + \cdots + n^k$. These polynomials can be determined recursively and may be written as follows:

$$(k+1)S_k(X) = X^{k+1} - \binom{k+1}{1}B_1 X^k + \binom{k+1}{2}B_2 X^{k-1} + \cdots + \binom{k+1}{k}B_k X.$$

The coefficients $B_1, B_2, \ldots, B_k$ had already been discovered by Bernoulli. In fact Euler had already studied these numbers and found that they can be generated by considering the formal inverse of the series

$$\frac{e^X - 1}{X} = 1 + \frac{1}{2!}X + \frac{1}{3!}X^2 + \frac{1}{4!}X^3 + \cdots ;$$

namely

$$\frac{X}{e^X - 1} = 1 + \frac{B_1}{1!}X + \frac{B_2}{2!}X^2 + \frac{B_3}{3!}X^3 + \cdots .$$

This series appears in the Taylor expansion of the cotangent function: $\cot x = i + (1/x) \cdot (2ix/e^{2ix} - 1)$.

It is easily seen that $B_k = 0$ for every odd $k$, $k \neq 1$. The first Bernoulli numbers are $B_1 = -1/2$, $B_2 = 1/6$, $B_4 = -1/30$, $B_6 = 1/42$, $B_8 = -1/30$, $B_{10} = 5/66$, $B_{12} = -691/2730$, $B_{14} = 7/6$, $B_{16} = -3617/510$, $B_{18} = 43867/798$. The numerators grow quickly, for example:

$$B_{34} = \frac{2577687858367}{6}.$$

Bernoulli numbers have fascinating arithmetical properties, but I have to refrain from describing them. I will just mention their relation with Riemann's zeta-function $\zeta(s) = \sum_{n=1}^{\infty} (1/n^s)$ (for $s > 1$). The following formula holds:

$$B_{2k} = (-1)^{k-1} \frac{2(2k)!}{(2\pi)^{2k}} \zeta(2k) \quad (\text{for } k \geq 1).$$

Through his studies of the class number formula, Kummer showed that a prime number $p$ is regular if and only if $p$ does not divide the numerators of the Bernoulli numbers $B_2, B_4, \ldots, B_{p-3}$.

From the data he acquired, it was reasonable to conjecture that there are infinitely many regular primes, at least they seemed to appear more frequently than the irregular primes. Yet, this has never been proved and appears to be extremely difficult. Paradoxically, Jensen proved in 1915, in a rather simple way that there are in fact infinitely many irregular primes.

This was the situation around 1850. The theorem was proved for regular primes, the Bernoulli numbers had entered the stage and the main question was how to proceed in the case of irregular primes.

## 5. Kummer's Work on Irregular Prime Exponents

In 1851 Kummer began examining the irregular prime exponents. Aiming to derive congruences which must be satisfied if the first case fails, he produced some of his deepest results on cyclotomic fields.

It is impossible to describe in a short space Kummer's highly technical considerations, but the main points, which we mention here, give at least some idea of his astonishing mastery. First, he carefully studied the periods of the cyclotomic polynomial

$$\Phi_p(X) = X^{p-1} + X^{p-2} + \cdots + X + 1.$$

Suppose $q$ is a prime number, $q \neq p$, $f$ is the order of $q$ modulo $p$, $p - 1 = fr$, and let $g$ be a primitive root modulo $p$, and $\zeta$ a primitive $p$th root of 1. Kummer considered the $r$ periods of $f$ terms each $\eta_0, \eta_1, \ldots, \eta_{r-1}$ (already defined and used by Gauss). For example $\eta_0 = \zeta + \zeta^{g^r} + \zeta^{g^{2r}} + \cdots + \zeta^{g^{(f-1)r}}$, the other periods being conjugate to $\eta_0$. If $A$ is the ring of cyclotomic integers, and $A'$ is the ring of integers of the field $K' = \mathbb{Q}(\eta_0) = \cdots = \mathbb{Q}(\eta_{r-1})$, Kummer showed that $A$ is a free module over $A'$, with basis $\{1, \zeta, \ldots, \zeta^{f-1}\}$, and $A' = \mathbb{Z}[\eta_0, \ldots, \eta_{r-1}]$ is a free abelian group with basis $\{\eta_0, \eta_1, \ldots, \eta_{r-1}\}$. He also studied the decomposition of the prime $q$ in the ring $A'$.

Then, Kummer gave his beautiful proof that the group of classes of ideals of the cyclotomic field is generated by the classes of the prime ideals with prime norm.

Another ingredient in his work was the use of the cyclotomic functions first introduced by Jacobi. If $q$ is an odd prime of the form $q = kp + 1$, if $h$ is a primitive root $\bmod q$, $\zeta$ a primitive $p$th root of 1 and $\eta$ a primitive $q$th root of 1, let

$$\langle \zeta, \eta \rangle = \sum_{t=1}^{q-1} \zeta^{\mathrm{ind}_h(t)} \eta^t$$

where $\mathrm{ind}_h(t)$, the index of $t$ (with respect to $h$, $q$) is the only integer $s$, $1 \le s \le q - 1$ such that $t \equiv h^s \pmod q$.

For every integer $d \in \mathbb{Z}$, let

$$\psi_d(\zeta) = \sum_{t=1}^{q-2} \zeta^{\mathrm{ind}_h(t) - (d+1)\,\mathrm{ind}_h(t+1)}.$$

If $Q$ is the ideal of $A$ generated by $q$ and $h^k - \zeta$ (where $q = kp + 1$) then of course $Q$ is a prime ideal of norm $q$, that is, $Aq = \prod_{i=0}^{p-2} \sigma^i(Q)$ (where $\sigma$ is a generator of the Galois group). The main results concern certain products of conjugates of $Q$ which are principal ideals:

$$A\langle \zeta, \eta \rangle^p = \prod_{i=0}^{p-2} \sigma^i(Q)^{g_{\pi - i}},$$

with $g_e \equiv g^e \pmod p$, $\pi = (p - 1)/2$ and if

$$I_d = \{i \mid 0 \le i \le p - 2, g_{\pi - i} + g_{\pi - i + \mathrm{ind}_g(d)} > p\},$$

then

$$A\psi_d(\zeta) = \prod_{i \in I_d} \sigma^i(Q).$$

All this was put together to give Kummer his congruences. If $x$, $y$, $z$ are pairwise relatively prime integers, not multiples of $p$, such that $x^p + y^p + z^p = 0$, then

$$(Az)^p = A(x^p + y^p) = A(x + y) \prod_{k=0}^{p-2} A(x + \zeta^{g^k} y),$$

where $g$ is a primitive root modulo $p$. The ideals $A(x + y)$, $A(x + \zeta^{g^k} y)$ are $p$th powers of ideals, say $A(x + y) = J_0^p$, $A(x + \zeta^{g^k} y) = J_k^p$ ($J_k$ being a conjugate of $J_0$). For every $d$, $1 \le d \le p - 2$, and $I_d$ defined as before, $\prod_{i \in I_d} \sigma^i(J_0)$ is a principal ideal, say $A\alpha$, where $\alpha = F(\zeta)$, $F(X)$ being a polynomial with coefficients in $\mathbb{Z}$ and degree at most $p - 2$. Then

$$\prod_{i \in I_d} (x + X^{g^i} y) = X^m (F(X))^p + \Phi_p(X) M(X),$$

where $M(X) \in \mathbb{Z}[X]$.

Considering these polynomials as functions of the real variable $t > 0$, letting $t = e^v$ and taking an appropriate branch of the logarithm we obtain:

$$\sum_{i \in I_d} \log(x + e^{vg^i} y) = mv + p \log F(e^v) + \log\left[ 1 + \frac{\Phi_p(e^v) M(e^v)}{e^{mv} (F(e^v))^p} \right].$$

Let $D^nG$ denote the $n$th derivative of $G(v)$, at $v = 0$. Kummer showed for $2s = 2, 4, \ldots, p - 3$ $(p \neq 2,3)$ that the following congruences are satisfied:

$$[D^{p-2s}\log(x + e^v y)]B_{2s} \equiv 0 \pmod{p},$$

where $B_{2s}$ is the Bernoulli number of index $2s$.

Since $D^j \log(x + e^v y) = R_j(x,y)/(x + y)^j$, where $R_j(X,Y)$ is a homogeneous polynomial of total degree $j$, multiple of $Y$, writing $R_j(X,Y) = X^j P_j(T)$, it follows that

$$P_{p-2s}(t)B_{2s} \equiv 0 \pmod{p}$$

for $2s = 2, 4, \ldots, p - 3$.

The polynomials $P_j(T)$ may be computed recursively. With these congruences, Kummer improved his previous result:

*If $p$ divides the numerator of at most one of the Bernoulli numbers $B_2, B_4, \ldots, B_{p-3}$, then the first case of Fermat's theorem holds for $p$.*

In 1905 Mirimanoff generalized this last result of Kummer, as follows:

*If $p$ does not divide the numerator of one of the four Bernoulli numbers $B_{p-3}, B_{p-5}, B_{p-7}, B_{p-9}$, then the first case holds for the prime $p$.*

This theorem is again a *tour de force*. However, due to the long computations involving large Bernoulli numbers, its applicability is limited.

It was becoming increasingly clear that new and significantly more powerful methods were necessary to provide any substantial progress.

Later, I shall describe the sensational work by Wieferich and Mirimanoff early this century, and how Furtwängler used class field theory (more specifically Eisenstein's reciprocity law for the power residue symbol) to improve and simplify these results. All this brought into the battle the newly created forces of class field theory.

# 6. Other Relevant Results

In 1856, Grünert considered the size of possible solutions of Fermat's equation.

He proved that if $x$, $y$, $z$ are nonzero integers such that $x^n + y^n = z^n$, with $0 < x < y < z$, then necessarily $x > n$. This was very easy to prove.

For example, if $p = 101$ the smallest nontrivial solution, if it exists, would involve numbers greater than $102^{101}$. This pointed to a fact which was becoming more and more apparent: In order to disprove Fermat's statement one has to deal with very large numbers.

In 1894, following the line of Sophie Germain, Wendt contributed an interesting theorem. He considered the determinant $W_n$ of the circulant matrix

$$\begin{pmatrix} 1 & \binom{n}{1} & \binom{n}{2} & \cdots & \binom{n}{n-1} \\ \binom{n}{n-1} & 1 & \binom{n}{1} & \cdots & \binom{n}{n-2} \\ \vdots & \vdots & \vdots & & \vdots \\ \binom{n}{1} & \binom{n}{2} & \binom{n}{3} & \cdots & 1 \end{pmatrix}$$

which is equal to $\prod_{j=0}^{n-1} [(1 + \xi_j)^n - 1]$, where $\xi_0 = 1, \xi_1, \ldots, \xi_{n-1}$ are the $n$th roots of 1.

Wendt proved:

*If $p$ is an odd prime, if there exists $h \geq 1$ such that $q = 2hp + 1$ is prime and 3th, if $q$ does not divide $W_{2h}$ and $p^{2h} \not\equiv 1 \pmod{q}$, then the first case of Fermat's conjecture holds for $p$.*

A first step in the proof is the following: if $x$, $y$, $z$ are integers not divisible by $q$ and if $x^p + y^p + z^p \equiv 0 \pmod{q}$ then $q$ divides $W_{2h}$.

This leads to the interesting and related problem: if $p, q$ are odd primes does the congruence

$$X^p + Y^p + Z^p \equiv 0 \pmod{q}$$

have a solution in integers $x$, $y$, $z$ not multiples of $q$? Of course this depends on $p, q$.

If, given $p$, there exist infinitely many primes $q$ such that the above congruence *does not* have a solution as indicated, then Fermat's theorem would hold for $p$.

But in 1909, Dickson showed that this hypothesis is false. More precisely, if $q \geq (p - 1)^2(p - 2)^2 + 6p - 2$ then the above congruence modulo $q$ has a solution. In the same year, Hurwitz generalized this theorem, in a very beautiful paper, by counting the number of solutions of

$$\alpha_1 X_1^p + \alpha_2 X_2^p + \cdots + \alpha_n X_n^p \equiv 0 \pmod{q}.$$

All these considerations led again to deep investigations of the number of zeros of polynomials over finite fields, eventually linking up with the Riemann hypothesis for function fields.

# 7. The Golden Medal and the Wolfskehl Prize

In 1816, and again in 1850, the Académie des Sciences de Paris offered a golden medal and a prize of 3000 Francs for the mathematician who would solve Fermat's problem. The judges in 1856 were Cauchy, Liouville, Lamé, Bertrand, and Chasles.

Cauchy wrote the following report:

> Eleven memoirs have been presented to the Secretary. But none has solved the proposed question. The Commissaries have nevertheless noted that the piece registered under number 2 contained a new solution of the problem in the special case developed by Fermat himself, namely when the exponent is equal to 4.
>
> Thus, after being many times put for a prize, the question remains at the point where M. Kummer left it. However, the mathematical sciences should congratulate themselves for the works which were undertaken by the geometers, with their desire to solve the question, specially by M. Kummer; and the Commissaries think that the Academy would make an honourable and useful decision if, by withdrawing the question from the competition, it would adjugate the medal to M. Kummer, for his beautiful researches on the complex numbers composed of roots of unity and integers.

In 1908 the very substantial Wolfskehl Prize, in the amount of 100,000 Mark, was offered with the same aim by the Königliche Gesellschaft der Wissenschaften, in Göttingen, Germany:

> By the power conferred on us, by Dr. Paul Wolfskehl, deceased in Darmstadt, hereby we fund a prize of one hundred thousand Marks, to be given to the person who will be the first to prove the great theorem of Fermat.
>
> In his will, Doctor Wolfskehl observed that Fermat (*Oeuvres*, Paris, 1891, volume I, p. 291, observation 2) asserted mutatis mutandis that the equation $x^\lambda + y^\lambda = z^\lambda$ has no integral solutions for any odd prime number $\lambda$. This theorem has to be proved, either following the ideas of Fermat, or completing the researches of Kummer (*Crelle's Journal*, vol. XL, page 130; Abhandlungen der Akademie der Wissenschaften zu Berlin, 1857), for all exponents $\lambda$, for which it has some meaning [consult Hilbert, *Theorie der algebraischen Zahlkörper*, 1894–1895, and *Enzyklopädie der mathematischen Wissenschaften*, (1900–1904), I C 4b, page 713].
>
> The following rules will be followed:
>
> The Königliche Gesellschaft der Wissenschaften in Göttingen will decide in entire freedom to whom the prize should be conferred. It will refuse to accept any manuscript written with the aim of entering the competition to obtain the Prize. It will only take in consideration those mathematical memoirs which have appeared in the form of a monograph in the periodicals, or which are for sale in the bookstores. The Society asks the authors of such memoirs to send at least five printed exemplars.
>
> Works which are published in a language which is not understood by the scholarly specialists chosen for the jury will be excluded from the competition. The authors of such works will be allowed to replace them by translations, of guaranteed faithfulness.
>
> The Society declines its responsibility for the examination of works not brought to its attention, as well as for the errors which might result from the fact that the author of a work, or part of a work, are unknown to the Society.
>
> The Society keeps the right of decision in the case where various persons would have dealt with the solution of the problem, or for the case where the solution is the result of the combined efforts of several scholars, in particular in what concerns the partition of the Prize, at its own discretion.
>
> The award of the Prize by the Society will take place not earlier than two years after the publication of the memoir to be crowned. The interval of time

is aimed to allow the German and foreign mathematicians to voice their opinion about the validity of the solution published.

As soon as the Prize will be conferred by the Society, the laureate will be informed by the secretary, on the name of the Society, and the result will be published everywhere the Prize would have been announced during the preceding year. The assignment of the Prize by the Society is not to be the subject of any further discussion.

The payment of the Prize will be made to the laureate, in the next three months after the award, by the Royal Cashier of Göttingen University, or, at the receivers own risk, at any other place he will have designated.

The capital may be delivered against receipt, at the Society's will, either in cash, or by the transfer of financial values. The payment of the Prize will be considered as accomplished by the transmission of these financial values, even though their total value at the day's course would not attain 100,000 Mark.

If the Prize is not awarded by September 13, 2007, no ulterior claim will be accepted.

The competition for the Prize Wolfskehl is open, as of today, under the above conditions.

Göttingen, June 27, 1908
Die Königliche Gesellschaft der Wissenschaften.

A memorandum dated 1958 states that the Prize of 100,000 DM has been reduced to approximately 7,600 DM, in virtue of the inflation and financial changes.

Dr. F. Schlichting, from the Mathematics Institute of the University of Göttingen, was kind enough to provide me with the following information on the Wolfskehl Prize:

Göttingen, March 23, 1974.

Dear Sir:

Please excuse the delay in answering your letter. I enclose a copy of the original announcement, which gives the main regulations, and a note of the "Akademie" which is usually sent to persons who are applying for the prize, now worth a little bit more than 10,000 DM. There is no count of the total number of "solutions" submitted so far. In the first year (1907–1908) 621 solutions were registered in the files of the Akademie, and today they have stored about 3 meters of correspondence concerning the Fermat problem. In recent decades it was handled in the following way: the secretary of the Akademie divides the arriving manuscripts into (1) complete nonsense, which is sent back immediately, and into (2) material which looks like mathematics. The second part is given to the mathematical department and there, the work of reading, finding mistakes and answering is delegated to one of the scientific assistants (at German universities these are graduated individuals working for Ph.D. or habilitation and helping the professors with teaching and supervision)—at the moment I am the victim. There are about 3 to 4 letters to answer per month, and there is a lot of funny and curious material arriving, e.g., like the one sending the first half of his solution and promising the second if we would pay 1000 DM in advance; or another one, who promised me 10 per cent of his profits from publications, radio and TV interviews after he got famous, if only I would support him now; if not, he threatened to send it to a Russian mathematics department to deprive us of the glory of discovering him. From time to time someone appears in Göttingen and insists on personal discussion.

Nearly all "solutions" are written on a very elementary level (using the notions of high school mathematics and perhaps some undigested papers in number theory), but can nevertheless be very complicated to understand. Socially, the senders are often persons with a technical education but a failed career who try to find success with a proof of the Fermat problem. I gave some of the manuscripts to physicians who diagnosed heavy schizophrenia.

One condition of Wolfskehl's last will was that the Akademie had to publish the announcement of the prize yearly in the main mathematical periodicals. But already after the first years the periodicals refused to print the announcement, because they were overflowed by letters and crazy manuscripts. So far, the best effect has been had by another regulation of the prize: namely, that the interest from the original 100,000 Mark could be used by the Akademie. For example, in the 1910s the heads of the Göttingen mathematics department (Klein, Hilbert, Minkowski) used this money to invite Poincaré to give six lectures in Göttingen.

Since 1948 however the remainder of the money has not been touched.

I hope that you can use this information and would be glad to answer any further questions.

Yours sincerely,<br>
F. Schlichting.

# Bibliography

I shall only refer here to items specifically connected with the historical aspects. The other references will be made later, as it will be appropriate.

?       Fermat, P.
Lettre à Mersenne, pour Sainte-Croix (Septembre 1636?, 1637?, June 1638?). *Oeuvres*, III, Gauthier-Villars, Paris, 1896, 286–292.

1640   Fermat, P.
Lettre à Mersenne (Mai ? 1640). *Oeuvres*, II, Gauthier-Villars, Paris, 1894, 194–195.

1640   Fermat, P.
Lettre à Frénicle de Bessy (18 Octobre 1640). *Oeuvres*, II, Gauthier-Villars, Paris, 1894, 206–212.

1654   Fermat, P.
Lettre à Pascal (29 Août 1654). *Oeuvres*, II, Gauthier-Villars, Paris, 1894, 307–310.

1657   Fermat, P.
Lettre à Digby (15 Août 1657). *Oeuvres*, II, Gauthier-Villars, Paris, 1894, 342–346.

1807   Gauss, C. F.
Journal des Savants de Göttingen, 10 Mars 1807.

1816   Gauss, C. F.
Letter to Olbers (March 21, 1816). *Werke*, $X_1$, 75–76, G. Teubner, Leipzig, 1917.

1839   Lamé, G.
Mémoire sur le dernier théorème de Fermat. *C. R. Acad. Sci. Paris*, **9**, 1839, 45–46.

1839  Cauchy, A. and Liouville, J.
Rapport sur un mémoire de M. Lamé relatif au dernier théorème de Fermat. *C. R. Acad. Sci. Paris*, **9**, 1839, 359–364. Reprinted in *Oeuvres Complètes*, (1), Gauthier-Villars, Paris, 1897, 499–504.

1840  Lamé, G.
Mémoire d'analyse indéterminée démontrant que l'équation $x^7 + y^7 = z^7$ est impossible en nombres entiers. *J. Math. Pures et Appl.*, **5**, 1840, 195–211.

1844  Eisenstein, F. G.
Letter to Stern (1844?). Reprinted in *Math. Werke*, II, Chelsea Publ. Co., New York, 1975, 791–795.

1844  Kummer, E. E.
De numeris complexis, qui radicibus unitatis et numeris integris realibus constant. Acad. Albert. Regiomont. gratulatur Acad. Vratislaviensis, 1844, 28 pages. Reprinted in *J. Math. Pures et Appl.*, **12**, 1847, 185–212. Reprinted in *Collected Papers*, vol. I, edited by A. Weil, Springer-Verlag, Berlin, 1975.

1847  Cauchy, A.
Various communications. *C. R. Acad. Sci. Paris*, **24**, 1847, 407–416, 469–483, 516–530, 578–585, 633–636, 661–667, 996–999, 1022–1030, 1117–1120, and **25**, 1847, 6, 37–46, 46–55, 93–99, 132–138, 177–183, 242–245. Reprinted in *Oeuvres Complètes*, (1), 10, Gauthier-Villars, Paris, 1897, 231–285, 290–311, 324–350, 354–368.

1847  Kummer, E. E.
Extrait d'une lettre de M. Kummer à M. Liouville. *J. Math. Pures et Appl.*, **12**, 1847, 136. Reprinted in *Collected Papers*, vol. I, edited by A. Weil, Spring-Verlag, Berlin, 1975.

1847  Lamé, G.
Démonstration générale du théorème de Fermat sur l'impossibilité en nombres entiers de l'équation $x^n + y^n = z^n$. *C. R. Acad. Sci. Paris*, **24**, 1847, 310–314.

1847  Liouville, J.
Remarques à l'occasion d'une communication de M. Lamé sur un théorème de Fermat. *C. R. Acad. Sci. Paris*, **24**, 1847, 315–316.

1847  Lamé, G.
Mémoire sur la résolution en nombres complexes de l'équation $A^5 + B^5 + C^5 = 0$. *J. Math. Pures et Appl.*, **12**, 1847, 137–171.

1847  Lamé, G.
Mémoire sur la résolution en nombres complexes de l'équation $A^n + B^n + C^n = 0$. *J. Math. Pures et Appl.*, **12**, 1847, 172–184.

1847  Lamé, G.
Note au sujet de la démonstration du théorème de Fermat. *C. R. Acad. Sci. Paris*, **24**, 1847, 352.

1847  Lamé, G.
Second mémoire sur le dernier théorème de Fermat. *C. R. Acad. Sci. Paris*, **24**, 1847, 569–572.

1847  Lamé, G.
Troisième mémoire sur le dernier théorème de Fermat. *C. R. Acad. Sci. Paris*, **24**, 1847, 888.

1856  Cauchy, A.
Rapport sur le concours relatif au théorème de Fermat (Commissaires MM. Bertrand, Liouville, Lamé, Chasles, Cauchy rapporteur). *C. R. Acad. Sci. Paris*, **44**, 1856, 208.

1860   Smith, H. J. S.
Report on the theory of numbers, Part II, Art. 61 "Application to the last theorem of Fermat", Report of the British Association for 1859, 228–267. *Collected Mathematical Works*, I, Clarendon Press, Oxford, 1894, 131–137. Reprinted by Chelsea Publ. Co., New York, 1965.

1883   Tannery, P.
Sur la date des principales découvertes de Fermat. *Bull. Sci. Math., Sér.* 2, 7, 1883, 116–128. Reprinted in *Sphinx–Oedipe*, 3, 1908, 169–182.

1910   Hensel, K.
Gedächtnisrede auf Ernst Eduard Kummer, *Festschrift zur Feier des* 100. *Geburtstages Eduard Kummers*, Teubner, Leipzig, 1910, 1–37. Reprinted in Kummer's *Collected Papers*, vol. I, edited by A. Weil, Springer-Verlag, Berlin, 1975.

1912   ———
Bekanntmachung (Wolfskehl Preis). *Math. Annalen*, 72, 1912, 1–2.

1929   Vandiver, H. S. and Wahlin, G.
Algebraic numbers, II. *Bull. Nat. Research Council*, 62, 1928. Reprinted by Chelsea Publ. Co., New York, 1967.

1937   Bell, E. T.
*Men of Mathematics*, Simon and Schuster, New York, 1937.

1943   Hofmann, J. E.
Neues über Fermats zahlentheoretische Herausforderungen von 1657. *Abhandl. Preuss. Akad. Wiss., Berlin*, No. 9, 1944.

1948   Itard, J.
Sur la date à attribuer à une lettre de Pierre Fermat. *Revue d'Histoire des Sciences et de leurs Applications*, 2, 1948, 95–98.

1959   Schinzel, A.
Sur quelques propositions fausses de P. Fermat. *C. R. Acad. Sci. Paris*, 249, 1959, 1604–1605.

1961   Bell, E. T.
*The Last Problem*, Simon and Schuster, New York, 1961.

1962   De Waard, C.
*Correspondence du Père Marin Mersenne*, vol. 7, Editions du Conseil National de la Recherche Scientifique, Paris, 1962, 272–283.

1966   Noguès, R.
*Théorème de Fermat, son Histoire*, A. Blanchard, Paris, 1966.

1975   Edwards, H. M.
The background of Kummer's proof of Fermat's last theorem for regular primes. *Arch. for History of Exact Sciences*, 14, 1975, 219–236. Postscript, *Arch. for History of Exact Sciences*, 17, 1977, 381–394.

1977   Mazur, B.
Review of Kummer's Papers, "Collected Works", Volumes I, II. *Bull. Amer. Math. Soc.*, 83, 1977, 976–988.

# LECTURE II

# Recent Results

Some of the most common questions I have been asked are:

a. For which exponents is Fermat's theorem true?
b. Is serious work still being done on the problem?
c. Will it be solved soon?

Anyone over 40, hearing my reply to the first question, will say: "When I was younger, we knew that it was true up to ..." and will then state some rather small exponent.

Below I will try to present whatever information I have gathered. I will not, however, attempt to answer the last question.

There has always been considerable work done on the subject—though of rather diverse quality—so it is necessary to be selective. My purpose is to show the various methods of attack, the different techniques involved, and to indicate important historical developments.

Here are 10 recent results which will later be discussed in more detail.

## 1. Stating the Results

1. Wagstaff (1976): Fermat's last theorem (FLT) holds for every prime exponent $p < 125000$.
2. Morishima and Gunderson (1948): The first case of FLT holds for every prime exponent $p < 57 \times 10^9$ (or, at worst, as I will explain, for every prime exponent $p < 3 \times 10^9$, according to Brillhart, Tonascia and Weinberger, 1971).

In fact the first case also holds for larger primes.

3. The first case of FLT holds for the largest prime known today.

The above results are on the optimistic side. But some mathematicians think that there might be a counterexample. How large would the smallest counterexample have to be for a given exponent $p$?

4. Inkeri (1953): If the first case fails for the exponent $p$, if $x$, $y$, $z$ are integers, $0 < x < y < z$, $p \nmid xyz$, $x^p + y^p = z^p$, then

$$x > \left(\frac{2p^3 + p}{\log(3p)}\right)^p.$$

And in the second case,

$$x > p^{3p-4} \quad \text{and} \quad y > \tfrac{1}{2}p^{3p-1}.$$

Moreover, Pérez Cacho proved in 1958 that in the first case, $y > \tfrac{1}{2}(p^2 P + 1)^p$, where $P$ is the product of all primes $q \neq p$ such that $q - 1$ divides $p - 1$.

There might also be only finitely many solutions. In this respect:

5. Inkeri and Hyyrö (1964): (a) Given $p$ and $M > 0$, there exist at most finitely many triples $(x,y,z)$, such that $0 < x < y < z$, $x^p + y^p = z^p$, and $y - x$, $z - y < M$.

   (b) Given $p$, there exist at most finitely many triples $(x,y,z)$ such that $0 < x < y < z$, $x^p + y^p = z^p$, and $x$ is a prime power.

For each such triple, cf. Inkeri (1976), we have the effective majoration (and this is a very important new feature):

$$x < y < \exp\exp\{[2^p(p - 1)^{10(p-1)}]^{(p-1)^2}\}.$$

Another sort of result, this time for even exponents is the following:

6. Terjanian (1977): If $x$, $y$, $z$ are nonzero integers, $p$ is an odd prime, and $x^{2p} + y^{2p} = z^{2p}$, then $2p$ divides $x$ or $y$. In other words, the first case of FLT is true for every even exponent.

The possibility that FLT (or even its first case) holds for infinitely many prime exponents is still open. In this respect we have:

7. Rotkiewicz (1965): If Schinzel's conjecture on Mersenne numbers is true, then there exist infinitely many primes $p$ such that the first case of FLT holds for $p$ (Schinzel conjectured that there exist infinitely many square-free Mersenne numbers).

The next results are intimately connected with the ideal class group of the cyclotomic fields $\mathbb{Q}(\zeta)$, where $\zeta$ is a primitive $p$th root of 1.

8. Vandiver (1929): If the second factor $h^+$ of the class number of $\mathbb{Q}(\zeta)$ is not a multiple of $p$ and if none of the Bernoulli numbers $B_{2np}$ ($n = 1,2,\ldots,(p - 3)/2$) is a multiple of $p^3$, then Fermat's last theorem holds for the exponent $p$.

9. Eichler (1965): If the first case fails for $p$, then $p^{[\sqrt{p}]-1}$ divides the first factor $h^*$ of the class number of $\mathbb{Q}(\zeta)$ and the $p$-rank of the ideal class group of $\mathbb{Q}(\zeta)$ is greater than $\sqrt{p} - 2$.
10. Brückner (1979): If the first case fails for $p$, then the irregularity index of $p$, $\mathrm{ii}(p) = \#\{k = 2,4,\ldots,p-3 \,|\, p$ divides the Bernoulli number $B_k\}$ satisfies

$$\mathrm{ii}(p) > \sqrt{p} - 2.$$

## 2. Explanations

Now, I shall explain the significance of these various theorems and computations.

Result (1). Wagstaff obtained his result with a computer, but what is the theory behind it?

Kummer's theorem asserts that FLT holds for the prime exponents $p$ which are regular. A prime $p$ is *regular* if $p$ does not divide the class number $h$ of the cyclotomic field $\mathbb{Q}(\zeta)$, where $\zeta$ is a primitive $p$th root of 1. Kummer showed that this is equivalent to $p$ not dividing the first factor $h^*$ of the class number. Since the computation of the class number, or even of its first factor, is rather involved, and even more because the class number grows so rapidly with $p$, it was imperative to find a more amenable criterion. Kummer characterized the regular primes $p$ by the condition:

$$p \nmid B_{2k} \quad \text{for } 2k = 2, 4, \ldots, p - 3.$$

Here $B_{2k}$ denotes the $2k$th Bernoulli number. These are defined by the formal power series expansion

$$\frac{X}{e^X - 1} = \sum_{n=0}^{\infty} B_n \frac{X^n}{n!}.$$

They may be obtained recursively; moreover if $n$ is odd, $n \geq 3$, then $B_n = 0$.

Vandiver gave a practical criterion to determine whether $p$ is irregular, by means of the congruence

$$\frac{4^{p-2k} + 3^{p-2k} - 6^{p-2k} - 1}{4k} B_{2k} \equiv \sum_{p/6 < a < p/4} a^{2k-1} \pmod{p}.$$

The advantage of this congruence is that it involves a sum of relatively few summands, contrary to the previous congruences. If both the right-hand side and the left-hand factor of the above congruence are multiples of $p$, then the above congruence does not decide the question and other similar congruences have to be used. Once it is known that $p$ is irregular, the following criterion is used (Vandiver, 1954 and Lehmer, Lehmer, and Vandiver, 1954):

Let $p$ be an irregular prime, let $P = rp + 1$ be a prime such that $P < p^2 - p$ and let $t$ be an integer such that $t^r \not\equiv 1 \pmod{P}$. If $p \,|\, B_{2k}$, with $2 \leq 2k \leq p - 3$

let

$$d = \sum_{n=1}^{(p-1)/2} n^{p-2k}$$

and

$$Q_{2k} = \frac{1}{t^{rd/2}} \prod_{a=1}^{(p-1)/2} (t^{ra} - 1)^{a^{p-1-2k}}.$$

If $Q_{2k}^r \not\equiv 1 \pmod{p}$ for all $2k$ such that $p \mid B_{2k}$, then FLT holds for the exponent $p$. This criterion is well suited to the computer.

During his extensive calculations, Wagstaff noted many facts about the irregular primes. The maximum irregularity index found was 5. Moreover,

$$\frac{\#\ (\text{irregular primes}\ p < 125000)}{\#\ (\text{primes}\ p < 125000)} \approx 0.39248 \approx 1 - \frac{1}{\sqrt{e}} = 0.39347.$$

This confirms a heuristic prediction of Siegel (1964).

Let me now recall various interesting results about regular and irregular primes.

It is suspected that there exist infinitely many regular primes, but this has never been proved. On the other hand, Jensen proved in 1915 that there exist infinitely many irregular primes. Actually they are abundant in the following sense. In 1975, Yokoi proved for $N$ an odd prime, and Metsänkylä (1976), for arbitrary $N \geq 3$, that if $H$ is a proper subgroup of the multiplicative group $(\mathbb{Z}/N\mathbb{Z})^*$, then there exist infinitely many irregular primes $p$ such that $p$ modulo $N$ is not in $H$.

Taking $N = 12$ and letting $H$ be the trivial subgroup, gives the following puzzling theorem previously obtained by Metsänkylä (1971): There exist infinitely many irregular primes $p$ which satisfy either one of the congruences $p \equiv 1 \pmod 3$, $p \equiv 1 \pmod 4$. But he couldn't decide which of these congruence classes must contain infinitely many irregular primes.

So it is rather startling that it is possible—and not too difficult—to show that there are infinitely many irregular primes, however, it is not known whether there are infinitely many regular ones, even though heuristic arguments seen to indicate that these are much more numerous.

Among the many conjectures—and all seem difficult to prove—let me mention:

1. There exist primes with arbitrarily large irregularity index.
2. There exist infinitely many primes with given irregularity index.
3. There exists a prime $p$ and some index $2k$ such that $p^2 \mid B_{2k}, 2 \leq 2k \leq p - 3$.

Result (2). The fact that the first case holds for all prime exponents less than $3 \times 10^9$ depends on the scarcity of primes $p$ satisfying the congruence $2^{p-1} \equiv 1 \pmod{p^2}$.

Fermat's little theorem says that if $p$ is a prime and $p \nmid m$, then $m^{p-1} \equiv 1 \pmod{p}$. Hence the quotient $q_p(m) = (m^{p-1} - 1)/p$ is an integer. It is called the *Fermat quotient* of $p$ with base $m$.

In 1909 Wieferich proved the following theorem:

*If the first case of FLT fails for the exponent p, then p satisfies the stringent condition that $2^{p-1} \equiv 1 \pmod{p^2}$; or equivalently $q_p(2) \equiv 0 \pmod{p}$.*

This theorem had a new feature, in that it gives a condition involving only the exponent $p$, and not a possible solution $(x,y,z)$ of Fermat's equation as in most of the previous results. The original proof of Wieferich's theorem was very technical, based on the so-called Kummer congruences for the first case:

If $p \nmid xyz$ and $x^p + y^p + z^p = 0$, then for $2k = 2, 4, \ldots, p-3$, we have the congruences (for a real variable $v$)

$$\left[\frac{d^{p-2k}\log(x + e^v y)}{dv^{p-2k}}\right]_{v=0} \times B_{2k} \equiv 0 \pmod{p}$$

(as well as the similar congruences for $(y,x)$, $(x,z)$, $(z,x)$, $(y,z)$, $(z,y)$). These congruences were obtained with intricate considerations involving the arithmetic of the cyclotomic field and transcendental methods (the latter, as a matter of fact, may be replaced by $p$-adic methods).

Thus, it suffices to show that $2^{p-1} \not\equiv 1 \pmod{p^2}$ to guarantee that the first case holds for $p$. For a few years no such $p$ was found. Only in 1913 Meissner showed that $p = 1093$ satisfies $2^{p-1} \equiv 1 \pmod{p^2}$. The next prime satisfying this congruence was discovered by Beeger in 1922; it is $p = 3511$. Since then, computations performed up to $3 \times 10^9$ by Brillhart, Tonascia and Weinberger (1971) have not found any other such prime. Thus, in the above range, the first case holds for all but these two primes.[1]

The handling of these exceptional primes was actually done by a similar criterion. Indeed, in 1910 Mirimanoff gave another proof of Wiefe.ich's theorem and showed also that if the first case fails for $p$ then $3^{p-1} \equiv 1 \pmod{p^2}$. The primes $p = 1093$ and $3511$ do not satisfy this congruence.

Several more criteria of a similar kind were successively obtained by various authors. In 1914 Frobenius and Vandiver showed independently that $q_p(5) \equiv 0 \pmod{p}$ and $q_p(11) \equiv 0 \pmod{p}$, if the first case fails for $p$. Successively, Pollaczek, Vandiver, Morishima proved that $q_p(m) \equiv 0 \pmod{p}$ must hold for all primes $m \leq 31$. Morishima proved the same criterion for $m = 37, 41, 43$ (except for finitely many primes $p$). The exceptions were ruled out by Rosser in 1940 and 1941. However, in 1948 Gunderson pointed out that Morishima's proof was incomplete. I have been assured by Agoh and Yamaguchi, who worked with Morishima and studied his papers, that the proofs are sound.

Rosser, Lehmer and Lehmer, using the above criteria (up to $m = 43$), and the Bernoulli polynomials to estimate the number of lattice points in a certain simplex in the real vector space of 14 dimensions, gave the following well-known bound:

If the first case fails for $p$, then $p > 252 \times 10^6$.

These computations have been superseded by the bound $6 \times 10^9$, obtained using a computer, as I have already indicated.

[1] In a last minute letter (Sept. 2, 1979) D. H. Lehmer informs me that he extended the search up to $4.7 \times 10^9$ without finding any further prime of that kind.

Furthermore, Gunderson devised, in 1948, another sharper method to bound the exponent. Assuming the Fermat quotient criteria up to 31, this gives the bound $p > 43 \times 10^8$, and up to 43, the bound is $p > 57 \times 10^9$.

Result (3). The largest prime known today[1] is the Mersenne number $M_q = 2^q - 1$ where $q = 19937$. It has 6002 digits. Its primeness was shown by Tuckerman in 1971, using the famous Lucas test: if $q > 2$, $M_q$ is prime if and only if $M_q$ divides $S_q$. The numbers $S_q$ are defined by recurrence: $S_2 = 4$, $S_{n+1} = S_n^2 - 2$, so the sequence is 4, 14, 194, . . . .

But how was it possible to show that the first case holds for such a large exponent? As a matter of fact, this is a consequence of Wieferich's and analogous criteria, and it is a special case of a result which was proved successively by Mirimanoff, Landau, Vandiver, Spunar, Gottschalk. Namely:

Suppose that there exists $m$ not divisible by $p$, such that $mp = a \pm b$, where the prime factors of $a$ and of $b$ are at most 43 (this depends on the Fermat quotient criteria). Then the first case holds for $p$. Therefore, it holds for all Mersenne primes $M_q = 2^q - 1$, as well as for many other numbers.

Do there exist infinitely many prime numbers $p$ satisfying the conditions of the preceding proposition? This is an open question. In 1968 Puccioni proved:

*If this set of primes is finite, then for all primes $l \le 43$, $l \not\equiv \pm 1 \pmod 8$ the set $\mathcal{M}_l = \{q | q \text{ is a pime and } l^{q-1} \equiv 1(q^3)\}$ is infinite.*

Primes in $\mathcal{M}_l$ are very hard to find, but this doesn't preclude these sets being infinite.

Result (4). The first lower bound for a counterexample to FLT was given by Grünert in 1856. He showed that if $0 < x < y < z$ and $x^n + y^n = z^n$ then $x > n$. So it is useless to try to find a counterexample with small numbers. For example, if $n = 101$ the numbers involved in any counterexample would be least $102^{101}$.

It was easy to improve this lower bound. Based on congruences of Carmichael (1913), if $x^p + y^p = z^p$, $0 < x < y < z$, then $x > 6p^3$.

But, with some clever manipulations Inkeri arrived at the lower bound already given. Taking into account that the first case holds for all prime exponents $p < 57 \times 10^9$, then

$$x > \left( \frac{2 \times 57^3 \times 10^{27} + 57 \times 10^9}{\log(171 \times 10^9)} \right)^{57 \times 10^9}$$

This is a very large number; it has more than $17 \times 10^{11}$ digits!

---

[1] Since this book was written, three larger Mersenne primes were discovered. Curt Noll and Laura Nickel, two 18-year old students at California State University at Hayward, announced in 1978 that if $q = 21701$ then $M_q$ is a prime. Again, in 1979, Noll found that $M_q$ is prime when $q = 23209$. Nelson and Slowinski went further and showing that $M_q$ is prime when $q = 44497$; this is the largest prime known today.

Similarly, for the second case we may take $p = 125000$, hence

$$x > (125 \times 10^3)^{3 \times 125 \times 10^3 - 4}.$$

This number has more than $19 \times 10^5$ digits.

To give some sense of the magnitudes involved, I have inquired about some physical constants, as they have been estimated by the physicists.

For example, the radius of the known universe is estimated to be $10^{28}$ cm. The radius of the atomic nucleus, about $10^{-13}$ cm. So the number of nuclei that may be packed in the universe, is just about $(10^{28+13})^3 = 10^{123}$—a very modest number indeed!

But I should add that the above is rather controversial, and I have quoted it only to stress the enormous disparity between the sizes of the candidates for a counterexample to FLT, and the reputedly largest physical constants.

Despite the monstrous size of the numbers involved, it is perhaps not quite safe to assert that no counterexample to the theorem will ever be available. Consider, for example, the equation

$$121^{10^{100}} - 1 = (11^{10^{100}} + 1)(11^{10^{100}} - 1)$$

which is easy to establish. Yet, the numbers involved have more than $10^{100}$ digits.

This being said, mathematicians had better try to prove FLT, or at least some weaker form of it, rather than look for a counterexample.

Result (5). For example, it might be possible to show that the Fermat equation has at most finitely many solutions. It might even be that the number of solutions is bounded by an effectively computable bound. I should warn however that this has not yet been proved.

It was only under a further restriction that a finiteness result was proved by Inkeri. He considered possible solutions $(x, y, z)$ such that the integers are not too far apart, more precisely $y - x < M$, and $z - y < M$, where $M > 0$ is given in advance. Then the problem becomes actually one of counting integer solutions of an equation involving only 2 variables. For this purpose there are the theorems of Siegel, or Landau, Roth, or similar ones. Actually Inkeri and Hyyrö used the following: Let $m$, $n$ be integers, $\max\{m,n\} \geq 3$. Let $f(X) = a_0 X^n + a_1 X^{n-1} + \cdots + a_n \in \mathbb{Z}[X]$, with distinct roots. If $a$ is an integer, $a \neq 0$, then the equation $f(X) = aY^m$ has at most finitely many solutions in integers.

Given this theorem they proved statement (a).

Concerning (b), I wish to mention that it partially answers a conjecture of Abel (1823). Abel conjectured that if $x^p + y^p + z^p = 0$ (with nonzero integers $x$, $y$, $z$) then, at any rate, $x$, $y$, $z$ are not prime powers. I suppose that Abel might have had in mind a procedure, which would produce from a nontrivial solution $(x, y, z)$ another one $(x_1, y_1, z_1)$, where the minimum number of prime factors of the integers $x_1$, $y_1$, $z_1$ is strictly smaller than it was for $x$, $y$, $z$. In this situation he would "descend" on this number, eventually finding a solution with some prime-power integer—and if this turned out to be impossible, he would have proved FLT.

To date Abel's conjecture has not been completely settled. Sauer in 1905, and Mileikowsky in 1932 obtained some partial results. In 1954 Möller proved:

If $x^n + y^n = z^n$, $0 < x < y < z$, and if $n$ has $r$ distinct odd prime factors then $z$, $y$ have at least $r + 1$ distinct prime factors, while $x$ has at least $r$ such factors. If $n = p$ is a prime, this tells that $y$, $z$ cannot be prime-powers. Moreover, if $p$ does not divide $xyz$, then $x$ also cannot be a prime-power (this was proved by Inkeri in 1946). It remains only to settle the case $p \mid xyz$, and to show that $x$ is not a prime-power.

Inkeri has succeeded in proving that there are at most finitely many triples $(x,y,z)$, as above, where $x$ is a prime-power. Using the methods of Baker, which give effective upper bounds for the integral solutions of certain diophantine equations, Inkeri showed (1976), that

$$x < y < \exp\exp[2^p(p-1)^{10(p-1)}]^{(p-1)^2}.$$

I pause now to indicate another very interesting use of Baker's estimates.

The famous Catalan problem is the following: to show that the only solution in natural numbers, $x$, $y$, $m > 1$, $n > 1$, of the equation $x^m - y^n = 1$ is $x = 3$, $m = 2$, $y = 2$, $n = 3$. This problem is still open. However, using Baker's methods, Tijdeman determined a number $C > 0$ such that if $(x,y,m,n)$ is a solution then $x$, $y$, $m$, $n$ are less than $C$. In particular, there are only finitely many solutions.

Closely related is the following conjecture, which is a generalization of a theorem of Landau (published in his last book of 1959):

Let $a_1 < a_2 < \cdots$ be the increasing sequence of all integers which are proper powers (i.e., squares, cubes, etc. . .). Then $\lim_{n \to \infty} (a_{n+1} - a_n) = \infty$.

In his result, Landau considered two fixed exponents $m$, $n$ and the sequence of $m$th powers and $n$th powers.

Result (6). Now I will turn to a more elementary result.

In his very first paper on Fermat's problem, published in 1837, Kummer considered Fermat's equation with exponent $2n$, where $n$ is odd. And he showed that if it has a nontrivial solution, $x^{2n} + y^{2n} = z^{2n}$, with $\gcd(n,xyz) = 1$ then $n \equiv 1 \pmod 8$.

So, there exist infinitely many primes $p$ such that the first case is true for the exponent $2p$.

Kummer's result was rediscovered several times. It has also been improved. For example, in 1960 Long showed that if $\gcd(n,xyz) = 1$, $x^{2n} + y^{2n} = z^{2n}$ then $n \equiv 1$ or $49 \pmod{120}$. Some more elementary manipulation shows that if $m \equiv 4$ or $6 \pmod{10}$ then $X^m + Y^m = Z^m$ cannot have a solution $(x,y,z)$ with $\gcd(m,xyz) = 1$. But the best possible result dealing with the first case, for an even exponent, was just obtained by Terjanian. It plainly states that the first case is true for any even exponent. The proof is ingenious, but elementary. This leads to the speculation that there might be an elementary

proof for the first case and arbitrary prime exponents. I think, however, that
it shows rather that the equation with prime exponents is far more difficult
to handle than with even exponents.

Result (7). Schinzel's conjecture has been supported by numerical evidence.
To date, no one has ever found a square factor of any Mersenne number.
Moreover if $p^2$ divides a Mersenne number, then $p > 9 \times 10^8$.

Rotkiewicz's theorem says that Schinzel's conjecture implies that there
exist infinitely many primes $p$ such that $2^{p-1} \not\equiv 1 \pmod{p^2}$. Hence by
Wieferich's theorem, there would exist infinitely many primes $p$ for which
the first case holds. I believe, however, that a proof of this last statement, and
a proof of Schinzel's conjecture are equally difficult.

Result (8). To better explain the meaning of Vandiver's result, it is necessary to return to Kummer's monumental theorem:

*If $p$ is a regular prime, then* FLT *holds for the exponent $p$.*

As I have already mentioned, Kummer was led to study the arithmetic of
cyclotomic fields, to take care of the phenomenon of nonunique factorization
into primes. To recover uniqueness Kummer created the concept of *ideal
numbers*. Later Dedekind interpreted these ideal numbers to be essentially
what we call today *ideals*. However, it should be said that Kummer's ideal
numbers were in fact today's divisors. Besides the ideal numbers, he considered of course the actual numbers, that is, the elements of the cyclotomic
field. For the ideal numbers unique factorization holds. Ideal numbers
were called *equivalent* when one was the product of the other by an actual
number. Kummer showed that the number of equivalence classes is finite—it
is called the *class number* of the cyclotomic field and usually denoted by $h$.

Moreover, Kummer indicated precise formulas for the computation of $h$.
He wrote $h = h^* h^+$, where

$$h^* = \frac{1}{(2p)^{(p-3)/2}} \left| G(\eta) G(\eta^3) \cdots G(\eta^{p-2}) \right|$$

$$h^+ = \frac{2^{(p-3)/2}}{R} \prod_{k=1}^{(p-3)/2} \left| \sum_{j=0}^{(p-3)/2} \eta^{2kj} \log\left|1 - \zeta^{g^j}\right| \right|.$$

In the above formulas, $\eta$ is a primitive $(p-1)$th root of 1; $g$ is a primitive
root modulo $p$; for each $j$, $g_j$ is defined by $1 \le g_j \le p - 1$ and $g_j \equiv g^j \pmod{p}$;
$G(X) = \sum_{j=0}^{p-2} g_j X^j$; and $R$ is the regulator of the cyclotomic field, which is
a certain invariant linked to the units of the field.

$h^*$ is called the *first factor*, while $h^+$ is the *second factor* of the class
number. Kummer proved that $h^*$, $h^+$ are integers—rather an unpredictable
fact, from the defining expressions. Actually, he recognized $h^+$ as being the
class number of the real cyclotomic field $\mathbb{Q}(\zeta + \zeta^{-1})$. He gave also the following interpretation of $h^+$. Let $U$ be the group of units of $\mathbb{Q}(\zeta)$, i.e., all
$\alpha \in \mathbb{Z}[\zeta]$ such that there exists $\beta \in \mathbb{Z}[\zeta]$ such that $\alpha\beta = 1$. Let $U^+$ denote the
set of those units which are real positive numbers. For every $k$, $2 \le k \le$

$(p - 1)/2$, let

$$\delta_k = \sqrt{\frac{1 - \zeta^k}{1 - \zeta} \times \frac{1 - \zeta^{-k}}{1 - \zeta^{-1}}},$$

so $\delta_k$ is a real positive unit of $\mathbb{Q}(\zeta)$. Let $V$ be the subgroup of $U^+$ generated by all these $(p - 3)/2$ "circular" units. Kummer showed that $h^+ = (U^+ : V)$, the index of $V$ in $U^+$.

Moreover, he proved that if $p$ does not divide $h^*$, then $p$ does not divide $h^+$. Therefore $p$ is a regular prime if and only if $p$ does not divide $h^*$. Then, he proceeded to compute $h^*$ for all primes $p \le 163$ and he found the following irregular primes $p = 37, 59, 67, 101, 103, 131, 149, 157$. Based on his computations, he conjectured that the first factor $h^* = h^*(p)$ of the class number is asymptotic to

$$h^*(p) \sim \gamma(p) = 2p\left(\frac{p}{4\pi^2}\right)^{(p-1)/4}$$

This conjecture, which agrees with recent numerical evidence, has yet to be proved. In 1951, Ankeny and Chowla proved that

$$\log h^*(p) = \log \gamma(p) + o(\log p) = \tfrac{1}{4}(p + 3)\log p - \tfrac{1}{2}p\log 2\pi + o(\log p).$$

Later, in 1964, Siegel published his weaker result:

$$\log h^*(p) = \log \gamma(p) + o(p \log \log p).$$

It follows that $\log h^*(p) \sim \tfrac{1}{4}p \log p$.

In 1976, Masley and Montgomery showed that if $p > 200$, then

$$(2\pi)^{-p/2}p^{(p-25)/4} \le h^*(p) \le (2\pi)^{-p/2}p^{(p+31)/4}$$

and Pajunen has shown also in 1976 that if $5 < p \le 641$, then

$$\tfrac{2}{3}\gamma(p) < h^*(p) < \tfrac{3}{2}\gamma(p).$$

Concerning the growth of the first factor, Ankeny and Chowla proved in 1951 that there exists $p_0$ such that $h^*(p)$ is monotonically increasing for $p \ge p_0$. It is conjectured by Lepistö that one may take $p_0 = 19$.

The second factor is much more difficult to handle, since it is tied to the structure of the group of units. It was Kummer who already found the first example, $p = 163$, for which $h^+(p)$ is even. However not many more examples were known before 1965, when Ankeny, Chowla, and Hasse, using a lemma of Davenport and class field theory, proved: if $q$ is a prime, $n > 1$, and $p = (2qn)^2 + 1$ is a prime, then $h^+(p) > 2$.

If $p | h^*(p)$ but $p \nmid h^+(p)$ the cyclotomic field is called *properly irregular*. It is *improperly irregular* if $p | h^+(p)$ and so $p | h^*(p)$. It is not known whether there are improperly irregular cyclotomic fields. At any rate, none has been found for $p < 125000$. Vandiver, Pollaczek, Dénes, and Morishima have studied irregular fields.

Vandiver's Result (8) finds its origin in Kummer's work of 1857, where Kummer considered in depth the first case for irregular primes. Vandiver analyzed the work of Kummer, corrected mistakes, filled gaps and was able to generalize it to include the second case also. A paper written by Vandiver in 1934 contains the claim that if $p$ does not divide the second factor, then the first case of FLT holds for $p$. However, his proof is now considered as questionable.

Result (9). The result of Eichler is beautiful and far reaching, and the method used relies on basic principles, rather than on the previous criteria. Actually, Eichler proved that if the first case fails for $p$ then $p^{[\sqrt{p}]-1}$ divides $h^*$. The other assertion may be proved in the same way.

Without knowing this variant of Eichler's theorem, Skula proved that if the $p$-rank of the ideal class group of $\mathbb{Q}(\zeta)$ is 1, that is, if the $p$-class group is cyclic, then the first case holds for $p$. A simpler proof was given by Brückner. But, this is contained already in Eichler's theorem.

Result (10). To explain the scope of the latest of Brückner's theorems, let $\Gamma$ be the ideal class group of $\mathbb{Q}(\zeta)$, $h$ the class number. If $p$ is irregular, then $\Gamma \neq p\Gamma$ so $\gamma_p = \dim(\Gamma/p\Gamma) \geq 1$ (where $\Gamma/p\Gamma$ is considered as a vector space over the field with $p$ elements). In 1965, Eichler proved that if the first case fails, then $\gamma_p > \sqrt{p} - 2$. However, the computation of $\gamma_p$ is difficult. Brückner succeeded in relating the above dimension $\gamma_p$ to the irregularity index ii$(p)$, and proved that if the first case fails for $p$, then more than $\sqrt{p} - 2$ Bernoulli numbers $B_{2k}$ (with $2 \leq 2k \leq p - 3$) are multiples of $p$.

This fits into a series of classical results. Cauchy (1847) and Genocchi (1852) proved that if the first case fails for $p$, then $B_{p-3}$ is a multiple of $p$. In 1857, Kummer showed that both $B_{p-3}$ and $B_{p-5}$ must be multiples of $p$. Later, Mirimanoff showed that $B_{p-7}$ and $B_{p-9}$ must also be multiples of $p$.

In 1934 Krasner proved quite an interesting result: there exists a prime $p_0$ (which could be effectively computed) such that if $p \geq p_0$ and if the first case of FLT fails for $p$, then the $k$ Bernoulli numbers $B_{p-3}, B_{p-5}, \cdots B_{p-(2k+1)}$ are all multiples of $p$; in this statement $k = [\sqrt[3]{\log p}]$. Thus, in the event of a solution in the first case of FLT a reasonably large number of successive Bernoulli numbers would be multiples of $p$. Even though this number is usually smaller than the one indicated by Brückner's theorem, in this case the Bernoulli numbers are consecutive. This is a most unlikely conclusion, perhaps pointing to the fact that the first case of Fermat's theorem may very well be true.

# Bibliography

1823 Abel, N. H.
Extraits de quelques lettres à Holmboe (1823). *Werke*, vol. 2. Grondahl, Christiania, 1881, 254–255.

1837   Kummer, E. E.*
De aequatione $x^{2\lambda} + y^{2\lambda} = z^{2\lambda}$ per numeros integros resolvenda. *J. reine u. angew. Math.*, **17**, 1837, 203–209.

1847   Cauchy, A.
Mémoire sur diverses propositions relatives à la théorie des nombres. *C. R. Acad. Sci. Paris*, **25**, 1847, 177–183. Also in *Oeuvres Complètes* (1), 10, Gauthier-Villars, Paris, 1897, 360–366.

1847   Kummer, E. E.
Beweis des Fermat'schen Satzes der Unmöglichkeit von $x^{\lambda} + y^{\lambda} = z^{\lambda}$ für eine unendliche Anzahl Primzahlen $\lambda$. *Monatsber. Akad. d. Wiss.*, *Berlin*, 1847, 132–139, 140–141, 305–319.

1850   Kummer, E. E.
Bestimmung der Anzahl nicht äquivalenter Classen für die aus $\lambda$-ten Wurzel der Einheit gebildeten complexen Zahlen und die Idealen factoren derselben. *J. reine u. angew. Math.*, **40**, 1850, 93–116.

1850   Kummer, E. E.
Zwei besondere Untersuchungen über die Classen-Anzahl und über die Einheiten der aus $\lambda$-ten Wurzeln der Einheit gebildeten complexen Zahlen. *J. reine u. angew Math.*, **40**, 1850, 117–129.

1850   Kummer, E. E.
Allgemeiner Beweis des Fermat'schen Satzes, dass die Gleichung $x^{\lambda} + y^{\lambda} = z^{\lambda}$ durch ganze Zahlen unlösbar ist, fur alle diejenigen Potenz-Exponenten $\lambda$, welche ungerade Primzahlen sind und in den Zählern der ersten $\frac{1}{2}(\lambda - 3)$ Bernoulli'schen Zahlen nicht vorkommen. *J. reine u. angew. Math.*, **40**, 1850, 130–138.

1852   Genocchi, A.
Intorno all' espressioni generali di numeri Bernoulliani. *Annali di scienze mat. e fisiche, compilate da Barnaba Tortolini*, **3**, 1852, 395–405.

1856   Grünert, J. A.
Wenn $n > 1$, so gibt es unter den ganzen Zahlen von 1 bis $n$ nicht zwei Werte von $x$ und $y$, für welche, wenn $z$ einen ganzen Wert bezeichnet, $x^n + y^n = z^n$ ist. *Archiv Math. Phys.*, **27**, 1856, 119–120.

1857   Kummer, E. E.
Einige Sätze über die aus den Wurzeln der Gleichung $\alpha^{\lambda} = 1$ gebildeten complexen Zahlen, für den Fall dass die Klassenzahl durch $\lambda$ theilbar ist, nebst Anwendungen derselben auf einen weiteren Beweis des letztes Fermatschen Lehrsatzes. *Math. Abhandl. d. Königl. Akad. d. Wiss.*, *Berlin*, 1857, 41–74.

1874   Kummer, E. E.
Über diejenigen Primzahlen $\lambda$ für welche die Klassenzahl der aus $\lambda$-ten Einheitswurzeln gebildeten complexen Zahlen durch $\lambda$ theilbar ist. Monatsber. *Königl. Preuss. Akad. d. Wiss. zu Berlin*, 1874, 239–248.

1897   Hilbert, D.
Die Theorie der algebraischen Zahlkörper. *Jahresbericht der Deutschen Mathematikervereinigung*, **4**, 1897, 175–546. Also in *Gesammelte Abhandlungen*, vol. I, Springer-Verlag, Berlin, 1932. Reprinted by Chelsea Publ. Co., New York, 1965.

1905   Mirimanoff, D.
L'équation indéterminée $x^l + y^l + z^l = 0$ et le critérium de Kummer. *J. reine u. angew. Math.*, **128**, 1905, 45–68.

---

\* The papers of Kummer are now easily accessible in volume I of his *Collected Papers*, edited by A. Weil, Springer-Verlag, Berlin, 1975.

†1905   Sauer, R.
Eine polynomische Verallgemeinerung des Fermatschen Satzes. Dissertation,
Giessen, 1905.

1909   Mirimanoff, D.
Sur le dernier théorème de Fermat et le critérium de M. A. Wieferich. Enseignement
Math., 1909, 455–459.

1909   Wieferich, A.
Zum letzten Fermat'schen Theorem. *J. reine u. angew. Math.*, **136**, 1909, 293–302.

1910   Bachmann, P.
*Niedere Zahlentheorie.* Teubner, Leipzig, 1910. Reprinted by Chelsea Publ. Co.,
New York, 1966.

1910   Mirimanoff, D.
Sur le dernier théorème de Fermat. *C. R. Acad. Sci. Paris*, **150**, 1910, 204–206.

1913   Carmichael, R. D.
Note on Fermat's last theorem. *Bull. Amer. Math. Soc.*, **19**, 1913, 233–236.

1913   Landau, E.
Réponse à une question de E. Dubouis. *L' Interm. des Math.*, **20**, 1913, 180.

1913   Meissner, W.
Über die Teilbarkeit von $2^p - 2$ durch das Quadrat der Primzahl $p = 1093$. Sitzungsberichte Akad. d. Wiss., Berlin, 1913, 663–667.

1914   Frobenius, G.
Über den Fermatschen Satz, III. Sitzungsberichte d. Berliner Akad. d. Wiss.,
1914, 129–131. Also in *Collected Works*, vol. 3, Springer-Verlag, Berlin, 1968,
648–676.

1914   Vandiver, H. S.
Extension of the criteria of Wieferich and Mirimanoff in connection with Fermat's
last theorem. *J. reine u. angew. Math.*, **144**, 1914, 314–318.

1914   Vandiver, H. S.
A note on Fermat's last theorem. *Trans. Amer. Math. Soc.*, **15**, 1914, 202–204.

1915   Jensen, K. L.
Om talteoretiske Egenskaber ved de Bernoulliske tal. *Nyt Tidsskrift f. Math.*, **26**,
B, 1915, 73–83.

1917   Pollaczek, F.
Über den grossen Fermat'schen Satz. Sitzungsber. Akad. d. Wiss. Wien, 126,
Abt. IIa, 1917, 45–59.

1919   Bachmann, P.
*Das Fermatproblem in seiner bisherigen Entwicklung*, Walter de Gruyter, Berlin,
1919.

1922   Beeger, N. G. W. H.
On a new case of the congruence $2^{p-1} \equiv 1(p^2)$. *Messenger of Mathematics*, **51**,
1922, 149–150.

1924   Pollaczek, F
Über die irregulären Kreiskörper der $l$-ten und $l^2$-ten Einheitswürzeln. *Math. Z.*,
**21**, 1924, 1–37.

1929   Vandiver, H. S.
On Fermat's last theorem. *Trans. Amer. Math. Soc.*, **31**, 1929, 613–642.

1930   Stafford, E. and Vandiver, H. S.
Determination of some properly irregular cyclotomic fields. *Proc. Nat. Acad. Sci.
U.S.A.*, **16**, 1930, 139–170.

1931   Morishima, T.
Über den Fermatschen Quotienten. *Jpn. J. Math.*, **8**, 1931, 159–173.

1931   Spunar, V. M.
On Fermat's last theorem, III. *J. Wash. Acad. Sci.*, **21**, 1931, 21–23.

1932   Mileikowsky, E. N.
Elementarer Beitrag zur Fermatschen Vermutung. *J. reine u. angew. Math.*, **166**, 1932, 116–117.

1933   Morishima, T.
Über die Einheiten und Idealklassen des Galoisschen Zahlkörpers und die Theorie des Kreiskörpers der $l^2$-ten Einheitswurzeln. *Jpn. J. of Math.*, **10**, 1933, 83–126.

1934   Krasner, M.
Sur le premier cas du théorème de Fermat. *C.R. Acad. Sci. Paris*, **199**, 1934, 256–258.

1934   Vandiver, H. S.
Fermat's last theorem and the second factor in the cyclotomic class number. *Bull. Amer. Math. Soc.* **40**, 1934, 118–126.

1938   Gottschalk, E.
Zum Fermatschen Problem. *Math. Annalen*, **115**, 1934, 157–158.

1939   Vandiver, H. S.
On basis systems for groups of ideal classes in a properly irregular cyclotomic field. *Proc. Nat. Acad. Sci. U.S.A.*, **25**, 1939, 586–591.

1939   Vandiver, H. S.
On the composition of the group of ideal classes in a properly irregular cyclotomic field. *Monatshefte f. Math. u. Phys.*, **48**, 1939, 369–380.

1940   Rosser, B.
A new lower bound for the exponent in the first case of Fermat's last theorem. *Bull. Amer. Math. Soc.* **46**, 1940, 299–304.

1941   Lehmer, D. H. and Lehmer, E.
On the first case of Fermat's last theorem. *Bull. Amer. Math. Soc.* **47**, 1941, 139–142.

1941   Rosser, B.
An additional criterion for the first case of Fermat's last theorem. *Bull. Amer. Math. Soc.*, **47**, 1941, 109–110.

1946   Inkeri, K.
Untersuchungen über die Fermatsche Vermutung. *Annales Acad. Sci. Fennicae*, A, I, No. 33, 1946, 1–60.

1948   Gunderson, N. G.
Derivation of Criteria for the First Case of Fermat's Last Theorem and the Combination of these Criteria to Produce a New Lower Bound for the Exponent. Thesis, Cornell University, 1948.

1951   Ankeny, N. C. and Chowla, S.
The class number of the cyclotomic field. *Can. J. Math.*, **3**, 1951, 486–494.

1953   Inkeri, K.
Abschätzungen für eventuelle Lösungen der Gleichung im Fermatschen Problem. *Ann. Univ. Turku*, A, I, 1953, No. 1, 3–9.

1954   Dénes, P.
Über irreguläre Kreiskörper. *Publ. Math. Debrecen*, **3**, 1954, 17–23.

1954   Lehmer, D. H., Lehmer, E., and Vandiver, H. S.
An application of high speed computing to Fermat's last theorem. *Proc. Nat. Acad. Sci. U.S.A.*, **40**, 1954, 25–33.

1954  Vandiver, H. S.
Examination of methods of attack of the second case of Fermat's last theorem. *Proc. Nat. Acad. Sci. U.S.A.*, **40**, 1954, 732–735.

1955  Möller, K.
Untere Schränke fur die Anzahl der Primzahlen, aus denen $x$, $y$, $z$ des Fermatschen Gleichung $x^n + y^n = z^n$ bestehen muss. *Math. Nachr.*, **14**, 1955, 25–28.

1958  Pérez-Cacho, L.
On some questions in the theory of numbers (in Spanish). *Rev. Mat. Hisp. Amer.*, (4), **18**, 1958, 10–27 and 113–124.

1959  Landau, E.
*Diophantische Gleichungen mit endlich vielen Lösungen* (new edition by A. Walfisz), V. E. B. Deutscher Verlag d. Wiss., Berlin, 1959.

1960  Long, L.
A note on Fermat's theorem. *Math. Gaz.*, **44**, 1960, 261–262.

1964  Inkeri, K. and Hyyrö, S.
Über die Anzahl der Lösungen einiger Diophantischer Gleichungen. *Annales Univ. Turku*, A, **I**, 1964, No. 78, 3–10.

1964  Siegel, C. L.
Zu zwei Bemerkungen Kummers. Nachr. Akad. Wiss. Göttingen, Math. Phys. K1. II, 1964, 51–57. *Gesammelte Abhandlugen*, vol. III, Springer-Verlag, New York, 1966, 436–442.

1964  Sierpiński, W.
*A Selection of Problems in the Theory of Numbers*, Macmillan, New York, 1964.

1965  Ankeny, N. C., Chowla, S. and Hasse, H.
On the class number of the maximal real subfield of a cyclotomic field. *J. reine u. angew. Math.*, **217**, 1965, 217–220.

1965  Eichler, M.
Eine Bemerkung zur Fermatschen Vermutung. *Acta Arithm.*, **11**, 1965, 129–131; Errata, 261.

1965  Montgomery, H. L.
Distribution of irregular primes. *Illinois J. Math.*, **9**, 1965, 553–558.

1965  Rotkiewicz, A.
Sur les nombres de Mersenne dépourvus de diviseurs carrés et sur les nombres naturels $n$ tels que $n^2 | 2^n - 2$. *Matematicky Vesnik*, **2**, (17), 1965, 78–80.

1966  Borevich, Z. I. and Shafarevich, I. R.
*Number Theory*, Academic Press, New York, 1966.

1968  Puccioni, S.
Un teorema per una resoluzioni parziali del famoso problema di Fermat. *Archimede*, **20**, 1968, 219–220.

1971  Brillhart, J., Tonascia, J. and Weinberger, P.
On the Fermat quotient, in *Computers in Number Theory*, Academic Press, New York, 1971, 213–222.

1971  Metsänkylä, T.
Note on the distribution of irregular primes. *Ann. Acad. Sci. Fenn.*, *A*, **I**, No. **492**, 1971, 7 pages.

1971  Tuckerman, B.
The 24th Mersenne prime. *Proc. Nat. Acad. Sci. U.S.A.*, **68**, 1971, 2319–2320.

1972   Brückner, H.
Zum Beweis des ersten Falles der Fermatschen Vermutung für pseudoreguläre
Primzahlen (Bemerkungen zur vorstehender Arbeit von L. Skula). *J. reine u. angew.
Math.*, **253**, 1972, 15–18.

1972   Skula, L.
Eine Bemerkung zu dem ersten Fall der Fermatschen Vermutung. *J. reine u.
angew. Math.*, **253**, 1972, 1–14.

1974   Lepistö, T.
On the growth of the first factor of the class number of the prime cyclotomic field.
*Annales Acad. Sci. Fennicae*, A, I, 1974, No. 577, 19 pages.

1975   Brückner, H.
Zum Ersten Fall der Fermatschen Vermutung. *J. reine u. angew Math.*, **274/5**,
1975, 21–26.

1975   Wagstaff, S. S.
Fermat's last theorem is true for any exponent less than 100000. *Notices Amer.
Math. Soc.*, **23**, 1975, A-53, abstract 731-10-35.

1975   Yokoi, H.
On the distribution of irregular primes. *J. Number Theory* **7**, 1975, 71–76.

1976   Inkeri, K.
A note on Fermat's conjecture. *Acta Arithm.*, **29**, 1976, 251–256.

1976   Masley, J. M. and Montgomery, H. L.
Cyclotomic fields with unique factorization. *J. reine. u. angew. Math.*, **286/7**, 1976,
248–256.

1976   Metsänkylä, T.
Distribution of irregular prime numbers. *J. reine. u. angew. Math.*, **282**, 1976,
126–130.

1976   Pajunen, S.
Computation of the growth of the first factor for prime cyclotomic fields. *BIT*,
**16**, 1976, 85–87.

1976   Tijdeman, R.
On the equation of Catalan. *Acta Arithm.*, **29**, 1976, 197–209.

1976   Wagstaff, S. S.
Fermat's last theorem is true for any exponent less than 125000. Communicated
by letter. See also: The irregular primes to 125000, *Math. Comp.*, **32**, 1978, 583–
591.

1977   Terjanian, G.
Sur l'équation $x^{2p} + y^{2p} = z^{2p}$. *C. R. Acad. Sci. Paris*, **285**, 1977, 973–975.

1978   Nickel, L. and Noll, C.
*Los Angeles Times*, November 16, 1978, part II, page 1. See also: Le dernier
premier, *Gazette Sci. Math. Québec*, **3**, 1979, 27.

1979   Brückner, H.
*Explizit Reziprozitatsgesetz und Anwendungen. Univ. Essen.*, 1979.

1979   Lehmer, D. H.
Letter to the author (Sept. 2, 1979), Berkeley, Cal.

1994   Hellegouarch, Y.
Vers une arithmétique nouvelle. *Revue Math. Spéc.*, **104**, no. 6, 1994.

1994   Hellegouarch, Y.
Fermat presque vain cu. Actes du Colloque *La Mémoire des Nombres*, Cherbourg,
1994 (to appear).

1995   Rubin, K. and Silverberg, A.
Report on Wiles' Cambridge lecture. *Bull Amer. Math. Soc.*, **31**, 1995, 15–38.

# LECTURE III

# B.K. = Before Kummer

In this lecture, I wish to report various early attempts to solve Fermat's problem. I begin by considering the case of exponent 2, which is much earlier than Fermat's time. As Zassenhaus kindly pointed out to me, 2 is the oddest of the primes. Among its special properties, this oddest of all the primes is even; it is also the only exponent for which it is known that the Fermat equation has a nontrivial solution.

Then, I will give Fermat's famous proof by infinite descent for the case of fourth powers. After that, I will present Euler's and Gauss's proof for the case of cubes, as well as sketch proofs for other exponents.

Other attempts were not restricted to specific exponents. Noteworthy—though not successful—were the contributions of Cauchy and Lamé. Barlow and Abel found interesting relations which must be satisfied by any possible solution of Fermat's equation. Sophie Germain proved a clever and beautiful theorem for the first case.

Since I have already given some of the early history of the problem in Lecture I, I will now limit myself to the technical details of the theorems.

All these methods, devised before Kummer, have a certain naiveté in common. Normally, they use only properties of the rational numbers. However elementary they may have been, they didn't lack ingenuity. On the contrary, they were often very tricky.

Before entering into the details, I wish once more to recall:

The *first case* of Fermat's theorem holds for the prime exponent $p > 2$ when there do not exist integers $x$, $y$, $z$, such that $p$ does not divide $xyz$ and $x^p + y^p + z^p = 0$.

The *second case* holds for $p > 2$ when there do not exist relatively prime integers $x$, $y$, $z$, such that $x^p + y^p + z^p = 0$ and $p$ divides $x$, $y$, or $z$.

In order to prove Fermat's theorem for all exponents $n \geq 3$, it is sufficient to prove it for the exponent 4 and for all prime exponents $p \geq 3$. Because of this, I will mainly be interested in prime exponents.

Another obvious though still general remark, which I will not repeat is the following. If there exists a solution for Fermat's equation,

$$x^n + y^n + z^n = 0,$$

we may assume without loss of generality that $\gcd(x,y,z) = 1$ and thus $x$, $y$, $z$ must be pairwise relatively prime.

# 1. The Pythagorean Equation

If $x$, $y$, $z$ are nonzero integers satisfying

$$X^2 + Y^2 = Z^2, \tag{1.1}$$

then so are $|x|$, $|y|$, $|z|$. These numbers are the lengths of the sides of a right-angled triangle.

To determine all nontrivial integer solutions of (1.1) it suffices to determine the so-called *primitive* (Pythagorean) *triples* $(x,y,z)$:

$$x, \; y, \; z > 0, \qquad \gcd(x,y,z) = 1, \qquad x \text{ even.}$$

All the other solutions are obtained by changing signs, permuting $x$, $y$, and by multiplication with some nonzero integer.

The following theorem gives a complete description of all the primitive triples:

**(1A)** *If $a$, $b$ are integers, $a > b > 0$, $\gcd(a,b) = 1$, $a$, $b$ not of the same parity, let*

$$
\begin{aligned}
x &= 2ab, \\
y &= a^2 - b^2, \\
z &= a^2 + b^2.
\end{aligned}
\tag{1.2}
$$

*Then $(x,y,z)$ is a primitive triple. And conversely, every primitive triple may be so obtained.*

Distinct pairs $(a,b)$ give rise to distinct primitive triples and these may all be obtained from some pair $(a,b)$. For example, the smallest primitive triples, ordered according to increasing values of $z$ are as follows:

$$(4,3,5), \quad (12,5,13), \quad (8,15,17),$$
$$(24,7,25), \quad (20,21,29), \quad (12,35,37).$$

Listing of the primitive triples amounts to determining the representations of odd positive integers as sums of two squares. Fermat proved, in this context, the well-known theorem:

**(1B)** $n > 0$ *is a sum of two squares of integers if and only if every prime factor* $p$ *of* $n$*, such that* $p \equiv 3 \pmod 4$*, appears to an even power in the factorization of* $n$ *into prime factors.*

There remains then the question of finding the number of representations as sums of two squares.

Let $r(n)$ denote the number of pairs $(a,b)$ of integers (*not* necessarily positive) such that $n = a^2 + b^2$. For example, $r(1) = 4$ and $r(5) = 8$. The determination of $r(n)$ was done by Jacobi and by Gauss, independently:

**(1C)** $r(n) = 4(d_1(n) - d_3(n))$,

*where*

$$d_1(n) = \#\{d \mid 1 \le d, d \mid n, d \equiv 1 \pmod 4\},$$
$$d_3(n) = \#\{d \mid 1 \le d, d \mid n, d \equiv 3 \pmod 4\}.$$

With this information, it is possible to determine all the primitive Pythagorean triples. Clear proofs of these beautiful theorems may be found in the book of Hardy and Wright.

# 2. The Biquadratic Equation

Here is Fermat's proof using the method of *infinite descent*. The idea is the following: assume that $(x_0, y_0, z_0)$ is one solution in nonzero integers, then there is another solution $(x_1, y_1, z_1)$ of the same kind, with $0 < |x_1| < |x_0|$. Since this procedure may be repeated indefinitely, one would obtain an infinite decreasing sequence of positive integers $|x_0| > |x_1| > |x_2| > \cdots > 0$— which is absurd. So, there couldn't be any solution in nonzero integers.

**(2A)** *The equation*

$$X^4 + Y^4 = Z^2 \tag{2.1}$$

*has no solution in nonzero integers. In particular, the same is true for the equation*

$$X^4 + Y^4 = Z^4. \tag{2.2}$$

PROOF. Let $(x,y,z)$ be a triple of positive integers satisfying (2.1). It is easy to see that we may assume, without loss of generality that $\gcd(x,y,z) = 1$. We may also assume that $x$ is even. Then $(x^2, y^2, z)$ is a primitive Pythagorean triple: $x^4 + y^4 = z^2$.

By (1A) there exist integers $a$, $b$ such that $a > b > 0$, $\gcd(a,b) = 1$, $a$, $b$ have different parity, and

$$x^2 = 2ab,$$
$$y^2 = a^2 - b^2, \qquad (2.3)$$
$$z = a^2 + b^2.$$

As easily seen, $b$ must be even.

Since $b^2 + y^2 = a^2$ and $\gcd(b,y,a) = 1$, by (1A) there exist integers $c$, $d$ such that $c > d > 0$, $\gcd(c,d) = 1$, $c$, $d$ of different parity, and

$$b = 2cd,$$
$$y = c^2 - d^2, \qquad (2.4)$$
$$a = c^2 + d^2.$$

Hence

$$x^2 = 2ab = 4cd(c^2 + d^2). \qquad (2.5)$$

Since $c$, $d$, $c^2 + d^2$ are pairwise relatively prime, from (2.5) using the uniqueness of factorization into primes, we conclude that $c$, $d$, $c^2 + d^2$ are squares of positive integers:

$$c = e^2,$$
$$d = f^2, \qquad (2.6)$$
$$c^2 + d^2 = g^2.$$

Hence

$$e^4 + f^4 = g^2, \qquad (2.7)$$

that is, the triple $(e, f, g)$ is a solution of (2.1).

But $z = a^2 + b^2 = (c^2 + d^2)^2 + 4c^2d^2 > g^4 > g > 0$. By infinite descent, this leads to a contradiction.  □

Various equations of 4th degree, similar to (2.1) may be treated with this methods. I quote the following theorems:

**(2B)** *The following equations have no solution in nonzero integers*:

$$X^4 - Y^4 = \pm Z^2, \qquad (2.8)$$
$$X^4 + 4Y^4 = Z^2, \qquad (Euler) \qquad (2.9)$$
$$X^4 - 4Y^4 = \pm Z^2. \qquad (2.10)$$

Also Legendre proved:

**(2C)** *If* $x$, $y$, $z$ *are nonzero integers*:

$$\text{If } x^4 + y^4 = 2z^2, \text{ then } x^2 = y^2 \text{ and } z^2 = x^4. \qquad (2.11)$$
$$\text{If } 2x^4 + 2y^4 = z^2, \text{ then } x^2 = y^2, \text{ and } z^2 = 4x^4. \qquad (2.12)$$

# 3. The Cubic Equation

In my first lecture, I mentioned that the first published proof of Fermat's theorem for the case of cubes is due to Euler. It appears in Euler's *Algebra* published in St. Petersburg in 1770. This book was posthumously translated into German in 1802 and into English in 1822. An important step in Euler's proof, which used divisibility properties of integers of the form $a^2 + 3b^2$, was done without sufficient justification. Legendre, who reproduced Euler's proof in his book (1830) did not give any further explanations. Since he was himself also an expert on such matters, he had certainly understood Euler's reasoning. However, later mathematicians were less comfortable about the possible gap. In 1894, Schumacher pointed it out explicitly. The gap was again the object of comments by Landau (1901), a paper by Holden (1906) and a note by Welsch (1910). Quite recently in 1966, Bergmann published a thorough analysis of Euler's proof, with historical considerations, throwing more light on this controversy. Indeed, already in 1760, in his paper Supplementum quorundam ... supponuntur, published in *Novi commentarii academiae scientiarum Petropolitanae* (also in *Opera Omnia*, series prima, II, pages 556–575), Euler rigorously proved that if $s$ is odd and $s^3 = a^2 + 3b^2$, with $\gcd(a,b) = 1$, then $s = u^2 + 3v^2$, with $u$, $v$ integers.

Another proof of Fermat's theorem for cubes was given by Gauss, and published posthumously. Both proofs use the method of infinite descent. However, while Euler worked with integers of the form $a^2 + 3b^2$, Gauss used complex algebraic numbers of the form $a + b\sqrt{-3}$.

**(3A)** *The equation*

$$X^3 + Y^3 + Z^3 = 0 \tag{3.1}$$

*has no solution in nonzero integers.*

PROOF. Suppose that $x$, $y$, $z$ are pairwise relatively prime integers such that $x^3 + y^3 + z^3 = 0$, with $x$, $y$ odd, $z$ even and $|z|$ is the smallest possible. This may be assumed without loss of generality. Then

$$\begin{aligned} x + y &= 2a, \\ x - y &= 2b, \end{aligned} \tag{3.2}$$

where $a$, $b$ are relatively prime nonzero integers of different parity. Therefore

$$-z^3 = x^3 + y^3 = (a + b)^3 + (a - b)^3 = 2a(a^2 + 3b^2). \tag{3.3}$$

It follows easily that $a^2 + 3b^2$ is odd, 8 divides $2a$, $b$ is odd and $\gcd(2a, a^2 + 3b^2) = 1$ or 3.

*Case 1.* $\gcd(2a, a^2 + 3b^2) = 1$.

From (3.3), $2a$ and $a^2 + 3b^2$ are cubes:

$$\begin{aligned} 2a &= r^3, \\ a^2 + 3b^2 &= s^3, \end{aligned} \tag{3.4}$$

where $s$ is odd. At this point, according to Euler, it is possible to write $s = u^2 + 3v^2$, with integers $u$, $v$ such that

$$
\begin{aligned}
a &= u(u^2 - 9v^2), \\
b &= 3v(u^2 - v^2).
\end{aligned}
\tag{3.5}
$$

Then $v$ is odd, $u \neq 0$, $u$ is even, $3 \nmid u$, $\gcd(u,v) = 1$, and

$$
r^3 = 2a = 2u(u - 3v)(u + 3v).
\tag{3.6}
$$

Note that $2u$, $u - 3v$, $u + 3v$ have to be pairwise relatively prime, so they are cubes of integers:

$$
\begin{aligned}
2u &= -l^3, \\
u - 3v &= m^3, \\
u + 3v &= n^3,
\end{aligned}
\tag{3.7}
$$

with $l$, $m$, $n$ not equal to zero (since 3 does not divide $u$). Thus

$$
l^3 + m^3 + n^3 = 0
\tag{3.8}
$$

with $l$ even. Moreover, since $b \neq 0$, $3 \nmid u$:

$$
|z^3| = |2a(a^2 + 3b^2)| = |l^3(u^2 - 9v^2)(a^2 + 3b^2)| \geq 3|l^3| > |l^3|.
$$

This contradicts the minimality of $|z|$.

*Case 2.* $\gcd(2a, a^2 + 3b^2) = 3$.

Let $a = 3c$, then $4 \mid c$, $3 \nmid b$ and

$$
-z^3 = 6c(9c^2 + 3b^2) = 18c(3c^2 + b^2),
\tag{3.9}
$$

where $18c$, $3c^2 + b^2$ are relatively prime, $3c^2 + b^2$ is odd, and not a multiple of 3.

Then $18c$, $3c^2 + b^2$ are cubes of integers:

$$
\begin{aligned}
18c &= r^3, \\
3c^2 + b^2 &= s^3,
\end{aligned}
\tag{3.10}
$$

with $s$ odd. By the same step as in Case 1, $s = u^2 + 3v^2$, with integers $u$, $v$ such that

$$
\begin{aligned}
b &= u(u^2 - 9v^2), \\
c &= 3v(u^2 - v^2).
\end{aligned}
\tag{3.11}
$$

Thus $u$ is odd, $v$ is even, $v \neq 0$, $\gcd(u,v) = 1$, and $2v$, $u + v$, $u - v$ are pairwise relatively prime. From

$$
\left(\frac{r}{3}\right)^3 = 2v(u + v)(u - v)
\tag{3.12}
$$

it follows that

$$
\begin{aligned}
2v &= -l^3, \\
u + v &= m^3, \\
u - v &= -n^3.
\end{aligned}
\tag{3.13}
$$

Hence $l^3 + m^3 + n^3 = 0$ with $l$, $m$, $n$ nonzero integers, $l$ even. Finally

$$|z|^3 = 18|c|(3c^2 + b^2) = 54|v(u^2 - v^2)|(3c^2 + b^2)$$
$$= 27|l|^3|u^2 - v^2|(3c^2 + b^2) \geq 27|l|^3 > |l|^3.$$

This contradicts the choice of $|z|$ minimal. $\square$

To justify the step concerning the expression of $s$ as $s = u^2 + 3v^2$, it is necessary to study more deeply the quadratic form $x^2 + 3y^2$. This is elementary but a bit too long to explain in detail. Therefore I will only summarize the main points; details were given, for example, by Carmichael in 1915.

Let $S = \{a^2 + 3b^2 \,|\, a, b \text{ integers}\}$. The following is a list of various properties of $S$:

(3B) *S is closed under multiplication:*

$$(a^2 + 3b^2)(c^2 + 3d^2) = (ac + 3bd)^2 + 3(ad - bc)^2$$
$$= (ac - 3bd)^2 + 3(ad + bc)^2. \qquad (3.14)$$

(3C) *Let $p$ be a prime and $n \geq 1$. If $p \in S$ and $pn \in S$, then $n \in S$.*

(3D) *A prime $p$ belongs to $S$ if and only if $p = 3$ or $p \equiv 1 \pmod 6$.*

(3E) *If $m = a^2 + 3b^2 \in S$, with $\gcd(a,b) = 1$ and if $n \,|\, m$, then $n \in S$.*

This gives a complete description of $S$:

(3F) *$m$ is in $S$ if and only if the following condition is satisfied: if $p$ is a prime, $p \equiv -1 \pmod 6$, or $p = 2$, then the exact power of $p$ dividing $m$ is even.*

Concerning the representations of integers $m \in S$, the first result is the following:

(3G) *If $p$ is a prime number, $p = a^2 + 3b^2 = c^2 + 3d^2$, then $a^2 = c^2$, $b^2 = d^2$.*

(3H) *Let $m_1$, $m_2$ be products of primes belonging to $S$, let $m = m_1 m_2$. For each representation $m = u^2 + 3v^2$ there are representations*

$$m_1 = a^2 + 3b^2,$$
$$m_2 = c^2 + 3d^2, \qquad (3.15)$$

*such that*

$$u = ac - 3bd,$$
$$v = ad + bc. \qquad (3.16)$$

And finally:

**(3I)** *Let s be an odd integer, such that* $s^3 = u^2 + 3v^2$, *with nonzero integers* $u$, $v$. *Then* $s = t^2 + 3w^2$, *with t, w integers and*

$$u = t(t^2 - 9w^2),$$
$$v = 3w(t^2 - w^2).$$
(3.17)

The above property was the one needed by Euler.

Now I turn to Gauss's proof. As a matter of fact, as I hope to explain, it does more than prove Fermat's last theorem for exponent 3.

Gauss worked with complex numbers of the form $a + b\zeta$, where $a$, $b$ are integers, and $\zeta = (-1 + \sqrt{-3})/2$ is a primitive cubic root of 1. Let $A$ be the set of all such numbers. Using modern language, $A$ is a ring, namely the ring of algebraic integers of the imaginary quadratic field $K = \mathbb{Q}(\sqrt{-3})$. Every (nonzero) element in $A$ whose inverse is again in $A$ is called a *unit* of $A$. The units of $A$ are known to be: $1, -1, \zeta, -\zeta, \zeta^2, -\zeta^2$; they are all roots of unity.

If $\alpha$, $\beta \in A$, $\alpha$ *divides* $\beta$ when there exists $\gamma \in A$ such that $\alpha\gamma = \beta$; this is written $\alpha \mid \beta$. If $\alpha \mid \beta$ and $\beta \mid \alpha$ then $\alpha$ and $\beta$ are *associated*; $\alpha \sim \beta$. This happens only when $\alpha = \beta\gamma$ where $\gamma$ is a unit.

$\alpha \in A$ is a *prime* element if the only elements of $A$ dividing $\alpha$ are either associated with $\alpha$ or units. There are "enough" prime elements, so that every $\alpha \in A$ may be written as a product of a unit and powers of prime elements. For the rings in question, such a factorization is essentially unique: if $\alpha = \omega\pi_1^{e_1} \cdots \pi_s^{e_s} = \omega'\pi_1'^{e_1'} \cdots \pi_{s'}'^{e_{s'}'}$ with $\omega$, $\omega'$ units, $s \geq 0$, $s' \geq 0$, $\pi_i$ distinct prime elements, $e_i \geq 1$, $\pi_j'$ distinct prime elements, $e_j' \geq 1$, then $s = s'$, and up to a permutation of indices $\pi_1 \sim \pi_1', \ldots, \pi_s \sim \pi_s'$, $e_1 = e_1', \ldots, e_s = e_s'$.

Since there is unique factorization into primes, it is possible to define the greatest common divisor of elements of $A$—they are unique, up to multiplication by units. Elements of $A$ are *relatively prime* if their greatest common divisors are units.

An element which plays an important role is $\lambda = 1 - \zeta = (3 - \sqrt{-3})/2$. $\lambda$ is a prime element and $3 \sim \lambda^2$.

If $\alpha$, $\beta$, $\gamma \in A$, then $\alpha \equiv \beta \pmod{\gamma}$ means that $\gamma \mid \alpha - \beta$. This congruence relation satisfies properties analogous to ordinary congruence.

I note that since $3 \sim \lambda^2$, if $\alpha \equiv \beta \pmod{\lambda}$, then $\alpha^3 \equiv \beta^3 \pmod{\lambda^3}$.

There are exactly three congruence classes modulo $\lambda$, namely the classes of 0, 1 and $-1$.

The following congruence was needed in Gauss's proof:

**Lemma.** *If* $\alpha \in A$ *and* $\lambda \nmid \alpha$, *then* $\alpha^3 \equiv \pm 1 \pmod{\lambda^4}$.

The proof is straightforward. And now, Gauss's theorem:

**(3J)** *The equation*

$$X^3 + Y^3 + Z^3 = 0$$
(3.18)

*has no solutions in algebraic integers* $\alpha$, $\beta$, $\gamma \in A$ *all different from 0.*

PROOF. Suppose the theorem is false. Dividing by a greatest common divisor, there is no loss of generality in assuming that there exists $\alpha, \beta, \gamma \in A$, relatively prime, such that $\alpha^3 + \beta^3 + \gamma^3 = 0$. It follows that $\alpha$, $\beta$, $\gamma$ are also pairwise relatively prime. Hence it may be assumed that $\lambda \nmid \alpha, \lambda \nmid \beta$.

*Case 1. $\lambda \nmid \gamma$.*

The congruence classes of $\alpha$, $\beta$, $\gamma$ are therefore those of 1 or $-1$. So $\alpha \equiv \pm 1$ (mod $\lambda$), hence $\alpha^3 \equiv \pm 1$ (mod $\lambda^3$). Similarly, $\beta^3 \equiv \pm 1$ (mod $\lambda^3$), $\gamma^3 \equiv \pm 1$ (mod $\lambda^3$). So

$$0 = \alpha^3 + \beta^3 + \gamma^3 \equiv \pm 1 \pm 1 \pm 1 \ (\text{mod } \lambda^3).$$

The combinations of signs give $\pm 1$ or $\pm 3$. Clearly $0 \not\equiv \pm 1$ (mod $\lambda^3$); also, if $0 \equiv \pm 3$ (mod $\lambda^3$), then $\lambda^3 | \pm 3 \sim \pm \lambda^2$ hence $\lambda | \pm 1$ and $\lambda$ would be a unit, a contradiction.

*Case 2. $\lambda | \gamma$.*
Let $\gamma = \lambda^n \delta$ with $n \geq 1$, $\lambda \nmid \delta$, $\delta \in A$ hence

$$\alpha^3 + \beta^3 + \lambda^{3n}\delta^3 = 0 \qquad (3.19)$$

with $\alpha, \beta, \delta \in A$, $n \geq 1$.
So the following property $(P_n)$ is satisfied.
$(P_n)$: *There exist $\alpha, \beta, \delta \in A$ such that $\lambda \nmid \alpha$, $\lambda \nmid \beta$, $\lambda \nmid \delta$, $\alpha$, $\beta$ are relatively prime and $\alpha$, $\beta$, $\delta$ are a solution of an equation of the form*

$$X^3 + Y^3 + \omega\lambda^{3n}Z^3 = 0, \qquad (3.20)$$

*where $\omega$ is a unit* (in (3.19), $\omega = 1$).
The idea of the proof is the following: to show that if $(P_n)$ is satisfied, then $n \geq 2$ and $(P_{n-1})$ is also satisfied. Repeating this procedure, eventually $(P_1)$ would be satisfied, which is a contradiction. This is nothing but a form of infinite descent (on the exponent $n$).
So there remain two steps in the proof.

*Step 1.* If $(P_n)$ is satisfied, then $n \geq 2$.
Since $\lambda \nmid \alpha$, $\lambda \nmid \beta$ by the lemma $\alpha^3 \equiv \pm 1$ (mod $\lambda^4$), $\beta^3 \equiv \pm 1$ (mod $\lambda^4$) and $\pm 1 \pm 1 \equiv -\omega\lambda^{3n}\delta^3$ (mod $\lambda^4$) with $\lambda \nmid \delta$. The left-hand side must be 0, since $\lambda \nmid \pm 2$, hence $3n \geq 4$ and $n \geq 2$.

*Step 2.* If $(P_n)$ is satisfied, then $(P_{n-1})$ is also satisfied.
By hypothesis:

$$-\omega\lambda^{3n}\delta^3 = \alpha^3 + \beta^3 = (\alpha + \beta)(\alpha + \zeta\beta)(\alpha + \zeta^2\beta). \qquad (3.21)$$

The prime element $\lambda$ must divide one of the factors in the right-hand side. Now $\alpha + \beta \equiv \alpha + \zeta\beta \equiv \alpha + \zeta^2\beta$ (mod $\lambda$) because $1 \equiv \zeta \equiv \zeta^2$ (mod $\lambda$), so $\lambda$ divides each factor. Hence

$$\frac{\alpha + \beta}{\lambda}, \frac{\alpha + \zeta\beta}{\lambda}, \frac{\alpha + \zeta^2\beta}{\lambda} \in A$$

and

$$-\omega\lambda^{3(n-1)}\delta^3 = \frac{\alpha+\beta}{\lambda}\cdot\frac{\alpha+\zeta\beta}{\lambda}\cdot\frac{\alpha+\zeta^2\beta}{\lambda}. \tag{3.22}$$

From $n \geq 2$ (by the first step) $\lambda$ divides one of the factors in the right-hand side. It is easily seen that $\alpha+\beta$, $\alpha+\zeta\beta$, $\alpha+\zeta^2\beta$ are pairwise incongruent modulo $\lambda^2$. Hence $\lambda$ divides only one of the factors of right-hand side of (3.22). For example $\lambda$ divides $(\alpha+\beta)/\lambda$ (the other cases are analogous, replacing $\beta$ by $\zeta\beta$ or $\zeta^2\beta$, which is permissible).

So $\lambda^{3(n-1)}$ divides $(\alpha+\beta)/\lambda$. Hence

$$\begin{aligned}
\alpha+\beta &= \lambda^{3n-2}\kappa_1,\\
\alpha+\zeta\beta &= \lambda\kappa_2,\\
\alpha+\zeta^2\beta &= \lambda\kappa_3,
\end{aligned} \tag{3.23}$$

with $\kappa_1, \kappa_2, \kappa_3 \in A$ and $\lambda$ does not divide $\kappa_1, \kappa_2, \kappa_3$.

Multiplying:

$$-\omega\delta^3 = \kappa_1\kappa_2\kappa_3. \tag{3.24}$$

It is easy to see that $\kappa_1, \kappa_2, \kappa_3$ are pairwise relatively prime. Since the ring $A$ has unique factorization, $\kappa_1, \kappa_2, \kappa_3$ are associated with cubes

$$\begin{aligned}
\kappa_1 &= \eta_1\varphi_1^3,\\
\kappa_2 &= \eta_2\varphi_2^3,\\
\kappa_3 &= \eta_3\varphi_3^3,
\end{aligned} \tag{3.25}$$

where $\eta_i$ are units, $\varphi_i \in A$ $(i=1,2,3)$, $\varphi_1, \varphi_2, \varphi_3$ are pairwise relatively prime, and $\lambda$ does not divide $\varphi_1, \varphi_2, \varphi_3$. Then

$$\begin{aligned}
\alpha+\beta &= \lambda^{3n-2}\eta_1\varphi_1^3,\\
\alpha+\zeta\beta &= \lambda\eta_2\varphi_2^3,\\
\alpha+\zeta^2\beta &= \lambda\eta_3\varphi_3^3.
\end{aligned} \tag{3.26}$$

From $1 + \zeta + \zeta^2 = 0$ it follows that

$$\begin{aligned}
0 &= (\alpha+\beta) + \zeta(\alpha+\zeta\beta) + \zeta^2(\alpha+\zeta^2\beta)\\
&= \lambda^{3n-2}\eta_1\varphi_1^3 + \zeta\lambda\eta_2\varphi_2^3 + \zeta^2\lambda\eta_3\varphi_3^3
\end{aligned}$$

so

$$\varphi_2^3 + \tau\varphi_3^3 + \tau'\lambda^{3(n-1)}\varphi_1^3 = 0, \tag{3.27}$$

where $\tau, \tau'$ are units, $\varphi_1, \varphi_2, \varphi_3 \in A$ are not multiples of $\lambda$, and $\varphi_2, \varphi_3$ are relatively prime.

If $\tau = 1$, then $\varphi_2, \varphi_3, \varphi_1$ is a solution of

$$X^3 + Y^3 + \tau'\lambda^{3(n-1)}Z^3 = 0. \tag{3.28}$$

If $\tau = -1$, then $\varphi_2, -\varphi_3, \varphi_1$ is such a solution.

It remains to show that $\tau \neq \pm\zeta, \pm\zeta^2$. Indeed, since $n \geq 2$,

$$\varphi_2^3 + \tau\varphi_3^3 \equiv 0 \pmod{\lambda^2}. \tag{3.29}$$

But by the lemma

$$\varphi_2^3 \equiv \pm 1 \ (\text{mod } \lambda^4), \qquad \varphi_3^3 \equiv \pm 1 \ (\text{mod } \lambda^4)$$

hence $\pm 1 \pm \tau \equiv 0 \ (\text{mod } \lambda^2)$.

However $\pm 1 \pm \zeta \not\equiv 0 \ (\text{mod } \lambda^2)$ and $\pm 1 \pm \zeta^2 \not\equiv 0 \ (\text{mod } \lambda^2)$. So $\tau \neq \pm\zeta$, $\pm\zeta^2$ and this establishes the propery $(P_{n-1})$, concluding the proof. $\square$

Various other cubic or sextic equations were treated by similar methods:

**(3K)** *The following equations have no solutions in nonzero integers:*

$$X^3 + 4Y^3 = 1, \tag{3.30}$$

$$X^6 - 27Y^6 = 2Z^3, \tag{3.31}$$

$$16X^6 - 27Y^6 = Z^3, \tag{3.32}$$

$$X^3 + Y^3 = 3Z^3. \tag{3.33}$$

# 4. The Quintic Equation

In 1825, Dirichlet read at the Academy of Sciences of Paris a paper where he claimed to have proved Fermat's theorem for the exponent 5. However, he neglected to consider one of the possible cases. In the meantime, Legendre independently found a complete proof, while Dirichlet was settling the remaining case to finish his proof.

Essentially, Dirichlet's proof uses facts about the arithmetic of the field $K = \mathbb{Q}(\sqrt{5})$. It would take too long to describe the proof in detail. It suffices to say that it proceeds by considering two cases separately. The first case is quite easy. The second case was treated by infinite descent.

The following lemma is the basic technical tool in the proof:

**Lemma.** *Let $a$, $b$ be relatively prime nonzero integers of different parity, $5 \nmid a, 5 \mid b$. If*

$$a^2 - 5b^2 = \pm \left( \frac{1 + \sqrt{5}}{2} \right)^e \left( \frac{f + g\sqrt{5}}{2} \right)^5 \tag{4.1}$$

*(with $e \geq 0$, $f$, $g$ integers of the same parity) then there exist relatively prime integers $c$, $d$, of different parity, $5 \nmid c$, such that*

$$
\begin{aligned}
a &= c(c^4 + 50c^2d^2 + 125d^4), \\
b &= 5d(c^4 + 10c^2d^2 + 5d^4).
\end{aligned}
\tag{4.2}
$$

A similar lemma is necessary if $(a^2 - 5b^2)/4$ is of the form (4.1). These lemmas reflect the unique factorization (up to units) of elements into products of prime elements, which is valid in $K = \mathbb{Q}(\sqrt{5})$.

## 5. Fermat's Equation of Degree Seven

In 1839, Lamé proved Fermat's theorem for the exponent 7. He had been preceded by Dirichlet, who proved the theorem for the exponent 14 in 1832. It should be emphasized that the proof for 14 was substantially easier than for 7. Obviously, Dirichlet would have been happier to discover a proof for the exponent 7.

In 1840, Lebesgue found a much simpler proof than Lamé's. It used the following polynomial identity.

$$(X + Y + Z)^7 - (X^7 + Y^7 + Z^7)$$
$$= 7(X + Y)(Y + Z)(Z + X)[(X^2 + Y^2 + Z^2 + XY + YZ + ZX)^2$$
$$+ XYZ(X + Y + Z)] \tag{5.1}$$

which was used already by Lamé.

Cauchy and Liouville, while reporting on Lamé's paper of 1839, indicated other general polynomial identities. If $p$ is a prime, $p > 3$, then

$$(X + 1)^p - X^p - 1 = pX(X + 1)(X^2 + X + 1)^\varepsilon G_p(X), \tag{5.2}$$

where

$$\varepsilon = \begin{cases} 1 & \text{when } p \equiv -1 \pmod 6, \\ 2 & \text{when } p \equiv 1 \pmod 6, \end{cases}$$

and $G_p(X)$ is a polynomial with integral coefficients, not a multiple of $X^2 + X + 1$.

The identity used by Lamé in 1840 was: If $m$ is odd, then

$$(X + Y + Z)^m - (X + Y - Z)^m - (X - Y + Z)^m - (-X + Y + Z)^m$$
$$= 4mXYZ \sum_{\substack{a,b,c \geq 0 \\ a+b+c=(m-3)/2}} \frac{(m-1)!}{(2a+1)!(2b+1)!(2c+1)!} X^{2a} Y^{2b} Z^{2c}. \tag{5.3}$$

With such complicated tools, Lamé published a "proof" of Fermat's theorem for arbitary exponent. As I explained in my first lecture, this was a fiasco, because Lamé made unjustified use of what amounts to unique factorization in the ring of cyclotomic integers (generated by the $n$th roots of 1)—and this is not generally valid.

For the exponent 7 the main steps in Lebesgue's proof are the following:

(a) If $x$, $y$, $z$ are pairwise relatively prime nonzero integers such that $x^7 + y^7 + z^7 = 0$ by (5.1):

$$s^7 = 7vt, \tag{5.4}$$

where

$$\begin{aligned} s &= x + y + z, \\ u &= x^2 + y^2 + z^2 + xy + xz + yz, \\ v &= (x + y)(y + z)(z + x), \\ t &= u^2 + xyzs. \end{aligned} \tag{5.5}$$

(b) Then $v \neq 0$, $s \neq 0$, $v$ and $s$ are even, $u$ is odd, $t \equiv 1 \pmod{4}$, $\gcd(t, xyz) = 1$ $\gcd(t, v) = 1$.

(c) $t$ is the 14th power of an integer and $7 \nmid t$. Let $t = q^{14}$, $q \mid u$ so $u = qr$.

(d) $v = 7^6 p^7$ with $p$ even, hence

$$(x + y)(x + z)(y + z) = 7^6 p^7, \tag{5.6}$$

$$(x^2 + y^2 + z^2 + xy + yz + zx)^2 + xyzs = q^{14}, \tag{5.7}$$

$$x + y + z = 7pq^2, \tag{5.8}$$

$$x^2 + y^2 + z^2 + xy + xz + yz = qr. \tag{5.9}$$

(e) Putting $r - \frac{1}{2}7^2 p^2 q^3 = a$, $q^3 = b$, $p^2 = 2^{m+1}c$ and manipulating with the above relations yields

$$a^2 = b^4 - 2^{2m} \times 3 \times 7^4 b^2 c^2 + 2^{4(m+1)} \times 7^7 c^4, \tag{5.10}$$

where $a$, $b$, $c$ are odd and relatively prime.

(f) The proof is concluded by showing that (5.10) is impossible. This is done by induction on $m$. As a matter of fact, this step is actually longer than the rest of the proof.

Summarizing, I have tried to describe, at least in part, the proof of Lebesgue (already simpler than Lamé's) so as to convey the idea of how involved the proofs were becoming, as the exponent increased.

Such elementary methods are not of the same level of difficulty as the problem. But, by this time, there were already many other attempts, which I'll describe in the following lectures.

## Bibliography

1760/1  Euler, L.
  Supplementum quorundam theorematum arithmeticorum quae in nonnullis demonstrationibus supponuntur. *Novi Comm. Acad. Sci. Petrop.*, **8**, (1760/1) 1763, 105–128. Also in *Opera Omnia*, Ser. I, vol. II, 556–575. Teubner, Leipzig-Berlin, 1915.

1770  Euler, L.
  *Vollständige Anleitung zur Algebra*, Royal Acad. of Sciences, St. Petersburg, 1770. Translated into English, 1822, Longman, Hurst, Rees, Orme & Co., London (see pages 449–454). See also *Opera Omnia*, Ser. I, vol. I, Teubner, Leipzig-Berlin, 1915, 484–489.

1808  Legendre, A. M.
  *Essais sur la Théorie des Nombres* (2e édition), Courcier, Paris, 1808.

1823  Legendre, A. M.
  Sur quelques objets d'analyse indéterminée et particulièrement sur le théorème de Fermat. *Mém. de l'Acad. des Sciences, Institut de France*, **6**, 1823, 1–60.

1825  Legendre, A. M.
  *Essais sur la Théorie des Nombres*, 1825 (2nd supplement). Reprinted by *Sphinx-Oedipe*, **4**, 1909, 97–128.

1828   Dirichlet, G. L.
Mémoire sur l'impossibilité de quelques équations indéterminées du $5^e$ degré. *J. reine u. angew Math.*, **3**, 1828, 354–375.

1830   Legendre, A. M.
*Théorie des Nombres*, $3^e$ édition, vol. II, Firmin Didot Frères, Paris, 1830. Reprinted by A. Blanchard, Paris, 1955.

1832   Dirichlet, G. L.
Démonstration du théorème de Fermat pour les $14^e$ puissances. *J. reine u. angew. Math.*, **9**, 1832, 390–393.

1839   Lamé, G.
Mémoire sur le dernier théorème de Fermat. *C. R. Acad. Sci. Paris*, **9**, 1839, 45–46.

1839   Cauchy, A. and Liouville, J.
Rapport sur un mémoire de M. Lamé relatif au dernier théorème de Fermat. *C. R. Acad. Sci. Paris*, **9**, 1839. Also appeared in *J. Math. Pures et Appl.*, **5**, 1840, 211–215.

1840   Lamé, G.
Mémoire d'analyse indéterminée démontrant que l'equation $x^7 + y^7 = z^7$ est impossible en nombres entiers. *J. Math. Pures et Appl.*, **5**, 1840, 195–211.

1840   Lebesgue, V. A.
Démonstration de l'impossibilité de résoudre l'équation $x^7 + y^7 + z^7 = 0$ en nombres entiers. *J. Math. Pures et Appl.*, **5**, 1840, 276–279.

1840   Lebesgue, V. A.
Addition à la note sur l'équation $x^7 + y^7 + z^7 = 0$. *J. Math. Pures et Appl.*, **5**, 1840, 348–349.

1847   Lamé, G.
Mémoire sur la résolution en nombres complexes de l'equation $A^5 + B^5 + C^5 = 0$. *J. Math. Pures et Appl.*, **12**, 1847, 137–171.

1876   Gauss, C. F.
*Collected Works*, vol. II. Königlichen Ges. d. Wiss., Göttingen, 1876, 387–391.

1894   Schumacher, J.
Nachtrag zu Nr. 1077, XXIII, 269. *Zeitschrift Math. Naturw. Unterricht*, **25**, 1894, 350.

1901   Landau, E.
Sur une démonstration d'Euler d'un théorème de Fermat. *L'Interm. des Math.*, **8**, 1901, 145–147.

1906   Holden, H.
On the complete solution in integers for certain values of $p$, of $a(a^2 + pb^2) = c(c^2 + pd^2)$. *Messenger of Math.*, **36**, 1906, 189–192.

1910   Welsch
Réponse à une question de E. Dubouis. *L'Interm. des Math.*, **17**, 1910, 179–180.

1915   Carmichael, R. D.
*Diophantine Analysis*, Wiley, New York, 1915.

1938   Hardy, G. H. and Wright, E. M.
*An Introduction to the Theory of Numbers*, Clarendon Press, Oxford, 1938 (reprinted 1968).

1962   Shanks, D.
*Solved and Unsolved Problems in Number Theory, I*, Spartan, Washington, 1962. Reprinted by Chelsea Publ. Co., New York, 1979.

1962   Sierpiński, W.
*Pythagorean Triangles*, Yeshiva Univ., New York, 1962.

1966 Bergmann, G.
Über Eulers Beweis des grossen Fermatschen Satzes für den Exponenten 3. *Math. Annalen*, **164**, 1966, 159–175.

1966 Grosswald, E.
*Topics from the Theory of Numbers*, Macmillan, New York, 1966.

1969 Mordell, L. J.
*Diophantine Equations*, Academic Press, New York, 1969.

# The Naïve Approach

In this lecture, I will relate what has been done with Fermat's problem without using any sophisticated methods. Let me say, that these attempts should not be looked down on. On the contrary, they show much ingenuity, and they have helped to understand the intrinsic difficulties of the problem. I'll point out, in various cases, how these attempts have brought to light quite a number of other interesting, perhaps more difficult problems than Fermat's.

If I have decided to group these various results under the heading of "the naive approach," it is only because Fermat's problem has proved itself to be at another level. In fact, it is possible that all other approaches tried as yet may someday be considered naive. Who knows?

I will group the various results according to the main ideas involved, rather than chronologically.

## 1. The Relations of Barlow and Abel

A natural thought in trying to prove Fermat's theorem is to assume that there exist integers $x$, $y$, $z$, different from 0 (not multiples of $p$ in the first case) and satisfying the Fermat equation

$$x^p + y^p + z^p = 0, \tag{1.1}$$

where $p$ is an odd prime, and then, though whatever manipulations our fancy dictates, to derive relations involving these numbers $x$, $y$, $z$, and $p$. In this way, the idea is to aim to reach a contradiction, thereby proving that (1.1) was impossible.

Since

$$-z^p = x^p + y^p = (x + y)(x^{p-1} - x^{p-2}y + x^{p-3}y^2 - \cdots - xy^{p-2} + y^{p-1}),$$
(1.2)

$(x^p + y^p)/(x + y)$ is an integer and it is certainly of importance to study this expression and its divisibility properties.

From the very beginning, attention was given to the study of

$$Q_n(a,b) = \sum_{k=0}^{n-1} a^k(-b)^{n-k-1}$$
(1.3)

for $n \geq 1$, $a$, $b$ nonzero integers.

If $a + b \neq 0$, then $Q_n(a,b) = (a^n - (-b)^n)/(a + b)$ and if moreover $n$ is odd then

$$Q_n(a,b) = \frac{a^n + b^n}{a + b}.$$
(1.4)

Results of this kind are scattered through the literature. I prefer to list them together here, even though some of the properties won't be required until much later. The proofs are exercises. To avoid repetition, in what follows, $a$, $b$ will be nonzero relatively prime integers, $n$, $m \geq 1$ and $p$ is a prime.

**Lemma 1.1** (Birkhoff and Vandiver 1904; also Inkeri, 1946). *If $p \mid a^n \pm b^n$ but $p \nmid a^m \pm b^m$ (for every proper divisor $m$ of $n$), then $p \equiv 1$ (mod $n$).*

**Lemma 1.2** (Möller, 1955; see also Inkeri, 1946 and Vivanti, 1947, for parts (4), (5), (6)).

1. If $\gcd(n,m) = d$, then $\gcd(Q_n(a,b), Q_m(a,b)) = Q_d(a,b)$.
2. $\prod_{p \mid n} Q_p(a,b)$ divides $Q_n(a,b)$.
3. If $n$ is odd, $n \geq 3$ and $a > b$, then $Q_n(a,b) \geq n$.

The equality holds only when: $a = 1$, $b = -1$; or $n = 3$, $a = 2$, $b = 1$; or $n = 3$, $a = -1$, $b = -2$.

4. $\gcd(Q_n(a,b), a + b) = \gcd(n, a + b)$.
5. If $p \neq 2$ and $p^e \| a + b$ (with $e \geq 1$), then $p^{e+1} \| a^p + b^p$.
6. If $p \neq 2$ and $p \mid a + b$, then $p^2 \nmid Q_p(a,b)$.
7. If $p \neq 2$, $p^e \| a + b$, $e \geq 1$ and $n = p^r m$, $p \nmid m$, $r \geq 0$, then $p^{e+r} \| a^n - (-b)^n$.
8. If every prime factor of $n$ divides $a + b$, then $n(a + b)$ divides $a^n - (-b)^n$.

In one form or another, these lemmas are the background of several results to be mentioned.

Barlow (famous for the Barlow tables), discovered in 1810 the following relations; these were found later in 1823 by Abel, who mentioned them in a letter to Holmboe.

**(1A)** *If $x$, $y$, $z$ are pairwise relatively prime (nonzero) integers, satisfying $x^p + y^p + z^p = 0$ and if $p \neq 2$ does not divide $z$, then there exist integers $t$, $t_1$ such that*

$$x + y = t^p, \qquad \frac{x^p + y^p}{x + y} = t_1^p, \qquad z = -tt_1. \tag{1.5}$$

*Moreover, $p \nmid tt_1$, $\gcd(t,t_1) = 1$, $t_1$ is odd and $t_1 > 1$.*

PROOF. Since $x + y + z \equiv x^p + y^p + z^p = 0 \pmod{p}$, it follows that $-z \equiv x + y \pmod{p}$, so $p \nmid x + y$.

Clearly

$$(-z)^p = x^p + y^p = Q_p(x,y) \cdot (x + y). \tag{1.6}$$

By Lemma 1.2, part (4), the two factors $Q_p(x,y)$, $x + y$ are relatively prime. By unique factorization, they are $p$th powers.

The rest of the proof is even easier. $\qquad \qquad \square$

So, if pairwise relatively prime integers $x$, $y$, $z$ are not multiples of $p$ and satisfy (1.1), then the Barlow–Abel relations

$$x + y = t^p, \qquad \frac{x^p + y^p}{x + y} = t_1^p, \qquad z = -tt_1,$$

$$y + z = r^p, \qquad \frac{y^p + z^p}{y + z} = r_1^p, \qquad x = -rr_1, \tag{1.7}$$

$$z + x = s^p, \qquad \frac{z^p + x^p}{z + x} = s_1^p, \qquad y = -ss_1,$$

are satisfied.

In the above expressions the integers $r$, $s$, $t$, $r_1$, $s_1$, $t_1$ are not multiples of $p$, $r_1$, $s_1$, $t_1$ are odd, $\gcd(t,t_1) = \gcd(r,r_1) = \gcd(s,s_1) = 1$, and $\gcd(r,s,t) = \gcd(r_1,s_1,t_1) = 1$.

(1.7) may be rewritten as

$$x = -r^p + \frac{r^p + s^p + t^p}{2} = \frac{-r^p + s^p + t^p}{2},$$

$$y = -s^p + \frac{r^p + s^p + t^p}{2} = \frac{r^p - s^p + t^p}{2}, \tag{1.8}$$

$$z = -t^p + \frac{r^p + s^p + t^p}{2} = \frac{r^p + s^p - t^p}{2}.$$

Abel had also indicated relations like the above, when $p \mid z$. In analogy with (1A), Abel stated (1823):

**(1B)** *If* $x$, $y$, $z$ *are pairwise relatively prime nonzero integers satisfying* (1.1), *if* $p \neq 2$ *divides* $z$, *then there exist integers* $n \geq 2$, $t$, $r$, $s$, $t_1$, $r_1$, $s_1$ *such that*

$$x + y = p^{pn-1}t^p, \qquad \frac{x^p + y^p}{x + y} = pt_1^p, \qquad z = -p^n t t_1,$$

$$y + z = r^p, \qquad \frac{y^p + z^p}{y + z} = r_1^p, \qquad x = -rr_1, \qquad (1.9)$$

$$z + x = s^p, \qquad \frac{z^p + x^p}{z + x} = s_1^p, \qquad y = -ss_1.$$

*As before,* $r$, $s$, $t$, $r_1$, $s_1$, $t_1$ *satisfy the same properties already indicated.*

PROOF. In view of (1A) the relations of the last two lines of (1.9) must hold.

By hypothesis $x + y \equiv -z \equiv 0 \pmod{p}$. Writing $u = x + y = p^{m-1}t'$ with $m \geq 2$, $p \nmid t'$ and writing

$$Q_p(x,y) = \frac{x^p + y^p}{x + y} = \frac{x^p + (u - x)^p}{u}$$

$$= u^{p-1} - pu^{p-2}x + \binom{p}{2}u^{p-3}x^2 - \cdots + px^{p-1},$$

then by Lemma 1.2, $\gcd(x + y, Q_p(x,y)) = p$. So $Q_p(x,y) = pt_1'$, $p \nmid t_1'$. From

$$(-z)^p = Q_p(x,y) \cdot (x + y) = p^m t' t_1'$$

it follows that $m$ is a multiple of $p$, $m = pn$, leading to the required relations.

The other assertions are immediate, except that $n \geq 2$ still needs to be proved.

From

$$y + z = r^p, \qquad z + x = s^p$$

it follows that

$$(r + s)^p \equiv r^p + s^p \equiv y + x + 2z \equiv 0 \pmod{p}.$$

So $r \equiv -s \pmod{p}$, hence $r^p \equiv -s^p \pmod{p^2}$.

By (1.9)

$$2z = r^p + s^p - p^{pn-1}t^p \equiv 0 \pmod{p^2}$$

showing that $n \geq 2$. $\qquad\qquad\qquad\qquad\qquad\qquad\qquad\qquad\qquad\qquad\qquad\qquad$ $\square$

## 2. Sophie Germain

Sophie Germain was a French mathematician, a contemporary of Cauchy and Legendre, with whom she corresponded. Her theorem, brought by Legendre to the attention of the illustrious members of the Institut de

France, was greeted with great admiration. I want to indicate how beautiful and neat is her result. Its core is the following proposition:

**(2A)** *Let $p$, $q$ be distinct odd primes, satisfying the following conditions:*

1. *$p$ is not congruent to the pth power of an integer modulo $q$.*
2. *If $x$, $y$, $z$ are integers and if*

$$x^p + y^p + z^p \equiv 0 \pmod{q}, \tag{2.1}$$

*then $q$ divides $x$, $y$ or $z$.*

*Then the first case of Fermat's theorem is true for the exponent $p$.*

PROOF. Assume the contrary: $x$, $y$, $z$ are pairwise relatively prime integers such that $x^p + y^p + z^p = 0$. A fortiori, from hypothesis (2), $q$ divides one (and only one) of the integers $x$, $y$, $z$. Say, $q \mid x$, $q \nmid yz$. By Relations (1.8)

$$-r^p + s^p + t^p = 2x \equiv 0 \pmod{q}.$$

By hypothesis (2) $q$ divides $r$, $s$, or $t$. As easily seen, $q \mid r$, $q \nmid st$. So the following congruences hold (by (1.7)):

$$y \equiv -z \pmod{q};$$
$$t_1^p \equiv y^{p-1} \pmod{q} \quad \text{because } (x+y)t_1^p = x^p + y^p;$$
$$r_1^p \equiv pt_1^p \pmod{q}, \quad \text{because } r_1^p = \frac{y^p + z^p}{y + z} = y^{p-1} + y^{p-2}(-z) + y^{p-3}(-z)^2 + \cdots$$
$$\equiv py^{p-1} \equiv pt_1^p \pmod{q}.$$

Noting that $q \nmid t_1$, and letting $t'$ be such that $t't_1 \equiv 1 \pmod{q}$, then $(t'r_1)^p \equiv p \pmod{q}$. This contradicts hypothesis (1). □

Sophie Germain's celebrated theorem of 1823 is the following:

**(2B)** *If $p$ is an odd prime such that $2p + 1$ is also a prime, then the first case of Fermat's theorem holds for $p$.*

PROOF. It suffices to check that the primes $p$ and $q = 2p + 1$ satisfy the hypotheses of (2A), which is quite easy.

If $p \equiv a^p \pmod{q}$, computing the Legendre symbol yields

$$\pm 1 = \left(\frac{a}{q}\right) \equiv a^{(q-1)/2} = a^p \equiv p \pmod{q}$$

so $p \equiv \pm 1 \pmod{q}$, and this is impossible.

Next, suppose $x^p + y^p + z^p \equiv 0 \pmod{q}$ and $q \nmid xyz$. Since $p = (q-1)/2$, the little Fermat theorem implies that

$$x^p \equiv \pm 1 \pmod{q},$$
$$y^p \equiv \pm 1 \pmod{q},$$
$$z^p \equiv \pm 1 \pmod{q}.$$

So $0 = x^p + y^p + z^p \equiv \pm 1 \pm 1 \pm 1 \pmod{q}$, which is again impossible.
And that is the proof!                                                    □

Various comments are in order.

Such a nice result inspired at once several generalizations. Indeed, using the same idea, though with a somewhat more detailed analysis, Legendre could prove the following result:

**(2C)** *The first case of Fermat's theorem holds for the odd prime exponent p, provided one of the following numbers is also a prime*: $4p + 1$, $8p + 1$, $10p + 1$, $14p + 1$, $16p + 1$.

With this theorem, Sophie Germain and Legendre covered all the primes $p < 100$ and thereby established the first case for these primes. Though it was only the first case, still it represented a considerable advance over previous attempts, even more so because it was proved as early as 1823.

The method however has its limitations. It is difficult to extend it for primes $p$ such that $2kp + 1$ is prime, when $k$ is large. This may be seen, examining the proofs given by Legendre. Secondly, the method does not work for the second case of Fermat's theorem.

More interesting is the problem it leads to. Given $k \geq 1$, what can be said about the set of primes $p$ such that $2hp + 1$ is also prime, for some $h$, $1 \leq h \leq k$?

More specifically, are there infinitely many Sophie Germain primes $p$ (those such that $2p + 1$ is also a prime)? This question is of the same order of difficulty as the well-known "twin prime" problem.

I wish to state that the answer is positive, provided one assumes the following *hypothesis* H of Schinzel (1958):

"Let $s \geq 1$, let $f_1(X)$, ..., $f_s(X)$ be irreducible polynomials with integral coefficients; assume that the leading coefficient of $f_1(X)$ is positive and that no integer $n > 1$ divides all the numbers $f_1(m)f_2(m)\cdots f_s(m)$. Then there exists one (and it may be proved, necessarily infinitely many) natural number(s) $m$ such that $f_1(m), f_2(m), \ldots, f_s(m)$ are all primes."

For example, taking $f_1(X) = X$, $f_2(X) = 2X + 1$ this hypothesis would imply that there exist infinitely many Sophie Germain primes. Taking $f_1(X) = X$, $f_2(X) = X + 2$, it yields infinitely many twin primes. As a matter of fact, when the polynomials are linear, the hypothesis was made by Dickson (1904).

For the case where $m = 1$, $f_1(X) = aX + b$, with relatively prime integers $a$, $b$, the answer is positive and given by the famous Dirichlet theorem on primes in arithmetic progression. This is in fact, the only case for which the answer is known.

It appears to be extremely difficult to prove hypothesis H, but just as difficult to disprove it—a matter for deep thought.

Bateman and Horn, in 1962, looked at Schinzel's hypothesis from the quantitative point of view. With a heuristic argument, they made plausible that

$$\lim_{N \to \infty} Q(N) = \infty,$$

where $Q(N)$ is the number of primes $p \le N$ such that $2p + 1$ is also prime.

Another result in this connection was given by Vaughan in 1973. Namely, the two statements $(H_1)$ and $(H_2)$ below cannot be simultaneously true:

$(H_1)$ there exist an infinity of primes $p$ such that $8p + 1$ is a prime.

$(H_2)$ for infinitely many $n$, $d(n) = d(n + 1)$ [where $d(n)$ denotes the number of positive divisors of $n$].

There have been more recent developments by Krasner (1940) and Dénes (1951). Krasner's approach, as well as Dénes's, are not completely naïve. Indeed, Krasner appeals to a theorem of Furtwängler (which requires class field theory), while Dénes uses some results about cyclotomic fields and estimates for the size of the least exponent $p$ for which the first case might be false. Here is Krasner's theorem:

**(2D)** *Assume that $p$ is an odd prime, and $h$ is an integer such that:*

1. $q = 2hp + 1$ *is a prime,*
2. $3 \nmid h$,
3. $3^{h/2} < 2hp + 1$,
4. $2^{2h} \not\equiv 1 \pmod{q}$.

*Then the first case of Fermat's theorem holds for $p$.*

In 1951 Dénes proved:

**(2E)** *If $p$ is an odd prime, if $h$ is an integer, not a multiple of $3$, $h \le 55$ and such that $q = 2hp + 1$ is a prime, then the first case of Fermat's theorem holds for $p$.*

## 3. Congruences

The idea now will be to derive congruences which must be verified by nonzero pairwise relatively prime integers $x$, $y$, $z$ satisfying

$$x^p + y^p + z^p = 0. \tag{3.1}$$

The first result is due to Fleck (1909). It was repeatedly rediscovered by Frobenius (1914), Vandiver (1914), and Pomey (1923). [To his credit, I must add that Vandiver also proved a corresponding result valid for the second case; however, since it uses the methods of class field theory, I will postpone giving it until later.]

**(3A)** *With the above assumptions on $x$, $y$, and $z$, if $p$ does not divide $x$, then*

$$x^{p-1} \equiv 1 \pmod{p^3}.$$

This is a rather stringent congruence, since by Fermat's little theorem, it is only expected that $x^{p-1} \equiv 1 \pmod{p}$. The following lemma, due to Sophie Germain is required for the proof (for notation, see §1).

**Lemma 3.1.** *If $p$ does not divide $x$, $y$, $z$, then $r_1 \equiv 1 \pmod{p^2}$, $s_1 \equiv 1 \pmod{p^2}$, $t_1 \equiv 1 \pmod{p^2}$.*

PROOF (following Pérez-Cacho, 1958). I prove that $r_1 \equiv 1 \pmod{p^2}$. It suffices to show that if $q$ is any prime dividing $r_1$, then $q \equiv 1 \pmod{p^2}$.

From $q|r_1$ it follows that $q|x$, $q \nmid yz$. Since $\gcd(r,r_1) = 1$, we have $q \nmid y + z$. $q|x$ implies that $s_1^p \equiv z^{p-1} \pmod{q}$ and $t_1^p \equiv y^{p-1} \pmod{q}$, so $0 \equiv y^p + z^p \equiv yt_1^p + zs_1^p \pmod{q}$, that is, $-yt_1^p \equiv zs_1^p \pmod{q}$. But $q|r_1$, so $q|y^p + z^p$, and $q \nmid y + z$. By Lemma 1.1, $q \equiv 1 \pmod{p}$. Hence, raising to the power $(q-1)/p$: $z^{(q-1)/p} \equiv -y^{(q-1)/p} \pmod{q}$.

But if $y'$ is such that $y'y \equiv -1 \pmod{q}$, then from $z^p \equiv -y^p \pmod{q}$, it follows that $(zy')^p \equiv 1 \pmod{q}$. So, from $z \not\equiv -y \pmod{q}$, the order of $zy'$ modulo $q$ is equal to $p$.

But $(zy')^{(q-1)/p} \equiv 1 \pmod{p}$ so $p$ divides $(q-1)/p$, that is, $q \equiv 1 \pmod{p^2}$, as it was required to show. □

Using this lemma, Theorem (3A) is proved as follows:

*Case 1. $p \nmid xyz$.*

Then $r_1 \equiv 1 \pmod{p^2}$. Hence $x = -rr_1 \equiv -r \pmod{p^2}$ and $x^p \equiv -r^p \pmod{p^3}$.

By symmetry

$$y^p \equiv -s^p \pmod{p^3} \quad \text{and} \quad z^p \equiv -t^p \pmod{p^3}.$$

Hence

$$r^p + s^p + t^p \equiv 0 \pmod{p^3}. \tag{3.2}$$

From (1.8), $x \equiv -r^p \pmod{p^3}$ so

$$x \equiv x^p \pmod{p^3} \quad \text{and} \quad x^{p-1} \equiv 1 \pmod{p^3}.$$

*Case 2. $p|z$, $p \nmid xy$.*

I omit the proof, since it is similar. It requires a lemma like 3.1, which asserts that if $p|z$ then $t_1 \equiv 1 \pmod{p^2}$. □

The quantities $r$, $s$, $t$ satisfy various divisibility properties. Vandiver (1925) and Inkeri (1946) are responsible for the following result:

**(3B)**

1. *If $p \nmid xyz$, then $x + y + z$ is a multiple of $rstp^3$ and $r + s + t$ is a multiple of $p^2$.*
2. *If $p|xyz$, then $x + y + z$ is a multiple of $rstp^2$ but $r + s + t$ is not a multiple of $p$.*

In particular, in the second case, $r + s + t \neq 0$.

But this is true in general, as shown by Spunar (1929) and again by James (1934), Segal (1938):

**(3C)** $r + s + t \neq 0$.

The next group of results involves congruences of the following type:

$$(1 + x)^{p^m} \equiv 1 + x^{p^m} \pmod{p^{m+1}}. \tag{3.3}$$

It is convenient before proceeding, to state the following lemma of Ferentinou-Nicolacopoulou (1965):

**Lemma 3.2.** *If $p$ is an odd prime and $p$ does not divide $a$, nor $a + 1$, if $m \geq 2$, then $(a + 1)^{p^m} \equiv a^{p^m} + 1 \pmod{p^{m+1}}$ if and only if $(a + 1)^{p^{m-1}} \equiv a^{p^{m-1}} + 1 \pmod{p^{m+1}}$.*

The special case where $m = 2$ had been noted by G. D. Birkhoff (see Carmichael, 1913).

The next lemma was proved by Carmichael for $m = 1$ in 1913; Klösgen proved it in general (1970).

**Lemma 3.3.** *Let $p$ be an odd prime, $m \geq 1$. Then the following conditions are equivalent:*

a. *There exist $x$, $y$, $z$, not multiples of $p$, such that*

$$x^{p^m} + y^{p^m} + z^{p^m} \equiv 0 \pmod{p^{m+1}}. \tag{3.4}$$

b. *There exists $a$, $1 \leq a \leq (p - 3)/2$ such that*

$$1 + a^{p^m} \equiv (1 + a)^{p^m} \pmod{p^{m+1}}. \tag{3.5}$$

I omit the proofs of these lemmas, since they don't offer much difficulty.

In particular, if the first case fails for the exponent $p$, there exist $x$, $y$, $z$, not multiples of $p$, satisfying (3.1). Then

$$x^p + y^p + z^p \equiv 0 \pmod{p^2},$$

hence

$$1 + a^p \equiv (1 + a)^p \pmod{p^2}. \tag{3.6}$$

But Carmichael and Meissner proved independently in 1913 the more precise congruences:

**(3D)** *Let (3.1) be satisfied. Then*

1. *If $p \nmid xyz$, then there exists $a$, $1 \leq a \leq (p - 3)/2$, such that*

$$1 + a^p \equiv (1 + a)^p \pmod{p^3}. \tag{3.7}$$

2. *If $p|xyz$, then there exists $a$, $1 \leq a \leq (p-3)/2$ such that*

$$1 + a^p \equiv (1 + a)^p \pmod{p^2}.\tag{3.8}$$

PROOF. I give only a proof of the first assertion. The other is handled in a similar way. By (3A) $x^p \equiv x \pmod{p^3}$, so $x^{p^2} \equiv x^p \equiv x \pmod{p^3}$. Similarly $y^{p^2} \equiv y \pmod{p^3}$, $z^{p^2} \equiv z \pmod{p^3}$. Hence by (3B):

$$x^{p^2} + y^{p^2} + z^{p^2} \equiv x + y + z \equiv 0 \pmod{p^3}.$$

By the preceding lemmas, there exists $a$, $1 \leq a \leq (p-3)/2$ such that $(1+a)^{p^2} \equiv 1 + a^{p^2} \pmod{p^3}$ hence $(1+a)^p \equiv 1 + a^p \pmod{p^3}$. ☐

The congruence (3.7) may be investigated by computers. Wagstaff verified (in 1975) that for every prime $p < 10^5$, $p \equiv -1 \pmod 6$, the congruence $(1 + X)^p \equiv 1 + X^p \pmod{p^3}$ has no solution $a$, such that $1 \leq a \leq (p-3)/2$. This shows that the first case holds for such exponents $p$ (I stated in Lecture II that the first case is known to hold for $p < 3 \times 10^9$).

In 1950, Trypanis announced without proof the following strengthening of Carmichael's result; this was rediscovered (and proved) by Ferentinou-Nicolacopoulou in 1965:

**(3E)** *If $p > 5$ and the first case fails for $p$, there exists $a$, $1 \leq a \leq (p-5)/2$ such that*

$$(1 + a)^{p^2} \equiv 1 + a^{p^2} \pmod{p^4}.\tag{3.9}$$

It is easily seen that (3.9) implies (3.7). A recent result of Gandhi (1976) follows at once from the above.

Wells Johnson (1977) investigated whether the congruences of Carmichael and Trypanis may be further extended, modulo every power $p^{n+2}$ $(n \geq 1)$. This is a situation best handled with $p$-adic methods.

Let $\alpha_j$ denote the unique $p$-adic integer such that $\alpha_j^{p-1} = 1$ and $\alpha_j \equiv j \pmod p$ (where $j = 1, 2, \ldots, p-1$). The existence and uniqueness of $\alpha_j$ follows from Hensel's lemma.

**(3F)** *If $p$ is a prime and $p \nmid a$, $p \nmid a + 1$ then the following conditions are equivalent:*

a. $(1 + a)^{p^n} \equiv 1 + a^{p^n} \pmod{p^{n+2}}$ for every $n \geq 1$.
b. $1 + \alpha_a = \alpha_{1+a}$.
c. $a^2 + a + 1 \equiv 0 \pmod p$.

It is worth noting that already in 1839 Cauchy had shown that (c) implies that $(1 + a)^p \equiv 1 + a^p \pmod{p^3}$. In essence this is his proof: Since $a^2 + a + 1 \equiv 0 \pmod p$, $-3$ is a square modulo $p$, so $p \equiv 1 \pmod 6$.

But Cauchy had established the identity, when $p \equiv 1 \pmod 6$:

$$(1 + X)^p - 1 - X^p = pX(1 + X)(1 + X + X^2)^2 f(X),\tag{3.10}$$

where $f(X)$ is a polynomial with integral coefficients. For $X = a$ this gives the required congruence.

It would be interesting to know whether the following statement is true:

**(S)** *If $a$ is an integer, $1 \leq a \leq p - 2$, and if $(1 + a)^p \equiv 1 + a^p \pmod{p^3}$, then $a^2 + a + 1 \equiv 0 \pmod{p}$.*

Assume that (S) is true. Then the first case of Fermat's theorem would hold for every prime exponent $p \equiv 5 \pmod 6$.

Otherwise, by (3D), there exists an $a$ such that $1 + a^p \equiv (1 + a)^p \pmod{p^3}$, hence by (S) $a^2 + a + 1 \equiv 0 \pmod p$ and therefore $p \equiv 1 \pmod 6$, a contradiction.

A strengthening of (S) is false, as shown by Arwin in 1920; there exist integers $a$ and a prime $p$ such that $(1 + a)^p \equiv 1 + a^p \pmod{p^2}$ and yet $a^2 + a + 1 \not\equiv 0 \pmod p$.

## 4. Wendt's Theorem

In 1894, Wendt gave a criterion for the first case, involving a certain matrix with binomial coefficients. But, because of the size of the matrix, his criterion is quite awkward. Even worse, it was later recognized, that it is essentially equivalent to Sophie Germain's theorem. Yet, I'll present it now, because of other interesting connections.

For $n \geq 2$ let $W_n$ be the determinant of the $n \times n$ matrix:

$$W_n = \det \begin{bmatrix} 1 & \binom{n}{1} & \binom{n}{2} & \cdots & \binom{n}{n-1} \\ \binom{n}{n-1} & 1 & \binom{n}{1} & \cdots & \binom{n}{n-2} \\ \vdots & \vdots & \vdots & & \vdots \\ \binom{n}{1} & \binom{n}{2} & \binom{n}{3} & \cdots & 1 \end{bmatrix}. \tag{4.1}$$

This is a circulant. For example, $W_2 = -3$, $W_3 = 2^2 \times 7$, $W_4 = -3 \times 5^3$, $W_8 = -3^7 \times 5^3 \times 17^3$. Stern explicitly computed this determinant in 1871. The general result is the following:

If $f(X) = a_0 + a_1 X + \cdots + a_{n-1} X^{n-1}$ the circulant with first row $(a_0, a_1, \ldots, a_{n-1})$ is equal to $\prod_{i=0}^{n-1} f(\zeta_i)$, where $\zeta_0 = 1, \zeta_1, \ldots, \zeta_{n-1}$ are the $n$th roots of 1.

Since $(1 + X)^n - X^n = 1 + \binom{n}{1}X + \binom{n}{2}X^2 + \cdots + \binom{n}{n-1}X^{n-1}$,

$$W_n = \prod_{i=0}^{n-1} [(1 + \zeta_i)^n - 1]. \tag{4.2}$$

$$W_n = 0 \Leftrightarrow 6 \mid n.$$

The first result leading towards Wendt's theorem is the following:

**(4A)** *Let $p \neq 2$, let $q = 2hp + 1$ ($h \geq 1$) be primes. Then there exist integers* $x, y, z$ *such that $q \nmid xyz$ and*

$$x^p + y^p + z^p \equiv 0 \pmod{q} \tag{4.3}$$

*if and only if $q \mid W_{2h}$.*

And now, Wendt's theorem:

**(4B)** *Let $p \neq 2$, let $q = 2hp + 1$ ($h \geq 1$) be primes and $3 \nmid h$. If $q \nmid W_{2h}$ and $p^{2h} \not\equiv 1 \pmod{q}$, then the first case of Fermat's theorem holds for the exponent $p$.*

I wish to say only a few words about the proof. The fact that $q \nmid W_{2h}$ implies the congruence has only the trivial solution, may be established with routine arguments using linear algebra over the field with $q$ elements.

Moreover, if $p \equiv r^p \pmod{q}$ then $p^{2h} \equiv r^{2ph} = r^{q-1} \equiv 1 \pmod{q}$, contrary to the hypothesis. Thus, the hypotheses of Sophie Germain's theorem (2A) are satisfied and so the first case holds for $p$.

Another corollary of Sophie Germain's theorem was given by Vandiver (1926):

**(4C)** *If $p$ is an odd prime, if $q = 2hp + 1$ is a prime, if $q \nmid W_{2h}$ and $2h = 2^v p^k$, where $p \nmid v$, $k \geq 0$, then the first case of Fermat's theorem holds for $p$.*

For example, this provides a new proof of Legendre's results for primes $p$ such that $4p + 1$, or $8p + 1$, or $16p + 1$ is also a prime.

It is not known whether for every prime $p$ there exists $h > 1$ such that $q = 2hp + 1$ is a prime and $q \nmid W_{2h}$. This is a difficult problem.

To be more specific, in 1909 Dickson proved that for every prime $p$ there exists an integer $q_0$ such that if $q$ is any prime, $q \geq q_0$, then there exist integers $x, y, z$, not multiplies of $q$, such that $x^p + y^p + z^p \equiv 0 \pmod{q}$. By (4A), this means that $q \mid W_{2h}$ for every prime $q$, $q \geq q_0$, $q = 2hp + 1$. So, there are at most finitely many primes $q$ of the form $q = 2hp + 1$ for which $q \nmid W_{2h}$. What is not known is whether *one* such prime $q$ actually exists.

I shall return to Dickson's theorem in one of the later lectures.

To conclude the discussion of Wendt's determinant, I present several of its divisibility properties.

**(4D)**

1. *If $n$ is even, then $W_n = -(2^n - 1)u^2$, where $u$ is an integer.*
2. *If $d \mid n$, then $W_d \mid W_n$.*

Less routine and more interesting are the next results.
Emma Lehmer proved in 1935:

**(4E)** *If p is an odd prime then* $W_{p-1}$ *is a multiple of* $p^{p-2} q_p(2)$, *where*

$$q_p(2) = \frac{2^{p-1} - 1}{p}. \tag{4.4}$$

The number $q_p(2)$ is an integer (by Fermat's little theorem) called the
*Fermat quotient* (with base 2). I'll return often to the Fermat quotient, in
connection with various important theorems.

In particular, Wieferich's theorem of 1909 (to which I referred in Lecture
II) states that if the first case fails for $p$, then $q_p(2) \equiv 0 \pmod{p}$. As a corollary:

**(4F)** *If* $p^{p-1} \nmid W_{p-1}$, *then the first case of Fermat's theorem is true for the*
*exponent p.*

It is important, therefore, to determine the residue of $W_{p-1}$ modulo $p^{p-1}$.
Carlitz proved in 1959:

**(4G)**

$$W_{p-1} \equiv \prod_{a=1}^{p-2} [(1 + a)^p - a^p - 1]$$

$$\equiv p^{p-2} \prod_{a=1}^{p-2} \left( \sum_{s=1}^{p-1} (-1)^{s-1} \frac{a^s}{s} \right) \pmod{p^{p-1}}.$$

From this congruence, he was able to improve E. Lehmer's result (1960):

**(4H)** *If* $p^{p+33} \nmid W_{p-1}$, *then the first case of Fermat's theorem is true for*
*the exponent p.*

# 5. Abel's Conjecture

In 1823, Abel conjectured that if $n > 2$ and if $x$, $y$, $z$ are nonzero pairwise
relatively prime integers such that

$$x^n + y^n = z^n, \tag{5.1}$$

then none of $x$, $y$, $z$ can be prime-powers.

I have explained in my second lecture what may have been the reason
for Abel's conjecture. Perhaps he had in mind a descent on the number of
prime factors of a would-be solution, but he left no written explanation.

This conjecture is still unsettled. To be more precise, only a very particular assertion still needs to be established. To fix notation, let $0 < x < y < z$.

In 1884, Jonquières proved that $y$ is not a prime, and if $x$ is prime, then $z - y = 1$.

Let me just pause to say that up to now the possibility that there exist numbers $x$, $y$, $n$ for which $0 < x < y$ and $x^n + y^n = (y + 1)^n$ cannot yet be ruled out.

Jonquières' result was rediscovered by Gambioli in 1901.

Lucas proved more in 1891: $y$, $z$ have at least two prime factors. He also had an alleged proof that $x$ is not a prime power, however in 1895, Markoff pointed out a gap. Again, in 1905, Sauer rediscovered the result of Lucas.

Mileikowsky gave a new proof of Lucas's theorem in 1932. He has also shown that if $n$ is not a prime, then $x$ cannot be a prime-power.

All these partial results are contained in the following recent result of Möller (1955).

**(5A)** *Let $m \geq 1$ be an odd integer with $r$ distinct prime factors, let $u \geq 0$. Let $0 < x < y$ be relatively prime integers and $a = y^{2^u m} + x^{2^u m}$, $b = y^{2^u m} - x^{2^u m}$. Then:*

1. *$a$ and $b$ have at least $r$ distinct prime factors.*
2. *If $a$ has exactly $r$ distinct prime factors, then $r = 1$ and $a = 2^3 + 1^3$ (so $u = 0, m = 3$).*
3. *If $b$ has exactly $r$ distinct prime factors, then $b = (x + 1)^m - x^m$ (so $u = 0$, $y = x + 1$).*

PROOF. I sketch the proof of the first assertion. This is a place where I need Lemma 1.2. Replacing $x$, $y$ by $x^{2^u}$, $y^{2^u}$, it may be assumed that $u = 0$. Let $p_1, \ldots, p_r$ be the distinct prime factors of $m$. Since $y \pm x \neq 0$, by Lemma 1.2

$$y^m \pm x^m = (y \pm x) \cdot Q_m(y, \pm x)$$

is a multiple of $(y \pm x) \prod_{i=1}^{r} Q_{p_i}(y, \pm x)$. The integers $Q_{p_i}(y, \pm x)$ are also pairwise relatively prime, so $a$, $b$ have at least $r$ distinct prime factors.

The other two assertions again need Lemma 1.2 for their proof.

As a corollary:

**(5B)** *If $n$ is an integer with $r$ distinct odd prime factors, if $0 < x < y < z$ are pairwise relatively prime integers such that (5.1) holds, then:*

1. *$z$, $y$ have at least $r + 1$ distinct prime factors.*
2. *$x$ has at least $r$ distinct prime factors; moreover, if $x$ has exactly $r$ distinct prime factors, then $n$ is odd and $z = y + 1$.*

PROOF. Let $n = 2^u m$ with $u \geq 0$. By (5A) $z^n = x^n + y^n$ and $y^n = z^n - x^n$ have at least $r + 1$ distinct prime factors (note that $n \neq 3$ and $z - x > 1$).

Similarly, from $x^n = z^n - y^n$, $x$ has at least $r + 1$ distinct prime factors, except when $z = y + 1$ and $n$ is odd, in which case, it can only be said that $x$ has at least $r$ distinct prime factors. □

And now, more explicitly.

**(5C)** *Let $n > 2$, let $0 < x < y < z$ be relatively prime integers satisfying (5.1). Then*

1. *$z$, $y$ are not prime powers.*
2. *If $x$ is a prime power, then $z = y + 1$ and $n$ is an odd prime.*

PROOF.

(1) If $z$ is a prime power, so is $z^n = x^n + y^n$. By the preceding result, $n$ is a power of 2, $n \geq 4$, and this contradicts Fermat's theorem, which is true for such exponents. The same argument is enough to show that $y$ is not a prime power.

(2) If $x$ is a prime power, by (5B), $n = p^e$, $e \geq 1$, $p$ an odd prime and $z = y + 1$. It remains to show that $e = 1$.

Assume $e > 1$ so $z^{p^{e-1}} - y^{p^{e-1}} > 1$ and

$$x^{p^e} = z^{p^e} - y^{p^e}$$
$$= (z^{p^{e-1}} - y^{p^{e-1}})(z^{(p-1)p^{e-1}} + z^{(p-2)p^{e-1}}y^{p^{e-1}} + \cdots + y^{(p-1)p^{e-1}}). \quad (5.2)$$

Hence $p$ does not divide

$$z^{p^e} - y^{p^e} = (y + 1)^{p^e} - y^{p^e} = \sum_{j=1}^{p^e} \binom{p^e}{j} y^{p^e - 1}$$

because $p$ divides all but the last summand. So $x^{p^e}$ is the power of a prime $q \neq p$. Hence $q$ divides both factors of (5.2), that is,

$$y^{p^{e-1}} \equiv z^{p^{e-1}} \pmod{q} \quad \text{and} \quad (5.3)$$
$$z^{(p-1)p^{e-1}} + z^{(p-2)p^{e-1}}y^{p^{e-1}} + \cdots + y^{(p-1)p^{e-1}} \equiv py^{(p-1)p^{e-1}} \equiv 0 \pmod{q}.$$

Thus $q$ divides $y$ and therefore by (5.3), $q$ divides $z = y + 1$, a contradiction. □

With more refined, but still elementary methods, Inkeri proved in 1946 that if $0 < x < y < z$, $x^p + y^p = z^p$, and $p \nmid xyz$, then $z - y > 1$ and so $x$ is not a prime-power. I'll return to this question in my lecture on estimates.

# 6. Fermat's Equation with Even Exponent

As I shall indicate, it is possible to prove the first case of Fermat's theorem for even exponents. Clearly, it suffices to consider the exponents $2p$, where $p$ is an odd prime.

The first result in this connection was obtained by Kummer, in 1837. It is his first paper on Fermat's equation and it is written in Latin. Later, this result was rediscovered many times (Niedermeier, 1943; Griselle, 1953; Oeconomu, 1956).

The best theorem concerning the exponent $2p$ was published by Terjanian, in December 1977. It is indeed quite surprising that his proof, which requires only very elementary considerations, was not found beforehand. I'll not jump to the conclusion that perhaps there is also a simple proof of Fermat's theorem awaiting to be discovered. I would rather say that Terjanian's result shows that the first case of Fermat's theorem for an even exponent is far easier than for a prime exponent.

I begin with Kummer's theorem:

**(6A)** *Let $n > 1$ be an odd integer. If there exist nonzero integers $x$, $y$, $z$ such that $x^{2n} + y^{2n} = z^{2n}$ and $\gcd(n,xyz) = 1$, then $n \equiv 1 \pmod 8$.*

PROOF. It is possible to take $x$, $y$, $z$ positive and relatively prime. A simple observation tells that $x$ may be assumed even, while $y$, $z$ are odd. Write

$$x^{2n} = z^{2n} - y^{2n} = (z^2 - y^2) \times \frac{z^{2n} - y^{2n}}{z^2 - y^2} \tag{6.1}$$

and observe that if the two factors on the right are relatively prime, then they are $2n$th powers.

$z^2 - y^2$ is even, and

$$\frac{z^{2n} - y^{2n}}{z^2 - y^2} = z^{2(n-1)} + z^{2(n-2)}y^2 + \cdots + y^{2(n-1)} \tag{6.2}$$

is a sum of $n$ odd summands, hence it is odd, therefore of the form $k^{2n}$, with $k$ odd.

Each summand on the right is of the form $(2a + 1)^2 = 4a(a + 1) + 1 \equiv 1 \pmod 8$.

Thus (6.2) becomes an equality of the form

$$8b + 1 = (8a_1 + 1) + \cdots + (8a_n + 1).$$

Therefore $n \equiv 1 \pmod 8$.                                                                 □

For example, the first case of Fermat's theorem holds for $2n = 14$.

Kummer's theorem was extended by Grey in 1954 and by Long in 1960. Just for the record, I quote one of Long's results:

**(6B)** *If $n$ is an integer whose last digit (in decimal notation) is 4 or 6, and if $x$, $y$, $z$ are nonzero integers such that $x^n + y^n = z^n$, then $\gcd(n,xyz) > 2$.*

Now I shall give the proof of Terjanian's theorem, which contains all the above results as corollaries. Once more, as in §1, it is question of the quotient

$$Q_n(z, -y) = \frac{z^n - y^n}{z - y}.$$

If $m$, $n$ are nonzero relatively prime integers, $n$ odd, let $\left(\frac{m}{n}\right) = (m|n)$ denote the Jacobi symbol defined as an extension of the Legendre symbol in the following manner: if $|n| = \prod_{p|n} p^{e_p}$, let $\left(\frac{m}{n}\right) = \left(\frac{m}{-n}\right) = \prod_{p|n} \left(\frac{m}{p}\right)^{e_p}$. Thus, if $m$ is a square modulo $|n|$, then $\left(\frac{m}{n}\right) = +1$.

**Lemma 6.1.** *Let $y$, $z$ be distinct nonzero integers.*

1. *If $m = nq + r$, $0 \le r < n < m$, then*

$$Q_m(z, -y) = z^r Q_q(z^n, -y^n) Q_n(z, -y) + y^{m-r} Q_r(z, -y).$$

2. *If $m = nq - r$, $0 \le r < n < m$, then*

$$Q_m(z, -y) = [z^{n-r} Q_{q-1}(z^n, -y^n) + y^{m-n}] Q_n(z, -y) - y^{m-n} z^{n-r} Q_r(z, -y).$$

3. *If $z$, $y$ are odd, relatively prime, $z \equiv y \pmod 4$ and $m$ is odd, then $Q_m(z, -y) \equiv m \pmod 4$, so $Q_m(z, -y)$ is odd.*

4. *If $z$, $y$ are odd, relatively prime, $z \equiv y \pmod 4$ and $m$ and $n$ are odd natural numbers, then*

$$\left(\frac{Q_m(z, -y)}{Q_n(z, -y)}\right) = \left(\frac{m}{n}\right).$$

PROOF. The assertions (1) and (2) follow at once from the definitions.

(3) Let $z = y + 4t$. Then

$$Q_m(z, -y) = \frac{(y + 4t)^m - y^m}{4t} = \binom{m}{1} y^{m-1} + \binom{m}{2} y^{m-2} 4t + \cdots$$

$$= my^{m-1} \equiv m \pmod 4,$$

because $m - 1$ is even, $y$ is odd, so $y^{m-1} \equiv 1 \pmod 4$.

(4) The assertion is proved by induction on $m + n$. It is trivial when $m = n = 1$. Let $m + n > 2$.

If $m > n$, then there exist an integer $r$, odd, $0 < r < n$, and $q$ such that $m = qn + r$ or $m = qn - r$.

If $m = qn + r$, then $m - r$ is even, so by (1) and induction

$$\left(\frac{Q_m(z, -y)}{Q_n(z, -y)}\right) = \left(\frac{y^{m-r} Q_r(z, -y)}{Q_n(z, -y)}\right) = \left(\frac{Q_r(z, -y)}{Q_n(z, -y)}\right) = \left(\frac{r}{n}\right) = \left(\frac{m}{n}\right).$$

If $m = qn - r$, then $m - n$ and $n - r$ are even, so by (2) and induction

$$\left(\frac{Q_m(z, -y)}{Q_n(z, -y)}\right) = \left(\frac{-y^{m-n} z^{n-r} Q_r(z, -y)}{Q_n(z, -y)}\right) = \left(\frac{-Q_r(z, -y)}{Q_n(z, -y)}\right)$$

$$= \left(\frac{-1}{Q_n(z, -y)}\right)\left(\frac{Q_r(z, -y)}{Q_n(z, -y)}\right) = \left(\frac{-1}{Q_n(z, -y)}\right)\left(\frac{r}{n}\right).$$

By (3), $Q_n(z, y) \equiv n \pmod 4$. If $n = \prod_{i=1}^{s} p_i^{e_i}$, then it is easy to check that $n - 1 \equiv \sum_{i=1}^{s} (p_i - 1)e_i \pmod 4$, so

$$\left(\frac{-1}{n}\right) = \prod_{i=1}^{s} \left(\frac{-1}{p_i}\right)^{e_i} = (-1)^{\sum_{i=1}^{s} ((p_i - 1)/2)e_i} = (-1)^{(n-1)/2}.$$

Since $Q_n(z,y) \equiv n \pmod 4$,

$$\frac{Q_n(z,-y)-1}{2} \equiv \frac{n-1}{2} \pmod 2$$

hence

$$\left(\frac{-1}{Q_n(z,-y)}\right) = \left(\frac{-1}{n}\right).$$

Thus

$$\left(\frac{Q_m(z,-y)}{Q_n(z,-y)}\right) = \left(\frac{-1}{n}\right)\left(\frac{r}{n}\right) = \left(\frac{m}{n}\right).$$

Now, if $m < n$, by Jacobi's reciprocity law and the above proof

$$\left(\frac{Q_m(z,-y)}{Q_n(z,-y)}\right) = (-1)^{\frac{1}{2}(Q_m(z,-y)-1)\frac{1}{2}(Q_n(z,-y)-1)}\left(\frac{Q_n(z,-y)}{Q_m(z,-y)}\right)$$

$$= (-1)^{\frac{1}{2}(m-1)\frac{1}{2}(n-1)}\left(\frac{n}{m}\right)$$

$$= \left(\frac{m}{n}\right). \qquad \square$$

After this lemma, Terjanian's result follows almost at once:

**(6C)** *Let $p$ be an odd prime. If $x, y, z$ are nonzero integers such that $x^{2p} + y^{2p} = z^{2p}$, then $2p$ divides $x$ or $y$.*

PROOF. There is no loss of generality in assuming that $x$, $y$, $z$ are pairwise relatively prime. Also $x$, $y$ cannot be both odd, since this would imply that $x^{2p} \equiv y^{2p} \equiv 1 \pmod 4$ and hence that $z^{2p} \equiv 2 \pmod 4$, which is impossible. Let $x$ be even, so $y$, $z$ are odd. Then

$$x^{2p} = z^{2p} - y^{2p} = (z^2 - y^2)\frac{z^{2p} - y^{2p}}{z^2 - y^2}.$$

By Lemma 1.2

$$\gcd\left(z^2 - y^2, \frac{z^{2p} - y^{2p}}{z^2 - y^2}\right) = p \text{ or } 1.$$

If the greatest common divisor is $p$, then $p$ divides $x^{2p}$, so $2p$ divides $x$.

I show now that it is not possible that $z^2 - y^2$ and $(z^{2p} - y^{2p})/(z^2 - y^2)$ are relatively prime. If they are, both must be squares. But $z^2 \equiv y^2 \pmod 4$. Since $p$ is not a square, there exists a prime $q$ such that $p$ is not a square modulo $q$. It follows from Lemma 6.1 that

$$-1 = \left(\frac{p}{q}\right) = \left(\frac{Q_p(z^2,-y^2)}{Q_q(z^2,-y^2)}\right),$$

which is absurd, because $Q_p(z^2,-y^2)$ is a square.

This concludes the proof. $\qquad \square$

# 7. Odds and Ends

In this final section, I have grouped results of many different kinds. I can see by these deductions, that their authors enjoyed toying with Fermat's equation and were fully aware that their contributions would not solve the problem. Yet, not only are the proofs sometimes rather elegant or simple, but some of these results have been rediscovered and used in more substantial ways.

I start with a few divisibility properties. Massoutié and Pomey proved in 1931 the following result:

**(7A)** *If $p \equiv -1 \pmod 6$ and $x$, $y$, $z$ are non-zero integers such that $x^p + y^p + z^p = 0$, then 3 divides $x$, $y$, or $z$.*

In 1934, Pomey claimed also to have proved a similar result for all $p > 2$ and also with 5, instead of 3. According to A. Brauer (1934) his argument was erroneous. Inkeri proved in 1946 the weaker statement for 5:

**(7B)** *If $p > 2$, $p \not\equiv 1, 9 \pmod{20}$ and if $x$, $y$, $z$ are nonzero integers such that $x^p + y^p + z^p = 0$, then 5 divides $x$, $y$, or $z$.*

There are various other partial results of this kind in the literature, but I shall refrain from giving them.

Other curiosities follow. Swistak proved in 1969:

**(7C)** *If $p > 2$, if $x$, $y$, $z$ are positive pairwise relatively prime integers, and $x^p + y^p = z^p$, then $p$ divides $\varphi(x)$, $\varphi(y)$, and $\varphi(z)$ (where $\varphi$ denotes Euler's totient function).*

Goldziher proved the next result in 1913. It was rediscovered by Mihaljinec in 1952 and Rameswar Rao in 1969:

**(7D)** *Positive integers $x$, $y$, $z$ in arithmetic progression cannot satisfy the equation $x^n + y^n = z^n$ (with $n > 2$).*

Rameswar Rao proved also:

**(7E)** *If $n > 2$ is an odd integer and if $x$, $y$, $z$ are positive integers such that $x^n + y^n = z^n$, then $\mu(x + y) = 0$ ($\mu$ denotes the Möbius function).*

Here are several statements which are equivalent to Fermat's theorem. They were given by Pérez-Cacho in 1946. The equivalence between conditions (1), (2) and (3) was first proved by Bendz in 1901, and was rediscovered by S. Chowla in 1978.

**(7F)** *Let $m \geq 2$, $n = 2m - 1$. Then the following statements are equivalent:*

1. *The equation $X^n + Y^n = Z^n$ has only the trivial solutions in $\mathbb{Z}$.*
2. *The equation $X(1 + X) = T^n$ has only the trivial solutions in $\mathbb{Q}$.*

3. *The equation $X^2 = 4Y^n + 1$ has only the trivial solutions in $\mathbb{Q}$.*
4. *The equation $X^2 = Y^{n+1} - 4Y$ has only the trivial solutions in $\mathbb{Q}$.*
5. *For every non-zero rational number $a$, the polynomial $Z^2 - a^m Z + a$ is irreducible over $\mathbb{Q}$.*
6. *The equation $(XY)^m = X + Y$ has only the trivial solutions in $\mathbb{Q}$.*
7. *The equation $X^m = (X/Y) + Y$ has only the trivial solutions in $\mathbb{Q}$.*
8. *If $u_1, r$ are nonzero rational numbers, if $u_1, u_2, \ldots$ is a geometric progression of ratio $r$, then $u_m^2 - u_1 + r \neq 0$.*
9. *If $\Delta$ is a triangle with vertices $A, B, C$, if the angle $\widehat{CAB} = 90°$, if $|\overline{AB}| = 2$, and if $|\overline{AB}| + |\overline{BC}|$ is a nth power of a rational number, then $|\overline{AC}|$ is not rational.*

*Moreover, these conditions imply:*

10. *The tangents to the parabola $Y^2 = 4X$ at every rational point distinct from the origin, cut the curve $Y = X^m$ at irrational points.*

In conclusion, I would like to mention a nice paper of Hurwitz (1908). He considered the diophantine equation

$$X^m Y^n + Y^m Z^n + Z^m X^n = 0,$$

where $m \geq n$, $\gcd(m,n) = 1$ (without loss of generality), and he proved:

**(7G)** *The above equation has the trivial solution alone if and only if Fermat's theorem is true for the exponent $m^2 - mn + n^2$.*

For example, letting $m = 3$, $n = 1$ it follows that

$$X^3 Y + Y^3 Z + Z^3 X = 0$$

has only the trivial solution.

# Bibliography

1810   Barlow, P.
       Demonstration of a curious numerical proposition. *J. Nat. Phil. Chem. and Arts*, **27**, 1810, 193–205 (this paper is referred to in Dickson's *History of the Theory of Numbers*, vol. 2, 733).
1811   Barlow, P.
       *An Elementary Investigation of Theory of Numbers*, Ed. J. Johnson and Co., St. Paul's Churchyard, London, 1811, 153–169.
1823   Abel, N.
       Extraits de quelques lettres à Holmboe, Copenhague, l'an $\sqrt[3]{6064321219}$ (en comptant la fraction décimale). *Oeuvres Complètes*, 2nd edition, vol. II, Grondahl, Christiania, 1881, 254–255.
1823   Legendre, A. M.
       Recherches sur quelques objets d'analyse indéterminée et particulièrement sur

le théorème de Fermat. *Mémoires de l'Acad. des Sciences, Institut de France*, **6**, 1823, 1–60.

1825  Legendre, A. M.
*Essai sur la théorie des nombres* (2nd supplement) Paris, 1825. Reprinted in *Sphinx-Oedipe*, **4**, 1909, 97–128.

1837  Kummer, E. E.
De aequatione $x^{2\lambda} + y^{2\lambda} = z^{2\lambda}$ per numeros integros resolvenda. *J. reine u. angew. Math.*, **17**, 1837, 203–209. Reprinted in *Collected Papers*, vol. I, edited by A. Weil, Springer-Verlag, Berlin, 1975, 135–142.

1839  Cauchy, A. and Liouville, J.
Rapport sur un mémoire de M. Lamé relatif au dernier théorème de Fermat. *C. R. Acad. Sci. Paris*, **9**, 1839, $2^e$ sem., 359–363. Reprinted in *J. Math. Pures et Appl.*, **5**, 1840, 211–215. Reprinted in Cauchy's *Oeuvres Complètes*, I, vol. 4, Gauthier-Villars, Paris, 1911, 499–504.

1841  Cauchy, A.
*Exercises d'Analyse et de Physique Mathématique*, **2**, 1841, 137–144 (Notes sur quelques Théorèmes d'Algèbre). *Oeuvres Complètes*, II, vol. 12, Gauthier-Villars, Paris, 1916, 157–166.

1871  Stern, M. A.
Einige Bemerkungen über eine Determinante. *J. reine u. angew. Math.*, **73**, 1871, 374–380.

1884  Jonquières, E.
Sur le dernier théorème de Fermat. *Atti Accad. Pont. Nuovi Lincei*, **37**, 1883/1884, 146–149. Reprinted in *Sphinx-Oedipe*, **5**, 1910, 29–32.

1891  Lucas, E.
*Théorie des Nombres*, Gauthier-Villars, Paris, 1891. Reprinted by A. Blanchard, Paris, 1961.

1894  Wendt, E.
Arithmetischen Studien über den letzten Fermatschen Satz, welcher aussagt dass die Gleichung $a^n = b^n + c^n$ fur $n > 2$, in ganzen Zahlen nicht auflösbar ist. *J. reine u. angew. Math.*, **113**, 1894, 335–346.

1895  Markoff, V.
Question 477. *L'Interm. des Math.*, I, **2**, 1895, 23.

1901  Bendz, T. R.
*Öfver diophantiska Ekvationen $x^n + y^n = z^n$* (in Swedish). Almqvist & Wyksells Boktryckeri, Uppsala, 1901, 35 pages.

1901  Gambioli, D.
Memoria bibliográfica sull'ultimo teorema di Fermat. *Periodico di Mat.*, **16**, 1901, 145–192.

1904  Birkhoff, G. D. and Vandiver, H. S.
On the integral divisors of $a^n - b^n$. *Annals of Math.*, (2), **5**, 1904, 173–180.

1904  Dickson, L. E.
A new extension of Dirichlet's theorem on prime numbers. *Messenger of Math.*, **33**, 1904, 155–161.

†1905  Sauer, R.
Eine polynomische Verallgemeinerung des Fermatschen Satzes. Dissertation, Giessen, 1905.

1908  Hurwitz, A.
Über die diophantische Gleichung $x^3 y + y^3 z + z^3 x = 0$. *Math. Annalen*, **65**, 1908, 428–430. Reprinted in *Mathematische Werke*, vol. II, Birkhäuser, Basel, 1963, 427–429.

1909   Dickson, L. E.
Lower limit for the number of sets of solutions of $x^e + y^e + z^e \equiv 0 \pmod{p}$. *J. reine u. angew. Math.*, **135**, 1909, 181–188.

1909   Dickson, L. E.
On the congruence $x^n + y^n + z^n \equiv 0 \pmod{p}$. *J. reine u. angew. Math.*, **135**, 1909, 134–141.

1909   Fleck, A.
Miszellen zum grossen Fermatschen Problem. *Sitzungsber. Berliner Math. Ges.*, **8**, 1909, 133–148.

1913   Carmichael, R. D.
Note on Fermat's last theorem. *Bull. Amer. Math. Soc.*, **19**, 1913, 233–236.

1913   Carmichael, R. D.
Second note on Fermat's last theorem. *Bull. Amer. Math. Soc.*, **19**, 1913, 402–403.

†1913   Goldziher, K.
Hatvǎnyszamok Telbontása hatvắnyszamok összegere. *Középiskolai Math. Lapok*, **21**, 1913, 177–184.

1913   Meissner, W.
Über die Teilbarkeit von $2^p - 2$ durch das Quadrat der Primzahl $p = 1093$. *Sitzungsber Alad. d. Wiss. zu Berlin*, 1913, pp. 663–667

1914   Meissner, W. Uber die Lösungen der Kongruenz $x^{p-1} \equiv 1 \pmod{p^m}$ und ihre Verwendung zur Periodenbestimmung mod $p$. *Sitzungsber d. Berliner Math. Ges.*, **13**, 1914, 96–107.

1914   Vandiver, H. S.
A note on Fermat's last theorem. *Trans. Amer. Math. Soc.*, **15**, 1914, 202–204.

1914   Frobenius, G.
Über den Fermatschen Satz, III. *Sitzungsber. Akad. d. Wiss. zu Berlin*, 1914, 653–681. Reprinted in *Collected Works*, vol. 3, Springer-Verlag, Berlin, 1968, 648–676.

1920   Arwin, A.
Über die Lösung der Kongruenz $(\lambda + 1)^p - \lambda^p - 1 \equiv 0 \pmod{p^2}$. *Acta Math.*, **42**, 1920, 173–190.

1923   Pomey, L.
Sur le dernier théorème de Fermat. *C. R. Acad. Sci. Paris*, **177**, 1923, p. 1187–1190.

1925   Vandiver, H. S.
A property of cyclotomic integers and its relation to Fermat's last theorem (2nd paper). *Annals of Math.*, **26**, 1925, 217–232.

1926   Vandiver, H. S.
Note on trinomial congruences and the first case of Fermat's last theorem. *Annals of Math.*, **27**, 1926, 54–56.

1929   Spunar, V. M.
On Fermat's last theorem II. *J. Washington Acad. Sci.*, **19**, 1929, 395–401.

1931   Massoutié, L.
Sur le dernier théorème de Fermat. *C. R. Acad. Sci. Paris*, **193**, 1931, 502–504.

1931   Pomey, L.
Nouvelles remarques relatives au dernier théorème de Fermat. *C. R. Acad. Sci. Paris*, **193**, 1931, 563–564.

1932   Mileikowsky, E. N.
Elementarer Beitrag zur Fermatschen Vermutung. *J. reine u. angew. Math.*, **166**, 1932, 116–117.

1934   Brauer, A.
Review of paper by L. Pomey "Sur le dernier théorème de Fermat (Divisibilité par 3 et par 5)". *Jahrbuch d. Fortschritte der Math.*, **60**, II, 1934, 928.

1934   James, G.
On Fermat's last theorem. *Amer. Math. Monthly*, **41**, 1934, 419–424.

1934   Pomey, L.
Sur le dernier théorème de Fermat (divisibilité par 3 et par 5). *C. R. Acad. Sci. Paris*, **199**, 1934, 1562–1564.

1935   Lehmer, E.
On a resultant connected with Fermat's last theorem. *Bull. Amer. Math. Soc.*, **41**, 1935, 864–867.

1938   Segal, D.
A note on Fermat's last theorem. *Amer. Math. Monthly*, **45**, 1938, 438–439.

1940   Krasner, M.
A propos du critère de Sophie Germain—Furtwängler pour le premier cas du théorème de Fermat. *Mathematica Cluj*, **16**, 1940, 109–114.

1943   Niedermeier, F.
Ein elementarer Beitrag zur Fermatschen Vermutung. *J. reine u. angew. Math.*, **185**, 1943, 111–112.

1946   Inkeri, K.
Untersuchungen über die Fermatsche Vermutung. *Annales Acad. Sci. Fennicae*, Ser. A, I, No. **33**, 1946, 60 pages.

1946   Pérez-Cacho, L.
Fermat's last theorem and the Mersenne numbers (in Spanish). *Rev. Acad. Ci. Madrid*, **40**, 1946, 39–57.

1947   Vivanti, G.
Un teorema di aritmetica e la sua relazione colla ipotesi di Fermat. *Ist. Lombardo Sei. Lett., Rend. Sci. Mat. Nat.* (3), **11** (80), 1947, 239–246.

1950   Trypanis, A.
On Fermat's last theorem. Proc. Int. Congress Math., Cambridge, Mass., 1950, vol. 1, 301–302.

1951   Dénes, P.
An extension of Legendre's criterion in connection with the first case of Fermat's last theorem. *Publ. Math. Debrecen*, **2**, 1951, 115–120.

1952   Mihaljinec, M.
Prilog Fermatovu Problemu (Une contribution au problème de Fermat). In Serbo-Croatian (French summary). Hrvatsko Prirodoslovno Društvo. *Glasnik Mat. Fiz. Astr.*, Ser. II, **7**, 1952, 12–18.

1953   Griselle, T.
Proof of Fermat's last theorem for $n = 2(8a + 1)$. *Math. Magazine*, **26**, 1953, 263.

1954   Grey, L. D.
A note on Fermat's last theorem. *Math. Magazine*, **27**, 1954, 274–277.

1955   Möller, K.
Untere Schranke für die Anzahl der Primzahlen, aus denen $x$, $y$, $z$ der Fermatschen Geichung $x^n + y^n = z^n$ bestehen muss. *Math. Nachr.*, **14**, 1955, 25–28.

1956   Oeconomu, G.
Sur le premier cas du théorème de Fermat pour les exposants pairs. *C. R. Acad. Sci. Paris*, **243**, 1956, 1588–1591.

1958   Pérez-Cacho, L.
On some questions in the theory of numbers (in Spanish). *Rev. Mat. Hisp. Amer.* (4), **18**, 1958, 10–27 and 113–124.

1958    Schinzel, A. and Sierpiński, W.
Sur certaines hypothèses concernant les nombres premiers. Remarque. *Acta Arithm.*, **4**, 1958, 185–208 and *Acta Arithm.*, **5**, 1959, 259.

1959    Carlitz, L.
A determinant connected with Fermat's last theorem. *Proc. Amer. Math. Soc.*, **10**, 1959, 686–690.

1960    Carlitz, L.
A determinant connected with Fermat's last theorem, II. *Proc. Amer. Math. Soc.*, **11**, 1960, 730–733.

1960    Long, L.
A note on Fermat's theorem. *Math. Gaz.*, **44**, 1960, 261–262.

1962    Bateman, P. T. and Horn, R. A.
A heuristic asymptotic formula concerning the distribution of prime numbers. *Math. Comp.*, **16**, 1962, 363–367.

1963    Schinzel, A.
A remark on a paper of Bateman and Horn. *Math. Comp.*, **17**, 1963, 445–447.

1965    Ferentinou-Nicolacopoulou, I.
A new necessary condition for the existence of a solution of the equation $x^p + y^p = z^p$ of Fermat (in Greek, French summary). *Bull. Soc. Math. Grèce* (N.S.), **6**, I, 1965, 222–236.

1965    Ferentinou-Nicolacopoulou, I.
Remarks on the article "A new necessary condition for the existence of a solution of the equation $x^p + y^p = z^p$ of Fermat" (in Greek, French summary). *Bull. Soc. Math. Grèce*, (N.S.), **6**, II, 1965, 356–357.

1969    Rameswar Rao, D.
Some theorems on Fermat's last theorem. *Math. Student*, **37**, 1969, 208–210.

1969    Swistak, J. M.
A note on Fermat's last theorem. *Amer. Math. Monthly*, **76**, 1969, 173–174.

1970    Klösgen, W.
Untersuchungen über Fermatsche Kongruenzen. *Gesells. f. Math. und Datenverarbeitung*, No. **36**, 124 pages, 1970, Bonn.

1973    Vaughan, R. C.
A remark on the divisor function $d(n)$. *Glasgow Math. J.*, **14**, 1973, 54–55.

1975    Everett, C. J. and Metropolis, N.
On the roots of $X^m \pm 1$ in the $p$-adic field $\mathbf{Q}_p$. *Notices Amer. Math. Soc.*, **22**, 1975, A-619. Preprint LA-UR-74-1835 Los Alamos Sci. Lab., Los Alamos, New Mexico.

1975    Wagstaff, S. S.
Letter to W. Johnson (unpublished).

1976    Gandhi, J. M.
On the first case of Fermat's last theorem. Preprint.

1977    Johnson, W.
On congruences related to the first case of Fermat's last theorem. *Math. Comp.*, **31**, 1977, 519–526.

1977    Terjanian, G.
Sur l'équation $x^{2p} + y^{2p} = z^{2p}$. *C. R. Acad. Sci. Paris*, **285**, 1977, 973–975.

1978    Chowla, S.
L-series and elliptic curves, part 4: On Fermat's last theorem. Number Theory Day. Springer Lect. Notes, No. 626, 1978, 19–24. Springer-Verlag, New York, 1978.

LECTURE V

# Kummer's Monument

This lecture is about Kummer's famous theorem: Fermat's last theorem is true for every exponent $p$ which is a regular prime. I will explain how Kummer arrived at the notion of a regular prime and show why his approach to the problem may be considered quite natural. In some sense, it forces itself on us.

## 1. A Justification of Kummer's Method

In my first Lecture, I already gave a historical survey of what was known about the problem circa 1835–1845.

To show the difficulties one might encounter and to show how natural Kummer's approach was, I wish to make some preliminary remarks. The first idea which has been used, time and time again, is to write

$$-z^p = x^p + y^p \tag{1.1}$$

then to express $x^p + y^p$ as a product of factors which are pairwise relatively prime and therefore must themselves be $p$th powers, since integers factor uniquely into primes.

For $p = 3$ this was done by Gauss (see Lecture III) using cubic roots of 1. It will work in general:

$$x^p + y^p = (x + y)(x + \zeta y)(x + \zeta^2 y)\cdots(x + \zeta^{p-1}y). \tag{1.2}$$

Here $\zeta$ represents a $p$th root of 1, namely, the complex number

$$\zeta = \cos\frac{2\pi}{p} + i\sin\frac{2\pi}{p}. \tag{1.3}$$

It is necessary to be sure that there are precisely $p$ $p$th roots of 1, all being powers of $\zeta$, and that

$$1 + \zeta + \zeta^2 + \cdots + \zeta^{p-1} = 0. \tag{1.4}$$

To follow the main idea, it is necessary to give some sense to the statement that the various factors $x + y$, $x + \zeta y$, $x + \zeta^2 y$, ... $x + \zeta^{p-1}y$ are pairwise relatively prime. This requires an a priori notion of divisibility. In such a theory, a quantity is an integer exactly when it is a multiple of 1. So "integers" must be defined. Since the only complex numbers which are relevant have to do with $\zeta$ and its powers, and since the ordinary operations of addition, subtraction, multiplication, and division have to be allowed, Kummer was led to consider those complex numbers which are obtained from $\zeta$ and the rational numbers using the above operations. These numbers have the form

$$a_0 + a_1\zeta + a_2\zeta^2 + \cdots + a_{p-2}\zeta^{p-2}. \tag{1.5}$$

A summand involving $\zeta^{p-1}$ is unnecessary, since it may be expressed by means of (1.4) in terms of lower powers of $\zeta$.

The totality of such numbers is the field $\mathbb{Q}(\zeta)$. Every element satisfies an algebraic equation over $\mathbb{Q}$, of degree at most (and in fact dividing) $p - 1$.

The elements of the form (1.5) where each $a_i$ is an integer, are called the *cyclotomic integers* of $\mathbb{Q}(\zeta)$. They form a ring $\mathbb{Z}[\zeta]$. Each ordinary integer is a cyclotomic integer.

If $\alpha, \beta \in \mathbb{Q}(\zeta)$, $\alpha$ *divides* $\beta$ when there exists $\gamma \in \mathbb{Z}[\zeta]$ such that $\alpha\gamma = \beta$. One writes $\alpha | \beta$.

The notions of divisibility of cyclotomic integers and ordinary integers share various properties.

However, and this is the main point, there are at least two major differences. First, there are cyclotomic integers (not equal to 1 or $-1$), which divide 1. For example, the numbers $1, -1, \zeta, -\zeta, \zeta^2, -\zeta^2, \ldots, \zeta^{p-1}, -\zeta^{p-1}$. In fact, I will show that there are many others. From the point of view of divisibility, they play no role, so they are called *units*, and their union is the multiplicative *group of units*.

Suppose two nonzero elements $\alpha, \beta$ are such that $\alpha$ divides $\beta$ and $\beta$ divides $\alpha$. In this case $\alpha, \beta$ are called *associated* elements and the notation $\alpha \sim \beta$ is used. It is obvious that $\alpha, \beta$ are associated if and only if $\alpha/\beta$ is a unit.

An example is the following:

$$1 - \zeta^k \sim 1 - \zeta^j \quad \text{when } 1 \leq k \neq j \leq p - 1.$$

Indeed, let $j'$ be such that $jj' \equiv k \pmod{p}$. Then

$$\frac{1 - \zeta^k}{1 - \zeta^j} = \frac{1 - (\zeta^j)^{j'}}{1 - \zeta^j} = 1 + \zeta^j + \zeta^{2j} + \cdots + \zeta^{(j'-1)j} \in \mathbb{Z}[\zeta].$$

In the same way $(1 - \zeta^j)/(1 - \zeta^k) \in \mathbb{Z}[\zeta]$.

The existence of units forces an indeterminacy in describing elements with given divisibility. Lumping together all the elements with the same behavior is therefore a reasonable idea.

Following the model of ordinary arithmetic, a cyclotomic integer is *prime* if the only cyclotomic integers which divide it are either associated to it, or units. So essentially primes have no proper divisor.

It is not at all difficult to show that every cyclotomic integer is a product of primes. However, it can be shown that it is false, in general, that if $\alpha$ is a prime and $\alpha$ divides $\beta\gamma$ then $\alpha|\beta$ or $\alpha|\gamma$. It is known that if this property were true, then the unique factorization theorem would hold. That is, every cyclotomic integer would be, in a unique way (up to units) equal to a product of primes. Conversely, the unique factorization theorem would imply the above property.

It took some time before mathematicians were convinced that in general unique factorization will not hold for cyclotomic integers. This is the second essential difference with ordinary arithmetic. Kummer himself found, sometime later, by an indirect method, the first example of this possibility—the field of 23rd roots of 1.

Now the crucial point in any proof of Fermat's theorem would have to be the knowledge that the factors in (1.2) are $p$th powers. This conclusion being unwarranted, Kummer invented certain "ideal numbers" such that the factors in (1.2) would become $p$th powers of these ideal numbers.

He had not only to invent these ideal numbers, but also to extend to them the language of divisibility and ensure that for these numbers, unique factorization holds. The next problem was to measure to what extent unique factorization for cyclotomic integers was lacking.

The success of Kummer's theory of cyclotomic ideal numbers prompted its generalization to other fields of algebraic numbers. This work was done by Dedekind, who reinterpreted Kummer's ideal numbers as being certain subsets in $\mathbb{Z}[\zeta]$ (or even in $\mathbb{Q}(\zeta)$), which constitute the Dedekind ideals (respectively, fractional ideals). Kummer's ideal numbers were closer to the modern-day notion of "divisors", which embodies all the features for the local and global study of number theory.

# 2. Basic Facts about the Arithmetic of Cyclotomic Fields

For the sake of fixing notations, I'll recall now some of the basic facts about cyclotomic fields, which we will need. All this, and much more may be found, with proofs, in any of the standard textbooks on algebraic numbers. I will adopt the current modern terminology.

Let $p \geq 3$ be a prime number. Let $\zeta = \zeta_p$ be a primitive $p$th root of 1.

$K = \mathbb{Q}(\zeta)$ is the cyclotomic field (corresponding to $p$). Its elements are the complex numbers $a_0 + a_1\zeta + \cdots + a_{p-2}\zeta^{p-2}$ (with $a_0, a_1, \ldots, a_{p-2} \in \mathbb{Q}$).
$A = \mathbb{Z}[\zeta]$ is the ring of cyclotomic integers: $a_0 + a_1\zeta + \cdots + a_{p-2}\zeta^{p-2}$ (with $a_0, a_1 \ldots, a_{p-2} \in \mathbb{Z}$).

The minimal polynomial of $\zeta$ over $\mathbb{Q}$ is the $p$th *cyclotomic polynomial*:

$$\Phi_p(X) = \frac{X^p - 1}{X - 1} = X^{p-1} + X^{p-2} + \cdots + X + 1.$$

It is an irreducible polynomial with roots $\{\zeta, \zeta^2, \ldots, \zeta^{p-1}\}$. The trace and norm of $\zeta$ are

$$\text{Tr}(\zeta) = \zeta + \zeta^2 + \cdots + \zeta^{p-1} = -1, \tag{2.1}$$

$$N(\zeta) = 1. \tag{2.2}$$

Moreover

$$\Phi_p(X) = \prod_{i=1}^{p-1} (X - \zeta^i) \tag{2.3}$$

and

$$\Phi_p(1) = \prod_{i=1}^{p-1} (1 - \zeta^i) = p. \tag{2.4}$$

Every $\alpha \in A$ may be written in unique way in the form indicated, so $\{1, \zeta, \ldots, \zeta^{p-2}\}$ is *an integral basis* of $A$. The elements $\alpha \in A$ are roots of monic irreducible polynomials with coefficients in $\mathbb{Z}$. The basis $\{1, \zeta, \ldots, \zeta^{p-2}\}$ has *discriminant*

$$d = \det(\text{Tr}(\zeta^{i+j}))_{i,j=0,1,\ldots,p-2}.$$

$d$ is also the discriminant of any other integral basis.

A computation yields

$$d = (-1)^{(p-1)/2} p^{p-2}. \tag{2.5}$$

A nonempty subset $I$ of $A$ is an *ideal* if it satisfies the following properties:

1. If $\alpha, \beta \in I$, then $\alpha + \beta, \alpha - \beta \in I$.
2. If $\alpha \in I$, $\beta \in A$, then $\alpha\beta \in I$.

Among the ideals there are $I = 0$ (consisting only of 0) and $I = A$. If $\alpha_1, \ldots, \alpha_n \in A$ the set $\{\sum_{i=1}^m \beta_i\alpha_i \,|\, \beta_1, \ldots, \beta_n \in A\}$ is an ideal of $A$. It is said to be *generated* by $\{\alpha_1, \ldots, \alpha_n\}$. If $\alpha \in A$, then the set $A\alpha = \{\beta\alpha \,|\, \beta \in A\}$ is called the *principal ideal* generated by $\alpha$.

Ideals of $A$ may be added and multiplied together. If $I, J$ are ideals:

$$I + J = \{\alpha + \beta \,|\, \alpha \in I, \beta \in J\}, \tag{2.6}$$

$$IJ = \left\{\sum_{i=1}^m \alpha_i\beta_i \,\big|\, m \geq 0, \alpha_i \in I, \beta_i \in J\right\}. \tag{2.7}$$

$I + J$, $IJ$ are again ideals. It is easy to establish the rules of operations with ideals.

If $I_1$, $I_2$, $I_3$ are ideals and $I_1 I_2 = I_3$ then $I_1$ and $I_2$ are said to divide $I_3$. If $P$ is an ideal, $P \neq 0$, $P \neq A$, and the only ideals dividing $P$ are $P$ and $A$, then $P$ is called a *prime ideal*. It may be shown that $I$ divides $J$ if and only if $I \supset J$ (as sets). Thus prime ideals are *maximal* among the ideals different from $A$ (as subsets of $A$), and conversely.

If $A\alpha = A\beta \neq 0$ then $\alpha/\beta$ and $\beta/\alpha$ are both in $A$, so $\beta$ divides $\alpha$ and $\alpha$ divides $\beta$. In the sense already indicated $\alpha$, $\beta$ are associated ($\alpha \sim \beta$) and $\alpha/\beta$, $\beta/\alpha$ are units of $A$. The group of units of $A$ is denoted by $U$; it is a subgroup of the multiplicative group $K^{\cdot}$ of nonzero elements of $K$.

The fundamental theorem for ideals is the following:

**(2A)** *Every nonzero ideal of $A$ is, in unique way, equal to the product of (not necessarily distinct) prime ideals.*

This theorem is also valid for any algebraic number field, as was shown by Dedekind.

It follows that the rules of divisibility of ordinary integers remain valid for the divisibility of ideals. In particular, it is possible to define the *greatest common divisor* $\gcd(I,J)$ and the *least common multiple* $\mathrm{lcm}(I,J)$ of ideals $I$, $J$. Since $I_1$ divides $I_2$ if and only if $I_1 \supset I_2$, then

$$\gcd(I,J) = I + J, \qquad \mathrm{lcm}(I,J) = I \cap J. \tag{2.8}$$

If $\alpha, \beta \in I$ the *congruence* $\alpha \equiv \beta \pmod{I}$ means that $\alpha - \beta \in I$, that is, $I$ divides the principal ideal $A(\alpha - \beta)$. Congruences modulo ideals satisfy properties analogous to ordinary congruences of integers. The set of congruence classes modulo $I$ forms the *residue ring* $A/I$; it is a field if and only if $I$ is a prime ideal.

It is of course important to describe the prime ideals of $A$. It is easy to see that if $Q$ is a prime ideal of $A$ it contains exactly one ordinary prime number $q$. Since $q \in Q$, $Q$ divides the principal ideal $Aq$, so $Q$ is among its prime ideal factors. This leads to determining the prime ideal factors of $Aq$ for any prime number $q$. There are two cases: $q = p$ and $q \neq p$.

**(2B)** *Let $\lambda = 1 - \zeta \in A$. Then $A\lambda$ is a prime ideal containing $p$, $Ap = (A\lambda)^{p-1}$ and $A/A\lambda$ is the field with $p$ elements.*

Because $Ap$ is the product of $p - 1$ ideals equal to $A\lambda$, the prime $p$ is said to be *ramified* in the cyclotomic field $K = \mathbb{Q}(\zeta)$.

For any prime $q \neq p$, Kummer also indicated how to proceed. Let $\mathbb{F}_q$ be the field with $q$ elements and let $\bar{\Phi}_p \in \mathbb{F}_q[X]$ be the polynomial obtained from $\Phi_p$ by reducing its coefficients modulo $q$. Let $\bar{\Phi}_p = {}'\bar{F}_0^{e_0} \cdot \ldots \cdot \bar{F}_{r-1}^{e_{r-1}}$ ($e_0 \geq 1, \ldots, e_{r-1} \geq 1, r \geq 0$), be the decomposition of $\bar{\Phi}_p$ as product of powers of irreducible polynomials of $\mathbb{F}_q[X]$. Then:

**(2C)** *With notation as above, $r$ divides $p - 1$, $e_0 = \cdots = e_{r-1} = 1$. If $f = (p - 1)/r$, then each polynomial $\bar{F}_i$ has degree $f$. The ideal $Aq$ is equal to* the product of $r$ distinct prime ideals, say $Aq = Q_0 \cdot Q_1 \cdot \ldots \cdot Q_{r-1}$. Each $A/Q_i$ is the unique field of degree $f$ over $\mathbb{F}_q$. The prime ideals $Q_i$ correspond to the factors $\bar{F}_i$ as follows: if $F_i \in \mathbb{Z}[X]$ and $F_i$ reduced modulo $q$ is $\bar{F}_i$, then $Q_i$ is generated by the two elements $q$, $F_i(\zeta)$. Moreover, $f$ is the order of $q$ modulo $p$, in the multiplicative group $\mathbb{F}_p^*$.

Thus, each prime $q \neq p$ is *unramified* in the cyclotomic field (since $e_0 = \cdots = e_{r-1} = 1$).

The cyclotomic field $K = \mathbb{Q}(\zeta)$ is a Galois extension of $\mathbb{Q}$; its Galois group is cyclic of order $p - 1$. If $g$ is any primitive root modulo $p$, then the automorphism $\sigma$ of $K$ such that $\sigma(\zeta) = \zeta^g$ is a generator of this Galois group.

The conjugates of $\zeta$, i.e., the roots of $\Phi_p(X)$ appear in pairs of complex-conjugate numbers (none are real numbers). It is convenient to label the conjugates of $\zeta$ as follows: $\zeta = \zeta^{(1)}, \zeta^{(2)}, \ldots, \zeta^{(p-1)}$, where $\zeta^{(i+\pi)}$ (with $\pi = (p-1)/2$) is the complex-conjugate of $\zeta^{(i)}$, for $i = 1, 2, \ldots, \pi$.

The Galois group transforms any ideal $I$ of $A$ into other ideals, which are the *conjugates* of $I$. If $I$ is a prime ideal so are its conjugates. Actually, it is possible that an ideal is equal to certain of its conjugates. In the decomposition of any prime $q$ into prime ideals as indicated in (2C) the prime ideals $Q_0, \ldots, Q_{r-1}$ are conjugate to each other.

The product of an ideal $I$ and all its conjugates gives a new ideal $N(I)$, called the *norm* of $I$. This is necessarily the principal ideal generated by an ordinary integer; in symbols $N(I) = An$. The number of elements of the residue ring $A/I$ is the absolute value $|n|$.

To measure the existence of nonprincipal ideals in $A$, it is convenient to introduce the *fractional ideals*. These are subsets $I$ of $K$ such that there exists $\alpha \in K$, $\alpha \neq 0$ for which $\alpha I = \{\alpha\beta \mid \beta \in I\}$ is an ideal of $A$. Thus, every ideal of $A$ is a fractional ideal; $K$ itself is not a fractional ideal. The set of nonzero fractional ideals of $K$ is a multiplicative abelian group under the operation:

$$IJ = \left\{ \sum_{i=1}^{m} \alpha_i\beta_i \mid \alpha_i \in I, \beta_i \in I \right\}.$$

This group is denoted by $\mathscr{I}d(K)$. Among the nonzero fractional ideals of $K$ are the principal fractional ideals, namely those of the form $A\alpha = \{\beta\alpha \mid \beta \in A\}$, where $\alpha \neq 0$. They constitute a subgroup of $\mathscr{I}d(K)$, denoted by $\mathscr{P}\mathscr{I}d(K)$.

Two nonzero fractional ideals $I$, $J$ are said to be *equivalent* when there exists a nonzero principal fractional ideal $A\alpha$ such that $I = (A\alpha)J$. This is indeed an equivalence relation; the nonzero principal fractional ideals are those equivalent to the unit ideal. Each equivalence class of ideals is called simply an *ideal class*. The class of the ideal $J$ shall be denoted $[J]$. The set of ideal classes forms a multiplicative abelian group, namely the quotient group $\mathscr{I}d(K)/\mathscr{P}\mathscr{I}d(K)$, which is called the *ideal class group* (or, simple, the class group) of $K$ and is denoted by $\mathscr{C}\ell(K)$.

(2D) *The group of ideal classes of $K$ is finite.*

For a proof of this theorem, which is true for any algebraic number field, see [Ri], page 123.

The number $h = h(p)$ of ideal classes of $K = \mathbb{Q}(\zeta_p)$ is called the *class number* of $K$. Thus $h = 1$ if and only if every ideal of $K$ is principal. This is equivalent to saying that every algebraic integer is equal, up to a unit, to the product of powers of prime algebraic integers (those whose only factors are units or associated numbers).

The structure of the abelian group $\mathscr{C}\ell(K)$ is easily described. For every prime $q$, let $\mathscr{C}\ell_q(K)$ be the $q$-primary component of the ideal class group $\mathscr{C}\ell(K)$. It consists of all ideal classes having order a power of $q$; in other words $\mathscr{C}\ell_q(K)$ is the $q$-Sylow subgroup of $\mathscr{C}\ell(K)$. By the structure theorem for finite abelian groups, $\mathscr{C}\ell(K)$ is isomorphic to the direct product of the groups $\mathscr{C}\ell_q(K)$. It should be noted that $\mathscr{C}\ell_q(K)$ is nontrivial if and only if $q$ divides the class number $h$ of $K$. Moreover, each $\mathscr{C}\ell_q(K)$ is isomorphic to the direct product of cyclic groups having order a power of $q$. If $q^t$ is the exponent of the group $\mathscr{C}\ell_q(K)$, that is, the maximum of the orders of the elements of $\mathscr{C}\ell_q(K)$, then the order of every element of $\mathscr{C}\ell_q(K)$ divides $q^t$ (this is a well-known property of finite abelian groups).

The following easy result is true in general for finite abelian groups:

**(2E)** *If $h = q^e u$, where $q$ is a prime not dividing $u$, $e \geq 1$, and if $q^t$ is the exponent of $\mathscr{C}\ell_q(K)$, then $t \leq e$.*

Some indications about how to compute the class number $h$ will be given later.

Turning to the group of units of the cyclotomic field, let $W$ be the group of roots of unity belonging to the cyclotomic field. Then

$$W = \{1, \zeta, \ldots, \zeta^{p-1}, -1, -\zeta, \ldots, -\zeta^{p-1}\}.$$

The structure of the group of units is as follows:

**(2F)** *$U \cong W \times U'$, where $U'$ is the free abelian multiplicative group with $r = (p-3)/2$ generators.*

That is, there exist units $\varepsilon_1, \ldots, \varepsilon_r$ in $K$ such that $\varepsilon_i^j \neq 1$ (for every $j \neq 0$) and such that every unit $\varepsilon \in U$ may be written in unique way in the form

$$\varepsilon = \omega \varepsilon_1^{e_1} \cdots \varepsilon_r^{e_r},$$

with $\omega \in W$, $e_i \in \mathbb{Z}$ $(i = 1, \ldots, r)$. This is a particular case of Dirichlet's general theorem (1846) on the structure of the group of units of any algebraic number field (see [Ri], page 148).

$\{\varepsilon_1, \ldots, \varepsilon_r\}$ is a *fundamental system* of units of $K$. Any two fundamental system of units have the same number of elements. Expressing the units of one fundamental system in terms of those of another, considering conjugates, their absolute value and logarithms, it may be seen that the following real number is independent of the system of units: $R = 2^{(p-3)/2}|\det(L)|$, where

$$L = \begin{pmatrix} \log|\varepsilon_1^{(1)}| & \cdots & \log|\varepsilon_1^{(r)}| \\ \vdots & & \vdots \\ \log|\varepsilon_r^{(1)}| & \cdots & \log|\varepsilon_r^{(r)}| \end{pmatrix}.$$

$R$ is called the *regulator* of the cyclotomic field $K$.

The explicit determination of a fundamental system of units turns out to be, in general, extremely difficult. I will indicate later its connection with the class number.

Among the units, I have already mentioned the roots of unity (they are 1, $-1$, $\zeta$, $-\zeta$, ... ,$\zeta^{p-1}$, $-\zeta^{p-1}$) and the elements $(1 - \zeta^j)/(1 - \zeta^k)$, where $1 \le j \ne k \le p - 1$.

Certain units are real numbers, for example the *Kummer units* (*Kreiseinheiten*):

$$\delta_k = \sqrt{\frac{1 - \zeta^k}{1 - \zeta} \cdot \frac{1 - \zeta^{-k}}{1 - \zeta^{-1}}} \quad \text{(positive square root)}$$

for $k = 2, \ldots, (p - 1)/2$. (It is necessary to prove that

$$\frac{1 - \zeta^k}{1 - \zeta} \cdot \frac{1 - \zeta^{-k}}{1 - \zeta^{-1}}$$

is a square in $K$; this follows from (2G) below).

Let $U^+$ be the totality of real units of $K$. It is a subgroup of the group $U$ of all units of $K$. Let $V$ denote the subgroup of $U^+$, generated by the $(p - 3)/2$ Kummer units. It is important to compare these three groups $V \subset U^+ \subset U$. First, Kummer proved:

**(2G)** *Every unit of $K$ is the product of a root of unity and a real positive unit.*

The Kummer units are multiplicatively independent: if

$$\prod_{k=2}^{(p-1)/2} \delta_k^{e_k} = 1, \text{ then } e_2 = \cdots = e_{(p-1)/2} = 0.$$

To say that they form a fundamental system of units amounts saying that $V = U^+$. Thus the index of $V$ in $U^+$ measures how far $V$ is from $U^+$ and it will be connected with the class number.

Later I will mention the deepest of Kummer's results about units, which is valid under a special hypothesis about the class number.

## 3. Kummer's Main Theorem

In 1847, Kummer announced his main theorem in a letter to Dirichlet, who then communicated it to the Academy of Sciences of Berlin.

In this proof, published in 1850, Kummer claims that Fermat's equation, for "regular" prime exponents, has only the trivial solution in the cyclotomic field $\mathbb{Q}(\zeta)$. Yet, his proof contains an "unaccountable lapse" (in Weil's expression; see Kummer's *Collected Papers*, vol. I, Notes by A. Weil, page 955). This concerns the unjustified assumption that a solution $\alpha$, $\beta$, $\gamma$ in the cyclotomic field may be chosen such that $\alpha$, $\beta$, $\gamma$ have no nontrivial common divisor. However, Kummer's argument did establish without any gaps, that there are no (nontrivial) solutions in ordinary integers when the exponent is a regular prime. Later, in 1897, Hilbert succeeded in adapting Kummer's proof, so as to exclude solutions from the cyclotomic field.

At the beginning, Kummer proved the theorem under two hypothesis about the exponent $p$:

**Hypothesis 1.** The $p$th power of a nonprincipal ideal is never a principal ideal.
**Hypothesis 2.** If $\alpha$ is a cyclotomic unit and there exists an ordinary integer $m$ such that $\alpha \equiv m \pmod{Ap}$, then there exists a unit $\beta$ such that $\alpha = \beta^p$.

In his second communication to the Academy, Kummer writes:

> My proof of Fermat's theorem, which Mr. Lejeune Dirichlet has communicated to the Royal Academy of Sciences, is based on two hypotheses. At that time, I could not in general decide which primes satisfied these hypothesis. For this purpose I was lacking the expression for the number of nonequivalent classes of complex ideal numbers, about which, already for a long time, Mr. Dirichlet has promised an article. After waiting for its publication, I have undertaken, with the help of some oral suggestions from Dirichlet, to derive the required expressions, and I have succeeded not only to discover them, but also to base, on these expressions, both hypotheses in my proof of Fermat's theorem.

Later, I will indicate how Kummer succeeded in characterizing the exponents $p$ for which these hypotheses hold. As a matter of fact, he actually proved that if the first hypothesis is satisfied then the second one follows automatically.

A role is played in the proof by the semi-primary integers. An element $\alpha \in A$ is said to be *semi-primary* if $\alpha \not\equiv 0 \pmod{A\lambda}$ but there exists an ordinary integer $m$ such that $\alpha \equiv m \pmod{A\lambda^2}$. The following lemma will be useful (it is easy to prove):

**Lemma 3.1.**
1. If $\alpha \equiv m + n\lambda \pmod{A\lambda^2}$, with $m, n$ ordinary integers, $m \not\equiv 0 \pmod{p}$ and if $l$ is an integer such that $lm \equiv n \pmod{p}$, then $\zeta^l\alpha$ is semi-primary.
2. If $\alpha$, $\beta \in A$ are semi-primary, there exists an integer $m$ such that $\alpha \equiv m\beta \pmod{A\lambda^2}$.

In the next lemma, I indicate what can be proved without making any hypothesis concerning the exponent.

**Lemma 3.2.** Let $p > 2$ be a prime. Assume that $\alpha$, $\beta$, $\gamma \in A$ and let

$$I = \gcd(A(\alpha + \zeta^k\beta)|k = 0,1,\ldots,p-1).$$

1. *If $\alpha^p + \beta^p + \gamma^p = 0$ and if $\lambda \nmid \gamma$, then for every $k = 0, 1, \ldots, p - 1$ there exists an ideal $J_k$ of $A$ such that*

$$A(\alpha + \zeta^k \beta) = J_k^p I. \tag{3.1}$$

*The ideals $J_0, J_1, \ldots, J_{p-1}$ are pairwise relatively prime and not multiples of $A\lambda$.*

2. *If $\alpha^p + \beta^p = \varepsilon \delta^p \lambda^{mp}$, where $\varepsilon$ is a unit of $A$, $\delta \in A$ and $\lambda$ does not divide $\alpha, \beta, \delta$, then $m \geq 2$.*

3. *If $\alpha^p + \beta^p + \gamma^p = 0$ and if $\gamma = \delta \lambda^m$, with $m \geq 1$, $\delta \in A$, $\lambda$ not dividing $\alpha, \beta, \delta$, then there exists $j_0$, $0 \leq j_0 \leq p - 1$, such that for every $k = 0, 1, \ldots, p - 1$ there exists an ideal $J_k$ of $A$, such that*

$$\begin{aligned} A(\alpha + \zeta^{j_0}\beta) &= (A\lambda)^{p(m-1)+1} I' J_{j_0}^p, \\ A(\alpha + \zeta^k \beta) &= (A\lambda) I' J_k^p \quad (\text{when } k \neq j_0), \end{aligned} \tag{3.2}$$

*where $I' = \gcd(A\alpha, A\beta)$. The ideals $J_0, J_1, \ldots, J_{p-1}$ are pairwise relatively prime and not multiples of $A\lambda$.*

PROOF.

(1) To begin

$$-\gamma^p = \alpha^p + \beta^p = \prod_{k=0}^{p-1} (\alpha + \zeta^k \beta). \tag{3.3}$$

First I note that if $1 \leq j < k \leq p - 1$, then $\gcd(A(\alpha + \zeta^j \beta), A(\alpha + \zeta^k \beta)) = I$. This is quite easy. Indeed, if $P$ is any prime ideal, $e \geq 1$, and $P^e$ divides both ideals $A(\alpha + \zeta^j \beta)$, $A(\alpha + \zeta^k \beta)$, then

$$\alpha + \zeta^j \beta \in P^e$$
$$\alpha + \zeta^k \beta \in P^e.$$

Then, taking the difference yields

$$(\zeta^j - \zeta^k)\beta = \zeta^j(1 - \zeta^{k-j})\beta \in P^e.$$

But $1 - \zeta^{k-j} \sim 1 - \zeta = \lambda$, so $\lambda\beta \in P^e$ and $P^e | A\lambda\beta$. Similarly $(\zeta^k - \zeta^j)\alpha = \zeta^j(\zeta^{k-j} - 1)\alpha \in P^e$ and $P^e | A\lambda\alpha$. But $P \nmid A\lambda$, otherwise $P | A\lambda$, so $P = A\lambda$ (both being prime ideals) and from (3.3) $A\lambda | A\gamma$ against the hypothesis. It follows that $P^e$ divides $A\alpha$, $A\beta$ and therefore $P^e | A(\alpha + \zeta^i \beta)$ for every $i = 0, 1, \ldots, p - 1$. This is enough to prove the assertion.

Let

$$J_k' = \frac{A(\alpha + \zeta^k \beta)}{I} \quad \text{for } k = 0, 1, \ldots, \quad p - 1, \tag{3.4}$$

so $J_0', J_1', \ldots, J_{p-1}'$ are pairwise relatively prime ideals, not multiples of $A\lambda$ and

$$\left(\frac{A\gamma}{I}\right)^p = \prod_{k=0}^{p-1} J_k'.$$

Then each $J'_k$ is the $p$th power of an ideal $J'_k = J^p_k$, in virtue of the unique factorization theorem for ideals.

(2) Multiplying $\alpha$, $\beta$ by any $p$th roots of 1 still gives algebraic integers satisfying the same equation $\alpha^p + \beta^p = \varepsilon \delta^p \lambda^{mp}$. Thus, by Lemma 3.1, it may be assumed that $\alpha$, $\beta$ are semi-primary integers.

Assume that $m = 1$. Since $\alpha$, $\beta$ are semi-primary, there exist $a, b \in \mathbb{Z}$ such that

$$\alpha \equiv a \,(\mathrm{mod}\, A\lambda^2) \quad \text{and} \quad \beta \equiv b \,(\mathrm{mod}\, A\lambda^2).$$

Since $Ap = A\lambda^{p-1}$,

$$\alpha^p \equiv a^p \,(\mathrm{mod}\, A\lambda^{p+1}) \quad \text{and} \quad \beta^p \equiv b^p \,(\mathrm{mod}\, A\lambda^{p+1}).$$

Since $m = 1$, $a^p + b^p = \alpha^p + \beta^p + \rho\lambda^{p+1} = \varepsilon\lambda^p(\delta^p + \varepsilon^{-1}\rho\lambda)$, where $\rho \in A$, $\lambda \nmid \rho$.

If $p^e$ is the exact power of $p$ dividing $a^p + b^p$ (with $e \geq 0$), then the exact power of $A\lambda$ dividing $A(a^p + b^p)$ is a multiple of $p - 1$, because $Ap = A\lambda^{p-1}$. Since $\lambda \nmid \delta$, $A\lambda \nmid A(\delta^p + \varepsilon^{-1}\rho\lambda)$, so the exact power of $A\lambda$ dividing the ideal $A(\varepsilon\lambda^p(\delta^p + \varepsilon^{-1}\rho\lambda))$ is $A\lambda^p$—a contradiction. This proves that $m \geq 2$.

(3) From

$$\varepsilon\delta^p\lambda^{mp} = \alpha^p + \beta^p = \prod_{k=0}^{p-1} (\alpha + \zeta^k\beta) \tag{3.5}$$

it follows that there exists $j$, $0 \leq j \leq p - 1$ such that $A\lambda \mid A(\alpha + \zeta^j\beta)$.

But then $A\lambda \mid A(\alpha + \zeta^k\beta)$ for every index $k$. Indeed, if $k \neq j$, then $\alpha + \zeta^k\beta = (\alpha + \zeta^j\beta) + \zeta^j(\zeta^{k-j} - 1)\beta$ and $\zeta^{k-j} - 1 \sim \lambda$. So

$$\frac{\alpha + \beta}{\lambda}, \frac{\alpha + \zeta\beta}{\lambda}, \ldots, \frac{\alpha + \zeta^{p-1}\beta}{\lambda} \in A.$$

These elements are pairwise incongruent modulo $A\lambda$. Otherwise,

$$\alpha + \zeta^j\beta \equiv \alpha + \zeta^k\beta \,(\mathrm{mod}\, A\lambda^2)$$

with $j \neq k$. So $\lambda^2 \mid (\zeta^j - \zeta^k)\beta \sim \lambda\beta$ and $\lambda \mid \beta$, against the hypothesis.

Since the number of congruence classes modulo $\lambda$ is $p$, there exists an unique index $j_0$, such that $(\alpha + \zeta^{j_0}\beta)/\lambda^p \equiv 0 \,(\mathrm{mod}\, A\lambda)$, that is, $\lambda^2 \mid \alpha + \zeta^{j_0}\beta$ and $\lambda^2 \nmid \alpha + \zeta^k\beta$ for $k \neq j_0$. So from (3.5)

$$(A\lambda)^{mp-(p-1)} = A\lambda^{p(m-1)+1} \quad \text{divides } A(\alpha + \zeta^{j_0}\beta)$$

and $p(m - 1) + 1 > 1$ since $m \geq 2$ (by part (2)).

If $I' = \gcd(A\alpha, A\beta)$, then $I' \mid A(\alpha + \zeta^k\beta)$ for every $k$ and $A\lambda \nmid I'$. Hence

$$\begin{aligned} A(\alpha + \zeta^{j_0}\beta) &= (A\lambda)^{p(m-1)+1}I'J'_{j_0}, \\ A(\alpha + \zeta^k\beta) &= (A\lambda)I'J'_k \quad \text{(for } k \neq j_0), \end{aligned} \tag{3.6}$$

where $J'_0, J'_1, \ldots, J'_{p-1}$ are ideals, not multiples of $A\lambda$.

These ideals are pairwise relatively prime: if $P$ is a prime ideal dividing $J'_j$, $J'_k$ $(k \neq j)$, then $(A\lambda)I'P$ divides $A(\alpha + \zeta^j\beta)$, $A(\alpha + \zeta^k\beta)$, hence also $A(\beta - \zeta^{j-k}\beta) = (A\lambda)A\beta$, so $I'P \mid A\beta$. Similarly $I'P \mid A\alpha$, hence $I'P$ divides $I' = \gcd(A\alpha, A\beta)$, which is impossible.

Multiplying the relations (3.6) yields

$$(A\delta)^p (A\lambda)^{mp} = \prod_{k=0}^{p-1} A(\alpha + \zeta^k \beta) = (A\lambda)^{mp} I'^p J_0' \cdots J_{p-1}'$$

so

$$(A\delta)^p = I'^p J_0' \cdots J_{p-1}'.$$

By the unique factorization of ideals, for each $k$ there exists an ideal $J_k$ such that $J_k' = J_k^p$ and clearly $A\lambda \nmid J_k$ moreover these ideals are pairwise relatively prime.

In view of (3.6) this concludes the proof.                                        □

A prime $p$ is said to be *regular* if $p$ does not divide the class number $h = h(p)$ of the cyclotomic field $K$.

**Lemma 3.3.** *The prime $p$ is regular if and only if the $p$th power of a non-principal ideal of $A$ is never a principal ideal.*

PROOF. By definition $p$ is regular when $p$ does not divide the order $h$ of the class group $\mathscr{C}\ell(K)$. This means that the $p$-Sylow subgroup of $\mathscr{C}\ell(K)$ is trivial. In other words, there is no element $[J]$ in $\mathscr{C}\ell(K)$ having order $p$: if $[J]^p = [A]$, then $[J] = [A]$. Or equivalently, if $J^p$ is a principal ideal, then $J$ is already a principal ideal.                                        □

I have indicated already that Kummer succeeded in proving that first hypothesis implies the second. This is the contents of his difficult and important *lemma on units*:

**Lemma 3.4.** *If $p$ is a regular prime, if $\alpha$ is a unit of $A$ such that there exists $m \in \mathbb{Z}$ satisfying the congruence $\alpha \equiv m \pmod{Ap}$, then $\alpha = \beta^p$, where $\beta$ is a unit of $A$.*

In order to prove this lemma, Kummer introduced $\lambda$-adic methods. I shall give a full proof of this lemma in my forthcoming other book on Fermat's theorem.

And now, finally, here is Kummer's main theorem:

**(3A)** *If $p$ is a regular prime, there exist no (nonzero) cyclotomic integers $\alpha$, $\beta$, $\gamma$ such that $\alpha^p + \beta^p + \gamma^p = 0$.*

PROOF. By Gauss's theorem for the exponent 3, the assertion is true for $p = 3$. I shall assume the theorem false (hence $p > 3$) and will consider the traditional two cases.

*First Case. $\alpha$, $\beta$, $\gamma$ are not multiples of $\lambda = 1 - \zeta$.*

I may assume $p \geq 7$. Indeed, if $p = 5$, then $\alpha$, $\beta$, $\gamma$ are congruent modulo $A\lambda$ to the integers $\pm 1$, or $\pm 2$, because $A/A\lambda = \mathbb{F}_5$ (the field with 5 elements).

Then $\alpha^5$, $\beta^5$, $\gamma^5$ are congruent to $\pm 1$, or $\pm 32$, modulo $A\lambda^5$ (note that $A5 = A\lambda^4$). In any case, $\alpha^5 + \beta^5 + \gamma^5 \not\equiv 0 \pmod{A\lambda^5}$, so the theorem would be true for the exponent 5.

Thus, let $p \geq 7$.

Multiplying $\alpha$, $\beta$ by any root of 1 still gives algebraic integers satisfying the same relation. So, by Lemma 3.1, it is possible to assume, without loss of generality, that $\alpha$, $\beta$ are semi-primary integers in $K$.

By Lemma 3.2

$$A(\alpha + \zeta^k \beta) = J_k^p I \qquad (k = 0, 1, \ldots, p - 1),$$

where the ideals $J_0, J_1, \ldots, J_{p-1}$ are pairwise relatively prime and not multiples of $A\lambda$, and $I = \gcd(A(\alpha + \beta), A(\alpha + \zeta\beta), \ldots, A(\alpha + \zeta^{p-1}\beta))$.

Since $\alpha + \zeta^{p-1}\beta \not\equiv 0 \pmod{A\lambda}$, because $A\lambda$ does not divide $A\gamma$, there exists a root of unity $\zeta'$ such that $\zeta'(\alpha + \zeta^{p-1}\beta)$ is a semi-primary integer. Let

$$\alpha' = \frac{\alpha}{\zeta'(\alpha + \zeta^{p-1}\beta)}, \qquad \beta' = \frac{\beta}{\zeta'(\alpha + \zeta^{p-1}\beta)}.$$

Hence $\alpha' + \zeta^{p-1}\beta' = \zeta'^{-1}$ and $A(\alpha' + \zeta^k\beta') = (J_k/J_{p-1})^p$ for $k = 0, 1, \ldots, p - 2$.

The fractional ideals $(J_k/J_{p-1})^p$ are principal, and $p$ is a regular prime. By Lemma 3.3, $J_k/J_{p-1}$ is also a principal ideal, say $J_k/J_{p-1} = A(\mu_k/n_k)$ (for $k = 0, 1, \ldots, p - 2$), where $\mu_k \in A$, $n_k \in \mathbb{Z}$ and $\mu_k$, $n_k$ have no common factor (not a unit) in $A$. We note also that $A\lambda$ does not divide $A\mu_k$, $An_k$.

Taking into account (2G), then

$$\alpha' + \zeta^k\beta' = \varepsilon_k \zeta^{c_k}\left(\frac{\mu_k}{n_k}\right) \qquad (0 \leq k \leq p - 2), \tag{3.7}$$

where $\varepsilon_k$ is a real unit of $K$, $0 \leq c_k \leq p - 1$. Since $A/A\lambda = \mathbb{F}_p$, $\mu_k \equiv m_k \pmod{A\lambda}$, where $m_k \in \mathbb{Z}$. Hence $\mu_k^p \equiv m_k^p \pmod{A\lambda^p}$ because $Ap = A\lambda^{p-1}$. Therefore

$$n_k(\alpha' + \zeta^k\beta') \equiv \varepsilon_k \zeta^{c_k} m_k^p \pmod{A\lambda^p}$$

(for $k = 0, 1, \ldots, p - 2$). Considering the complex-conjugates, we have

$$n_k(\overline{\alpha'} + \zeta^{-k}\overline{\beta'}) \equiv \varepsilon_k \zeta^{-c_k} m_k^p \pmod{A\lambda^p}$$

(because $\overline{\lambda} = 1 - \zeta^{-1}$ is associate to $\lambda$). Hence $\varepsilon_k m_k^p \equiv \zeta^{-c_k}n_k(\alpha' + \zeta^k\beta') \equiv \zeta^{c_k}n_k(\overline{\alpha'} + \zeta^{-k}\overline{\beta'}) \pmod{A\lambda^p}$ and since $A\lambda$ does not divide $An_k$,

$$\alpha' + \zeta^k\beta' \equiv \zeta^{2c_k}(\overline{\alpha'} + \zeta^{-k}\overline{\beta'}) \pmod{A\lambda^p}. \tag{3.8}$$

We now evaluate the exponents $c_k$. Since $\alpha$, $\beta$, $\zeta'(\alpha + \zeta^{p-1}\beta)$ are semi-primary integers, it follows from Lemma 3.1 that there exist rational integers $a$, $b$ such that

$$\alpha \equiv a\zeta'(a + \zeta^{p-1}\beta) \pmod{A\lambda^2},$$
$$\beta \equiv b\zeta'(a + \zeta^{p-1}\beta) \pmod{A\lambda^2}.$$

Since $A\lambda \nmid A(\alpha + \zeta^{p-1}\beta)$,

$$\alpha' + \zeta^k\beta' \equiv a + \zeta^k b \pmod{A\lambda^2}.$$

Similarly
$$\bar{\alpha}' + \zeta^{-k}\bar{\beta}' \equiv a + \zeta^{-k}b \ (\mathrm{mod}\ A\lambda^2).$$

But $\zeta^j = (1 - \lambda)^j \equiv 1 - j\lambda \ (\mathrm{mod}\ A\lambda^2)$ for every $j \in \mathbb{Z}$. Hence
$$a + b - kb\lambda \equiv (1 - 2c_k\lambda)(a + b + kb\lambda) \ (\mathrm{mod}\ A\lambda^2)$$
so
$$2c_k(a + b)\lambda \equiv 2kb\lambda \ (\mathrm{mod}\ A\lambda^2)$$
hence
$$c_k(a + b) \equiv kb \ (\mathrm{mod}\ A\lambda).$$

Since $a, b, c_k, k \in \mathbb{Z}$,
$$c_k(a + b) \equiv kb \ (\mathrm{mod}\ p).$$

From $\alpha + \zeta^{p-1}\beta \equiv (a + \zeta^{p-1}b)\zeta'(\alpha + \zeta^{p-1}\beta) \ (\mathrm{mod}\ A\lambda^2)$ it follows that
$$1 \equiv (a + \zeta^{p-1}b)\zeta' \ (\mathrm{mod}\ A\lambda^2).$$

But $\zeta \equiv 1 \ (\mathrm{mod}\ A\lambda)$, hence $\zeta^{p-1} \equiv \zeta' \equiv 1 \ (\mathrm{mod}\ A\lambda)$, so
$$a + b \equiv 1 \ (\mathrm{mod}\ A\lambda),$$

hence $a + b \equiv 1 \ (\mathrm{mod}\ p)$, so $c_k \equiv kb \ (\mathrm{mod}\ p)$ for $k = 0, 1, \ldots, p - 2$.

In particular
$$c_0 \equiv 0 \ (\mathrm{mod}\ p), \qquad c_1 \equiv b \ (\mathrm{mod}\ p), \qquad c_2 \equiv 2b \ (\mathrm{mod}\ p), \qquad c_3 \equiv 3b \ (\mathrm{mod}\ p).$$

Thus we may rewrite (3.8) as follows (for $k = 0,1,2,3$):
$$\begin{aligned}
\alpha' + \beta' - \bar{\alpha}' - \bar{\beta}' &= \rho_0\lambda^p, \\
\alpha' + \zeta\beta' - \zeta^{2b}\bar{\alpha}' - \zeta^{2b-1}\bar{\beta}' &= \rho_1\lambda^p, \\
\alpha' + \zeta^2\beta' - \zeta^{4b}\bar{\alpha}' - \zeta^{4b-2}\bar{\beta}' &= \rho_2\lambda^p, \\
\alpha' + \zeta^3\beta' - \zeta^{6b}\bar{\alpha}' - \zeta^{6b-3}\bar{\beta}' &= \rho_3\lambda^p,
\end{aligned} \tag{3.9}$$

where $\rho_0, \rho_1, \rho_2, \rho_3 \in A$.

Consider the matrix of coefficients
$$M = \begin{pmatrix} 1 & 1 & -1 & -1 \\ 1 & \zeta & -\zeta^{2b} & -\zeta^{2b-1} \\ 1 & \zeta^2 & -\zeta^{4b} & -\zeta^{4b-2} \\ 1 & \zeta^3 & -\zeta^{6b} & -\zeta^{6b-3} \end{pmatrix}.$$

By Cramer's rule
$$\alpha' = \frac{\det(M_1)}{\det(M)}, \qquad \beta' = \frac{\det(M_2)}{\det(M)}, \qquad \bar{\alpha}' = \frac{\det(M_3)}{\det(M)}, \qquad \bar{\beta}' = \frac{\det(M_4)}{\det(M)},$$

where $M_i$ is the matrix obtained from $M$ by replacing the $i$th column by the right-hand side of the relations (3.9). Thus $\det(M_i) \in A\lambda^p$ and since $A\lambda$ does not divide $A\alpha'$, $A\beta'$, $A\bar{\alpha}'$, $A\bar{\beta}'$, then necessarily $A\lambda^p$ divides $\det(M)$, that is

$$(1 - \zeta)(1 - \zeta^{2b})(1 - \zeta^{2b-1})(\zeta - \zeta^{2b})(\zeta - \zeta^{2b-1})(\zeta^{2b} - \zeta^{2b-1}) \equiv 0 \ (\mathrm{mod}\ A\lambda^p). \tag{3.10}$$

We discuss several possibilities.

1. $b \equiv 0 \pmod{p}$: then $\beta \equiv 0 \pmod{A\lambda}$, against the hypothesis.
2. $b \equiv 1 \pmod{p}$: then $\beta \equiv \alpha + \beta \pmod{A\lambda}$, hence $\alpha \equiv 0 \pmod{A\lambda}$, against the hypothesis.
3. $n \not\equiv 0, 1 \pmod{p}$ and $2b \not\equiv 1 \pmod{p}$: then all the factors in (3.10) are associated with $\lambda$, thus $\lambda^p$ divides $\lambda^6$, so $p \leq 6$, against the hypothesis.
4. $b \not\equiv 0, 1 \pmod{p}$ and $2b \equiv 1 \pmod{p}$: then $2\beta \equiv \alpha + \beta \pmod{A\lambda}$ so $\alpha \equiv \beta \pmod{A\lambda}$.

In view of the symmetry of the relation $\alpha^p + \beta^p + \gamma^p = 0$, we must also have $\alpha \equiv \gamma \pmod{A\lambda}$. But $\alpha^p \equiv \alpha$, $\beta^p \equiv \beta$, $\gamma^p \equiv \gamma \pmod{A\lambda}$, because $A/A\lambda = \mathbb{F}_p$. It follows that $\alpha + \beta + \gamma \equiv 3\alpha \equiv 0 \pmod{A\lambda}$ so $\alpha \equiv 0 \pmod{A\lambda}$ contradicting the hypothesis. So, we have again reached a contradiction, concluding the proof for the first case.

*Second Case.* $\lambda$ divides $\alpha$, $\beta$, or $\gamma$.

We assume without loss of generality that $\lambda$ divides $\gamma$.

We may write $\gamma = \delta\lambda^m$, where $\delta \in A$, $\lambda$ does not divide $\delta$, and

$$\alpha^p + \beta^p = -\delta^p \lambda^{mp}.$$

So there is a relation of the form

$$\alpha^p + \beta^p = \varepsilon \delta^p \lambda^{mp}, \tag{3.11}$$

where $\varepsilon$ is a unit, $\lambda$ does not divide $\delta$ and $m$ is minimal. By Lemma 3.2, $m \geq 2$. It follows that $\lambda$ does not divide $\alpha$ otherwise $\lambda$ also divides $\beta$ and from (3.11). after dividing by $\lambda^p$, we would obtain a similar relation with exponent $(m-1)p$ for $\lambda$, against the minimality of $m$. Similarly $\lambda$ does not divide $\beta$.

Our purpose will be to derive a relation analogous to (3.11) with smaller exponent of $\lambda$. By Lemma 3.2, after changing $\beta$ into $\zeta^{j_0}\beta$ (without loss of generality), we may write

$$\begin{aligned} A(\alpha + \beta) &= (A\lambda)^{p(m-1)+1} I' J_0^p, \\ A(\alpha + \zeta^k \beta) &= (A\lambda) I' J_k^p \quad \text{(for } 1 \leq k \leq p-1), \end{aligned} \tag{3.12}$$

where $I' = \gcd(A\alpha, A\beta)$, and the ideals $J_0, J_1, \ldots, J_{p-1}$ are pairwise relatively prime and not multiples of $A\lambda$. Then

$$A\left(\frac{\alpha + \zeta^k \beta}{\alpha + \beta}\right) = (A\lambda)^{-p(m-1)}\left(\frac{J_k}{J_0}\right)^p \quad (1 \leq k \leq p-1). \tag{3.13}$$

This shows that the fractional ideals $(J_k/J_0)^p$ are principal and since $p$ is regular, $J_k/J_0$ is also principal. Thus there exist elements $\mu_k \in A$, $n_k \in \mathbb{Z}$ such that $J_k/J_0 = A(\mu_k/n_k)$. Since $A\lambda$ does not divide $J_k$ ($0 \leq k \leq p-1$), we may assume that $\lambda$ does not divide the elements $\mu_k$, $n_k$. So there exist units $\varepsilon_k$ of $A$ such that

$$(\alpha + \zeta^k \beta)\lambda^{p(m-1)} = \varepsilon_k(\alpha + \beta)\left(\frac{\mu_k}{n_k}\right)^p \quad (1 \leq k \leq p-1).$$

In particular, if $k = 1, 2$, then

$$(\alpha + \zeta\beta)\lambda^{p(m-1)} = \varepsilon_1(\alpha + \beta)\left(\frac{\mu_1}{n_1}\right)^p,$$

$$(\alpha + \zeta^2\beta)\lambda^{p(m-1)} = \varepsilon_2(\alpha + \beta)\left(\frac{\mu_2}{n_2}\right)^p.$$

Multiplying the first relation by $1 + \zeta$ and then subtracting the second relation gives

$$\zeta(\alpha + \beta)\lambda^{p(m-1)} = (\alpha + \beta)\left[\left(\frac{\mu_1}{n_1}\right)^p \varepsilon_1(1 + \zeta) - \left(\frac{\mu_2}{n_2}\right)^p \varepsilon_2\right].$$

Hence

$$(\mu_1 n_2)^p - \frac{(\mu_2 n_1)^p \varepsilon_2}{\varepsilon_1(1 + \zeta)} = \frac{\zeta}{\varepsilon_1(1 + \zeta)} \lambda^{p(m-1)}(n_1 n_2)^p.$$

But $1 + \zeta$ is a unit of $A$ and the above relation has the form:

$$(\alpha')^p + \varepsilon'(\beta')^p = \varepsilon''(\delta')^p\lambda^{(m-1)p}, \tag{3.14}$$

where $\alpha' = \mu_1 n_2$, $\beta' = \mu_2 n_1$, $\delta' = n_1 n_2$, $\lambda \nmid \delta'$, and $\varepsilon'$, $\varepsilon''$ are units. This is not yet like relation (3.11), but we shall transform it into a relation of that type.

Since $p(m - 1) \geq p$, $\lambda^p$ divides $(\alpha')^p + \varepsilon'(\beta')^p$. But $A\beta' = A\mu_2 n_1$ is relatively prime to $A\lambda$, i.e., $A\beta' + A\lambda = A$, so there exists an element $\kappa \in A$ such that $\kappa\beta' \equiv 1 \pmod{A\lambda}$. Then $\kappa^p\beta'^p \equiv 1 \pmod{A\lambda^p}$ and $(\kappa\alpha')^p + \varepsilon' \equiv 0 \pmod{A\lambda^p}$. So there exists $\rho \in A$ such that $\varepsilon' \equiv \rho^p \pmod{A\lambda^p}$. But $A/A\lambda = \mathbb{F}_p$ so there exists $r \in \mathbb{Z}$ such that $\rho \equiv r \pmod{A\lambda}$. Then $\varepsilon' \equiv \rho^p \equiv r^p \pmod{A\lambda^p}$. By Kummer's Lemma 3.4 on units, $\varepsilon'$ is the $p$th power of a unit $\varepsilon_1'$ in $A : \varepsilon' = (\varepsilon_1')^p$ and we may rewrite (3.14) as follows:

$$(\alpha')^p + (\varepsilon_1'\beta')^p = \varepsilon''(\delta')^p\lambda^{(m-1)p}. \tag{3.15}$$

This is now a relation like (3.11), with $m - 1$ instead of $m$. This contradicts the choice of a minimal $m$, and the proof is concluded.     □

In particular, if $p$ is a regular prime, there exist no nonzero integers $x, y, z \in \mathbb{Z}$ such that $x^p + y^p + z^p = 0$.

In order to understand the force of Kummer's theorem, it will be necessary to find out which primes are regular. Is there any neat characterization of regularity? This, and other questions, will be considered in the next lecture.

## Bibliography

1846   Dirichlet, G. L.
    Zur Theorie der Complexen Einheiten. Bericht Akad. d. Wiss., Berlin, 1846, 103–107. Also in *Werke*, Vol. I, G. Reimer Verlag, Berlin, 1889, 640–643. (Reprinted by Chelsea Publ. Co., New York, 1969).

1847   Kummer, E. E.*
Zur Theorie der complexen Zahlen. *J. reine u. angew. Math.*, **35**, 1847, 319–326.

1847   Kummer, E. E.
Beweis des Fermatschen Satzes der Unmöglichkeit von $x^\lambda + y^\lambda = z^\lambda$ für eine unendliche Anzahl Primzahlen $\lambda$. *Monatsber. Akad. d. Wiss.*, *Berlin*, 1847, 132–139, 140–141, 305–319.

1847   Kummer, E. E.
Über die Zerlegung der aus Wurzeln der Einheit Gebildeten complexen Zahlen in ihren Primfactoren. *J. reine u. angew Math.*, **35**, 1847, 327–367.

1847   Dirichlet, G. L.
Bemerkungen zu Kummer's Beweis des Fermat schen Satzes, die Unmöglichkeit von $x^\lambda + y^\lambda = z^\lambda$ für eine unendliche Anzahl von Primzahlen $\lambda$ betreffend. *Monatsber. Akad. d. Wiss.*, 1847, 139–141. Also in *Werke*, vol. II, G. Reimer Verlag, Berlin, 1889, 254–255. Reprinted by Chelsea Publ. Co., New York, 1969.

1850   Kummer, E. E.
Allgemeiner Beweis des Fermat'schen Satzes dass die Gleichung $x^\lambda + y^\lambda = z^\lambda$ durch ganze Zahlen unlösbar ist, für alle diejenigen Potenz-Exponenten $\lambda$, welche ungerade Primzahlen sind und in die Zählern der ersten $\frac{1}{2}(\lambda - 3)$ Bernoulli'schen Zahlen als Factoren nicht Vorkommen. *J. reine u. angew. Math.*, **40**, 1850, 130–138.

1897   Hilbert, D.
Die Theorie der algebraischen Zahlkörper. *Jahresber. d. Deutschen Math. Verein.* **4**, 1897, 175–546. Also in *Gesammelte Abhandlungen*, vol. I. Springer-Verlag, Berlin, 1932. Reprinted by Chelsea Publ. Co., New York, 1965.

* See also *Collected Papers*, vol. I, edited by A. Weil, Springer-Verlag, Berlin, 1975.

# LECTURE VI

# Regular Primes

To utilize the force of Kummer's main theorem, it is imperative to determine when a prime $p$ is regular. In theory, at least, this is simple. All one needs to do is to compute the class number of the cyclotomic field $K_p = \mathbb{Q}(\zeta_p)$, where $\zeta_p$ is a primitive $p$th root of 1. And so Kummer began by discovering formulas for the class number. However, it soon became apparent that the computations involved were much too difficult, and other more suitable criteria for regularity were needed. These were also discovered by Kummer.

I begin this lecture by giving all these classical results of Kummer and then I will describe more recent developments in this area.

## 1. The Class Number of Cyclotomic Fields

Kummer established formulas for the class number of cyclotomic fields in a series of very important papers (1850, 1851). A proof of these formulas may be found, for example, in the book by Borevich and Shafarevich, or in Hilbert's *Zahlbericht*.

One formula involving the class number and other invariants of the cyclotomic field may be attributed to Dirichlet.

I recall the notation. Let $p$ be an odd prime, let $\zeta = \zeta_p$ be a primitive $p$th root of 1, let $K = K_p = \mathbb{Q}(\zeta)$ and $A = A_p = \mathbb{Z}[\zeta]$, the ring of integers of the cyclotomic field $K$.

For every real number $t \geq 1$, let $\sigma(t)$ be the number of ideals $I$ of $A$ having norm $N(I) \leq t$. Then

$$\lim_{t \to \infty} \frac{\sigma(t)}{t} = \frac{2^{(p-3)/2} \pi^{(p-1)/2} R}{p^{p/2}} \quad h \neq 0, \tag{1.1}$$

where $h = h(p)$ is the class number of $K$ and $R$ is its regulator (defined in the preceding lecture).

There is an important connection between the class number and the Dedekind zeta-function of the field $K$.

It is well-known that the Riemann zeta-series $\sum_{n=1}^{\infty} 1/n^s$ defined for $s$ real, $s > 1$, converges uniformly and absolutely, for every $\delta > 0$, in the interval $[1 + \delta, \infty)$. It defines therefore a continuous function for $1 < s < \infty$:

$$\zeta(s) = \sum_{n=1}^{\infty} \frac{1}{n^s} \tag{1.2}$$

$\zeta(s)$ is the *Riemann zeta-function*. $\zeta(s)$ admits the following representation as an infinite convergent product (Euler):

$$\zeta(s) = \prod \frac{1}{1 - \dfrac{1}{p^s}} \quad \text{(for } s > 1\text{)}, \tag{1.3}$$

(where the product extends over the set of all primes $p$).

The difference

$$\zeta(s) - \frac{1}{s - 1} \tag{1.4}$$

remains bounded when $s$ tends to 1 ($s > 1$). In particular,

$$\lim_{s \to 1 + 0} (s - 1)\zeta(s) = 1. \tag{1.5}$$

With every algebraic number field $F$, it is possible to associate its Dedekind zeta-function (when $F = \mathbb{Q}$ this gives the Riemann zeta-function). To do this, for every integer $n \geq 1$ let $v(n)$ be the number of ideals $I$ of the ring of algebraic integers of $F$ such that $N(I) = n$. For every real number $s > 1$ the series

$$\sum_I \frac{1}{N(I)^s} = \sum_{n=1}^{\infty} \frac{v(n)}{n^s} \tag{1.6}$$

(where the sum is over the set of all ideals $I$ of the ring of integers of $F$) converges uniformly and absolutely, for every $\delta > 0$, in the interval $[1 + \delta, \infty)$. Therefore it defines a continuous function of $s$, $1 < s < \infty$, called the *Dedekind zeta-function* of $F$:

$$\zeta_F(s) = \sum_I \frac{1}{N(I)^s}. \tag{1.7}$$

This zeta-function also admits a Euler product expansion:

$$\zeta_F(s) = \prod_P \frac{1}{1 - \dfrac{1}{N(P)^s}} \quad \text{(for } s > 1\text{)} \tag{1.8}$$

(where this product extends over the set of all prime ideals $P$).

It may be shown that if $K = \mathbb{Q}(\zeta)$, then

$$\lim_{s \to 1+0} (s - 1)\zeta_K(s) = \frac{2^{(p-3)/2}\pi^{(p-1)/2}R}{p^{p/2}} \quad h \neq 0. \tag{1.9}$$

Formulas (1.1) and (1.9) are clearly not suitable for the computation of the class number. They have to be transformed into something more appropriate.

Let $m \geq 1$ be an integer. A mapping $\chi: \mathbb{Z} \to \mathbb{C}$ (complex numbers) is a *modular character*, belonging to the modulus $m$, when the following conditions are satisfied:

1. $\chi(a) = 0$ if and only if $\gcd(a,m) \neq 1$.
2. If $a \equiv b \pmod{m}$, then $\chi(a) = \chi(b)$.
3. $\chi(ab) = \chi(a)\chi(b)$.

It follows that $|\chi(a)| = 1$ for every integer $a$, $\gcd(a,m) = 1$.

Among the modular characters with modulus $m$ there is the *principal character* modulo $m$, denoted $\chi_0$, namely: $\chi_0(a) = 1$ when $\gcd(a,m) = 1$, $\chi_0(a) = 0$ when $\gcd(a,m) \neq 1$.

If $m$ divides $m'$ and $\chi$ is a character with modulus $m$, it induces a character $\chi'$ with modulus $m'$, which is defined as follows: if $\gcd(a,m') \neq 1$, then $\chi'(a) = 0$; if $\gcd(a,m') = 1$, then $\chi'(a) = \chi(a)$.

Let $\chi$ be a character with modulus $m$, If there exists a divisor $d$ of $m$ and a character $\chi'$ with modulus $d$ which induces $\chi$, then $\chi$ is said to be *imprimitive*; otherwise $\chi$ is a *primitive character*. If $\chi$ is a primitive character with modulus $m$, then its modulus is called the *conductor of* $\chi$, denoted by $f_\chi$. In particular, the principal character of modulus 1 is primitive and has conductor equal to 1.

If $\chi$ is a character with modulus $m$, the series $\sum_{n=1}^{\infty} \chi(n)/n^s$ is called the *L-series of* $\chi$. For every $\delta > 0$ and every character $\chi$ with modulus $m$, the L-series of $\chi$ converges uniformly and absolutely in the interval $[1 + \delta, \infty)$. Hence it defines a continuous function $L(s|\chi)$ of $s$ on the interval $(1,\infty)$:

$$L(s|\chi) = \sum_{n=1}^{\infty} \frac{\chi(n)}{n^s} \quad \text{(for } 1 < s). \tag{1.10}$$

The function $L(s|\chi)$ admits the product representation

$$L(s|\chi) = \prod_p \frac{1}{1 - \dfrac{\chi(p)}{p^s}} \quad \text{(for } 1 < s). \tag{1.11}$$

For the principal character $\chi_0$, with modulus $m$:

$$L(s|\chi_0) = \prod_{p|m} \left(1 - \frac{1}{p^s}\right)\zeta(s) \quad \text{(for } 1 < s). \tag{1.12}$$

Actually, if $\chi \neq \chi_0$, the domain of convergence of the L-series of $\chi$ is larger: for every $\delta > 0$ it converges uniformly on the interval $[\delta,\infty)$. Hence it defines a continuous function on $(0,\infty)$, still denoted $L(s|\chi)$.

If $\chi$ is a character with modulus $m$, if $\xi = \zeta_m$ is a primitive $m$th root of 1, say $\xi = \cos(2\pi/m) + i\sin(2\pi/m)$, the sums

$$\tau_k(\chi) = \sum_{\substack{a=1 \\ \gcd(a,m)=1}}^{m} \chi(a)\xi^{ak} \qquad (k = 0,1,\ldots,m-1) \tag{1.13}$$

are called the *Gauss sums* belonging to the character $\chi$ (and to $\xi$). In particular

$$\tau_0(\chi) = \sum_{\substack{a=1 \\ \gcd(a,m)=1}}^{m} \chi(a) = \begin{cases} \varphi(m) & \text{when } \chi = \chi_0, \\ 0 & \text{when } \chi \neq \chi_0. \end{cases} \tag{1.14}$$

The *principal Gauss sum* is

$$\tau(\chi) = \tau_1(\chi) = \sum_{\substack{a=1 \\ \gcd(a,m)=1}}^{m} \chi(a)\xi^{a}. \tag{1.15}$$

If $\chi$ is a primitive character with conductor $f$, then $|\tau(\chi)|^2 = f$.

Now if $K = \mathbb{Q}(\zeta)$, then its Dedekind zeta-function is expressible as a product involving the $L$-series of the characters modulo $p$:

$$\zeta_K(s) = \prod_{\chi} L(s|\chi) \cdot \frac{1}{1 - \dfrac{1}{p^s}} \qquad (\text{for } s > 1) \tag{1.16}$$

and therefore

$$\zeta_K(s) = \prod_{\chi \neq \chi_0} L(s|\chi) \cdot \zeta(s) \quad (\text{for } s > 1). \tag{1.17}$$

From this may be deduced the following expression for the class number $h(p)$ of $\mathbb{Q}(\zeta_p)$:

$$h(p) = \frac{p^{p/2}}{2^{(p-3)/2}\pi^{(p-1)/2}R} \times \prod_{\chi \neq \chi_0} L(1|\chi). \tag{1.18}$$

In order to compute $h = h(p)$ it is still necessary to evaluate $L(1|\chi)$ for $\chi \neq \chi_0$. It may be shown that

$$L(s|\chi) = \frac{1}{p}\sum_{k=0}^{p-1} \overline{\tau_k(\chi)}\left[\sum_{n=1}^{\infty} \frac{\zeta^{-nk}}{n^s}\right] \qquad (\text{for } s > 1). \tag{1.19}$$

This gives

$$h(p) = \frac{p^{p/2}}{2^{(p-3)/2}\pi^{(p-1)/2}R} \times \frac{1}{p^{p-2}} \prod_{\chi \neq \chi_0} \left\{\sum_{k=1}^{p-1} \tau_k(\chi)\log\frac{1}{1-\zeta^{-k}}\right\}. \tag{1.20}$$

This last formula has an advantage over the previous one. It does not involve any infinite product or series. However it is still not suitable for direct computation, since the integer $h$ is expressed in terms of logarithms and complex numbers. To put it in appropriate form it is necessary to evaluate $L(1|\chi)$ when $\chi$ is an even character, that is, $\chi(a) = \chi(-a)$, and when $\chi$ is an odd character, that is, $\chi(a) = -\chi(-a)$.

If $\chi$ is even, then

$$L(1\,|\,\chi) = -\frac{\tau_1(\chi)}{p} \sum_{k=1}^{p-1} \overline{\chi(k)} \log\left(\sin\frac{k\pi}{p}\right). \tag{1.21}$$

If $\chi$ is odd, then

$$L(1\,|\,\chi) = \frac{\pi i \tau_1(\chi)}{p^2} \sum_{k=1}^{p-1} \overline{\chi(k)}k. \tag{1.22}$$

This gives the important formulas:

**(1A)** *The class number of* $K = \mathbb{Q}(\zeta)$ *is* $h(p) = h_1(p)h_2(p)$ *where*

$$h_1(p) = \frac{1}{(2p)^{(p-3)/2}} |G(\eta)G(\eta^3) \cdots G(\eta^{p-2})| \tag{1.23}$$

*and*

$$h_2(p) = \frac{2^{(p-3)/2}}{R} \times \prod_{k=1}^{(p-3)/2} \left| \sum_{j=0}^{(p-3)/2} \eta^{2kj} \log|1 - \zeta^{g^j}| \right|, \tag{1.24}$$

*where* $\eta$ *is a primitive* $(p-1)$th *root of* $1$, $g$ *is a primitive root modulo* $p$, $g_j \equiv g^j \pmod{p}$ *with* $1 \le g_j \le p-1$ *and*

$$G(X) = \sum_{j=0}^{p-2} g_j X^j.$$

The numbers $h_1(p)$, $h_2(p)$ are called respectively the *first factor* and the *second factor* of the class number of $K = \mathbb{Q}(\zeta)$.

The formula for the first factor may be also written in the following ways. First:

$$h_1(p) = \gamma(p) \cdot \prod_{\chi \in S} L(1\,|\,\chi), \tag{1.25}$$

where $S$ is the set of odd characters modulo $p$ and

$$\gamma(p) = 2p\left(\frac{p}{4\pi^2}\right)^{(p-1)/4} = \frac{p^{(p+3)/4}}{2^{(p-3)/2}\pi^{(p-1)/2}}. \tag{1.26}$$

Secondly:

$$h_1(p) = \frac{1}{(2p)^{(p-3)/2}} \left| \prod_{\chi \in S} \sum_{k=1}^{p-1} \chi(k)k \right|, \tag{1.27}$$

where $S$ was defined above.

At this stage of the theory nothing yet may be said about the nature of the numbers $h_1(p)$, $h_2(p)$, but it will soon be seen that they are integers and in fact, they admit an arithmetical interpretation. For this purpose it is necessary to study the field $K^+ = \mathbb{Q}(\zeta + \zeta^{-1})$ which is equal to $K \cap \mathbb{R}$. $K^+$ is called the *real cyclotomic field*.

The degree of the extension $K^+\,|\,\mathbb{Q}$ is $(p-1)/2$. The ring of integers $A^+$ of $K^+$ is equal to $A^+ = \mathbb{Z}[\zeta + \zeta^{-1}]$. The regulator of $K^+$ is $R^+ = R/2^{(p-3)/2}$, where $R$ is the regulator of $K$.

Comparing the Dedekind zeta-series for $K$ and $K^+$ leads eventually to the following result:

**(1B)** *The second factor $h_2(p)$ of the class number of $K$ is equal to the class number of $K^+ = \mathbb{Q}(\zeta + \zeta^{-1})$. In particular, $h_2(p)$ is a positive integer.*

For this reason, the second factor is also called the *real class number* of $\mathbb{Q}(\zeta)$ and denoted by $h^+ = h^+(p)$. From now on, I shall use this notation, since it is more suggestive.

Another interpretation of $h^+(p)$ is the following:

**(1C)** $h^+(p)$ *is the index of the subgroup $V$ in the group $U^+$.*

Recall that $U^+$ denotes the group of positive real units of $K$ and $V$ is the subgroup generated by the Kummer circular units

$$\delta_k = \sqrt{\frac{1 - \zeta^k}{1 - \zeta} \cdot \frac{1 - \zeta^{-k}}{1 - \zeta^{-1}}} \tag{1.28}$$

(it suffices to take $k = 2, \ldots, (p - 1)/2$).

If $g$ is a primitive root modulo $p$, let $\delta = \delta_g$. If $\sigma$ is the generator of the Galois group $G$ of $\mathbb{Q}(\zeta)|\mathbb{Q}$ such that $\sigma(\zeta) = \zeta^g$, let

$$\Delta = \left|\det(\log \sigma^{i+j}(\delta))_{i,j=0,1,\ldots,(p-5)/2}\right| \tag{1.29}$$

Then the second factor may be written as

$$h^+(p) = \frac{2^{(p-3)/2}\Delta}{R}. \tag{1.30}$$

The Kummer units are independent, since the index $(U^+ : V)$ is finite. They constitute a fundamental system of units if and only if $h^+(p) = 1$.

The number $h(p)/h^+(p) = h_1(p)$ is also called the *relative class number* of $\mathbb{Q}(\zeta)$ and denoted by $h^* = h^*(p)$. From now on I shall use this notation.

Kummer also showed:

**(1D)** $h^*$ *is a positive integer.*

In principle, formula (1.23) allows us to compute the first factor of the class number—at least when $p$ is not too large. As I shall soon indicate, Kummer made extensive computations. On the other hand, formulas (1.24) and (1.30) are still unwieldy, the point being that the regulator is extremely difficult to compute. Indeed, it would require the knowledge of a fundamental system of units for $K$. This is the crucial difficulty.

Recall however that all that is needed, is to know whether the prime $p$ divides the class number.

In 1850 Kummer succeeded in proving the important:

**(1E)** $p$ *divides the class number $h$ if and only if $p$ divides the first factor $h^*$.*

To establish this theorem, Kummer introduced transcendental methods. More precisely, for the first time he made use of $p$-adic methods (without calling them by this name). Such methods appeared explicitly only with Hensel (1908). So, once more, Kummer was a pioneer of genius.

## 2. Bernoulli Numbers and Kummer's Regularity Criterion

In 1850, Kummer proved the following beautiful condition for regularity:

**(2A)** $p$ *divides* $h^*$ *if and only if there exists an integer* $k$, $1 \le k \le (p-3)/2$, *such that* $p^2$ *divides the sum* $\sum_{j=1}^{p-1} j^{2k}$.

PROOF. I sketch the main points in the proof. By (1.23), $h^* = |\mu|/(2p)^{t-1}$ where $t = (p-1)/2$, $\mu = G(\eta)G(\eta^3) \cdots G(\eta^{p-2})$.

From $|\mu| = h^*(2p)^{t-1}$, it follows that $\mu$ is independent of the choice of $g$ (a primitive root modulo $p$) and of $\eta$ (a $(p-1)$th root of 1). It is convenient to choose $g$ such that $g^{p-1} \equiv 1 \pmod{p^2}$, which is always possible.

Let $B = \mathbb{Z}[\eta]$ be the ring of integers of the field $\mathbb{Q}(\eta)$. To show that $p$ divides $h^*$ is equivalent to proving that if $P$ is any prime ideal of $B$ dividing $Bp$, then $P$ divides $B(\mu/p^{t-1})$.

Let $P$ be such a prime ideal. By changing $\eta$ appropriately into some other $(p-1)$th root of 1, it is possible to insure that $P|B(1 - g\eta^{-1})$. This may be seen by considering the decomposition of $Bp$ into a product of prime ideals.

Since $\mu = G(\eta)G(\eta^3) \cdots G(\eta^{p-2})$, $P$ divides $B(\mu/p^{t-1})$ if an only if there exists $l$, odd, $1 \le l \le p - 2$, such that $P$ divides $B(G(\eta^l)/p)$, that is, $P^2$ divides $BG(\eta^l)$. It may be seen that $P$ does not divide $BG(\eta^{p-2})$ (by the choice of $\eta$). Since

$$1 - g^{p-1} = \prod_{j=0}^{p-2} (1 - g\eta^j) \equiv 0 \pmod{p^2},$$

$P^2$ divides $1 - g\eta^{p-2}$, so $g \equiv \eta \pmod{P^2}$. Therefore

$$G(\eta^l) \equiv \sum_{j=0}^{p-2} g_j g^{jl} \pmod{P^2}.$$

Since $p$ is unramified in $B$, $P^2$ divides $BG(\eta^l)$ exactly when $p^2$ divides $\sum_{j=0}^{p-2} g_j g^{jl}$. Putting $g_j \equiv g^j + a_j p \pmod{p^2}$, with integers $a_j$, yields

$$g_j^{l+1} \equiv (l+1)g_j g^{jl} - lg^{j(l+1)} \pmod{p^2}$$

(for $j = 0, 1, \ldots, p - 2$). Hence

$$\sum_{j=0}^{p-2} g_j^{l+1} \equiv (l+1)\left( \sum_{j=0}^{p-2} g_j g^{jl} \right) \pmod{p^2}.$$

Since $p \nmid l + 1$, $p^2$ divides $\sum_{j=0}^{p-2} g_j g^{jl}$ if and only if $p^2$ divides $\sum_{j=0}^{p-2} g_j^{l+1} = \sum_{j=1}^{p-1} j^{l+1}$. Putting $2k = l + 1$, this gives the stated condition.   □

The sums $\sum_{k=1}^{n} j^k$ have been studied by Fermat, Pascal and Jakob Bernoulli.

**(2B)** *For every integer $k \geq 0$ there exists a polynomial $S_k(X) \in \mathbb{Q}[X]$ with the following properties:*

1. *$S_k(X)$ has degree $k + 1$ and leading coefficient $1/(k + 1)$.*
2. *$(k + 1)! S_k(X) \in \mathbb{Z}[X]$.*
3. *for every $n \geq 1$, $S_k(n) = \sum_{j=1}^{n} j^k$.*

*These polynomials are determined recursively as follows:*

$$S_0 = X$$

*and if $k \geq 1$, then*

$$\binom{k + 1}{1} S_k(X) + \binom{k + 1}{2} S_{k-1}(X) + \cdots + \binom{k + 1}{k} S_1(X) + S_0(X)$$
$$= (X + 1)^{k+1} - 1. \quad (2.1)$$

*Moreover, $S_k(X)$ has no constant term.*

PROOF. This is an easy induction on $k$.   □

For example:

$$S_0(X) = X,$$
$$S_1(X) = \tfrac{1}{2}X^2 + \tfrac{1}{2}X,$$
$$S_2(X) = \tfrac{1}{3}X^3 + \tfrac{1}{2}X^2 + \tfrac{1}{6}X,$$
$$S_3(X) = \tfrac{1}{4}X^4 + \tfrac{1}{2}X^3 + \tfrac{1}{4}X^2.$$

In view of (2B), each polynomial $S_k(X)$ may be written in the form

$$(k + 1)S_k(X) = X^{k+1} + \binom{k + 1}{1} b_{k1} X^k + \binom{k + 1}{2} b_{k2} X^{k-1}$$
$$+ \cdots + \binom{k + 1}{k} b_{kk} X, \quad (2.2)$$

where the coefficients $b_{kj}$ are rational numbers.

Euler discovered the noteworthy fact that $b_{jk} = b_{kk}$ for all $j \geq k$. This remarkably simplifies the computation of the successive polynomials $S_k(X)$. But, even more is true, the coefficients $b_{jk}$ satisfy a recurrence relation and are also intimately connected with a power-series development.

To be more explicit, the formal power series

$$\frac{e^X - 1}{X} = 1 + \frac{1}{2!} X + \frac{1}{3!} X^2 + \frac{1}{4!} X^3 + \cdots \quad (2.3)$$

is invertible, since its constant term is 1. Let the inverse be written as follows:

$$\frac{X}{e^X - 1} = B_0 + \frac{B_1}{1!} X + \frac{B_2}{2!} X^2 + \frac{B_3}{3!} X^3 + \cdots, \tag{2.4}$$

where $B_i$ are rational numbers ($i \geq 0$). For example, a simple computation gives $B_0 = 1$, $B_1 = -\frac{1}{2}$, $B_2 = \frac{1}{6}$.

**(2C)** *The numbers $B_i$ satisfy the following recurrence relation:*

$$\binom{k+1}{1} B_k + \binom{k+1}{2} B_{k-1} + \cdots + \binom{k+1}{k} B_1 + 1 = 0 \tag{2.5}$$

*and for every $k \geq 1$:*

$$(k+1)S_k(X) = X^{k+1} - \binom{k+1}{1} B_1 X^k + \binom{k+1}{2} B_2 X^{k-1}$$

$$+ \cdots + \binom{k+1}{k} B_k X. \tag{2.6}$$

PROOF. Formula (2.5) is easily obtained by considering the coefficients of the powers of $X$ in the product of $(e^X - 1)/X$ and $X/(e^X - 1)$.

For each $n \geq 2$ let

$$U(T) = k! T(1 + e^T + e^{2T} + \cdots + e^{(n-1)T}),$$

where $T$ is an indeterminate.

The coefficient of $T^{k+1}$ is equal to

$$k! \left( \frac{1}{k!} + \frac{2^k}{k!} + \cdots + \frac{(n-1)^k}{k!} \right) = S_k(n-1). \tag{2.7}$$

On the other hand,

$$U(T) = k! T \frac{e^{nT} - 1}{e^T - 1} = k! \times \frac{T}{e^T - 1} \times (e^{nT} - 1)$$

$$= k! \left( \sum_{i=0}^{\infty} \frac{B_i}{i!} T^i \right) \left( \sum_{i=1}^{\infty} \frac{n^i T^i}{i!} \right). \tag{2.8}$$

Comparing the coefficient of $T^{k+1}$ in (2.8) with (2.7):

$$S_k(n-1) = k! \left( \sum_{i=0}^{k} \frac{B_i}{i!} \frac{n^{k+1-i}}{(k+1-i)!} \right),$$

hence

$$(k+1)S_k(n-1) = n^{k+1} + \binom{k+1}{1} B_1 n^k + \binom{k+1}{2} B_2 n^{k-1}$$

$$+ \cdots + \binom{k+1}{k} B_k n.$$

Since this holds for every $n \geq 2$ and $S_k(n) = n^k + S_k(n - 1)$,

$$(k + 1)S_k(X) = X^{k+1} - \binom{k+1}{1}B_1 X^k + \binom{k+1}{2}B_2 X^{k-1}$$

$$+ \cdots + \binom{k+1}{1}B_k X. \qquad \square$$

It is easily seen from

$$\frac{-X}{e^{-X} - 1} - \frac{X}{2} = \frac{X}{e^X - 1} + \frac{X}{2} = 1 + \sum_{k=2}^{\infty} \frac{B_k}{k!} X^k$$

that

$$B_{2k+1} = 0 \quad \text{for } k \geq 1. \tag{2.9}$$

The numbers $B_i$ are called the *Bernoulli numbers*. They appear for the first time in Jakob Bernoulli's posthumous work *Ars Conjectandi* (1713, page 97). Bernoulli computed $B_k$ for $k \leq 10$.

Euler rediscovered the Bernoulli numbers, while calculating the formula for $S_k(n)$. He computed $B_k$ for $k \leq 30$.

As will be seen, the Bernoulli numbers play a fundamental role in connection with regularity of primes, and so I'll return to them repeatedly. The algebraic and arithmetic properties of the Bernoulli numbers are quite fascinating and the literature about them is very extensive.

One of the main theorems about Bernoulli numbers was communicated by von Staudt to Gauss. The proof was published by von Staudt, in 1840, just after Clausen published the statement of the theorem. This theorem gives the exact value of the denominator of any Bernoulli number.

**(2D)** *If $k \geq 1$, then*

$$B_{2k} + \sum_{\substack{p \text{ prime} \\ p-1 \mid 2k}} \frac{1}{p} \in \mathbb{Z}. \tag{2.10}$$

There are numerous proofs of this theorem, for example: Hardy and Wright give the proof by Rado (1934).

Write

$$B_{2k} = \frac{N_{2k}}{D_{2k}} \quad \text{with } D_{2k} > 0, \ \gcd(N_{2k}, D_{2k}) = 1. \tag{2.11}$$

Then $D_{2k}$ is square-free and a prime $p$ divides $D_{2k}$ if and only if $p - 1 \mid 2k$. In particular, $6 \mid D_{2k}$ (for every $k \geq 1$).

Moreover, it follows also that if $p - 1 \mid 2k$, then:

$$pB_{2k} \equiv -1 \pmod{p}. \tag{2.12}$$

Another congruence which may be established is the following:

**(2E)** *For every integer $m \geq 2$,*

$$mN_{2k} \equiv D_{2k}S_{2k}(m - 1) \pmod{m^2}. \tag{2.13}$$

To make this precise, if $p$ is any prime number, a rational number $a/b$ (with $a$, $b$ integers, $b \neq 0$, $\gcd(a,b) = 1$) is said to be *p-integral* if $p$ does not divide $b$. If moreover, $p$ divides $a$, then $p$ is said to divide $a/b$. If $a/b$, $c/d$ are rational numbers, then $a/b \equiv c/d \pmod{p}$ when $p$ divides $(a/b) - (c/d)$. The properties of such congruences are analogous to those of the ordinary congruences of integers.

So to say that $p$ divides the Bernoulli number $B_{2k} = N_{2k}/D_{2k}$ means that $p \mid N_{2k}$ (and of course $p \nmid D_{2k}$, hence $p - 1 \nmid 2k$).

Though it is completely known which primes divide the denominator of Bernoulli numbers, it remains almost a complete mystery to predict which primes divide $N_{2k}$. Yet, according to the next regularity criterion of Kummer (1850) this is precisely what matters.

**(2F)** *$p$ divides $h^*$ (that is, $p$ is not a regular prime) if and only if $p$ divides the numerator of one of the Bernoulli numbers $B_2, B_4, \ldots, B_{p-3}$.*

PROOF. By (2.13) (with $m = p$) and by (2A), $p$ divides $h^*$ if and only if $p^2$ divides $pN_{2k}$, (where $2 \leq k \leq p - 3$), that is, if and only if $p \mid N_{2k}$.  □

The situation now has been much improved. It suffices to determine the Bernoulli numbers with indices up to $p - 3$, by means of the recurrence relation (2.5), and to determine whether their numerators are multiples of $p$. The only problem lies in the fact that as the index grows, so does the numerator of the corresponding Bernoulli number—in a rather dizzying way.

For example

$$B_{34} = \frac{2577687858367}{6}$$

which is still manageable. However, the numerator of $B_{220}$ already has 250 digits. And in conjunction with Fermat's problem, the computations have been pushed much further ahead, requiring further developments of the theory of Bernoulli numbers.

Right now, I only wish to say that Kummer discovered that 37, 59, 67 are the only irregular primes less than 100 (in 1850, 1851). Later in 1874, he showed that 101, 103, 131, 149, 157 are the irregular primes less than 164. For example $37 \mid B_{32}$, $59 \mid B_{44}$. As for $p = 157$ it divides both $B_{62}$ and $B_{110}$.

# 3. Various Arithmetic Properties of Bernoulli Numbers

As I have already intimated, a deeper study of the divisibility properties of Bernoulli numbers is essential, because the absolute value of their numerators grows very rapidly.

This follows from a celebrated formula by Euler, connecting the Bernoulli numbers and the values of Riemann's zeta-function at even positive integers.

First, Euler established the series development of the cotangent function (for $z$ a complex number, $|z| < \pi$):

$$\cot z = \frac{1}{z} - 2z \sum_{n=1}^{\infty} \frac{1}{n^2\pi^2 - z^2}$$

$$= \frac{1}{z} - 2z \sum_{n=1}^{\infty} \sum_{k=0}^{\infty} \frac{z^{2k}}{(n\pi)^{2k+2}}. \tag{3.1}$$

Because of the absolute convergence of the series, the order of summation may be interchanged, yielding:

$$\cot z = \frac{1}{z} - 2 \sum_{k=1}^{\infty} \zeta(2k) \frac{z^{2k-1}}{\pi^{2k}}. \tag{3.2}$$

From this, it follows:

$$B_{2k} = (-1)^{k-1} \frac{2(2k)!}{(2\pi)^{2k}} \zeta(2k) \quad \text{(for } k \geq 1\text{).} \tag{3.3}$$

Thus, knowing the Bernoulli numbers is essentially equivalent to knowing the values of the Riemann zeta-function at the positive even integers—a point which is worth stressing.

Often quoted are the special cases:

$$\sum_{n=1}^{\infty} \frac{1}{n^2} = \frac{\pi^2}{6}, \qquad \sum_{n=1}^{\infty} \frac{1}{n^4} = \frac{\pi^4}{90}.$$

For $k$ sufficiently large, $\zeta(2k)$ may be easily computed to a high degree of accuracy. Since the fractional part of $B_{2k}$ is known, by the theorem of von Staudt and Clausen, $B_{2k}$ may be determined with an error less than $\frac{1}{2}$. This method has been used in the past to build tables of Bernoulli numbers, but has its own limitations, since the series for $\zeta(2k)$ converges very slowly.

From (3.3) it is seen that $B_{2k} > 0$ if and only if $k$ is odd. It is also seen without difficulty that the sequence $|B_{2k}|$ is strictly increasing (for $k \geq 4$).

By means of Stirling's formula:

$$n! \sim \sqrt{2\pi} \, e^{-n} n^{n+\frac{1}{2}} \tag{3.4}$$

it follows that

$$|B_{2k}| \sim 4\sqrt{\pi k} \left(\frac{k}{\pi e}\right)^{2k} \tag{3.5}$$

and also that for every $M \geq 1$

$$\lim_{k \to \infty} \frac{|B_{2k}|}{(2k)^M} = \infty. \tag{3.6}$$

All this shows how hopeless is the exact determination of the numerator $N_{2k}$, except for rather small values of $2k$.

One of the most useful congruence properties of Bernoulli numbers was already discovered in 1851 by Kummer himself. He had the idea of considering the function

$$f(2k) = \frac{B_{2k}}{2k} \text{ modulo } p,$$

defined for integers $2k$ such that $p - 1 \nmid 2k$. Note that in this case, by the theorem of von Staudt and Clausen, $B_{2k}/2k$ is $p$-integral and it is possible to consider its residue class modulo $p$.

Kummer showed that the function $f$ has period $p - 1$. Explicitly:

$$\frac{B_{2k+p-1}}{2k + p - 1} \equiv \frac{B_{2k}}{2k} \pmod{p}. \tag{3.7}$$

This is nothing but a special case of the following more general congruence, also proved by Kummer:

**(3A)** *Let $n \geq 1$ and $k$ be integers such that $2k \geq n + 1$. If $p$ is an odd prime and if $a$ is an integer, $p \nmid a$, then*

$$\sum_{j=0}^{n} (-1)^j \binom{n}{j} \frac{(a^{2k+j(p-1)} - 1)B_{2k+j(p-1)}}{2k + j(p - 1)} \equiv 0 \pmod{p^n}. \tag{3.8}$$

*In particular, if $p - 1$ does not divide $2k$*

$$\sum_{j=0}^{n} (-1)^j \binom{n}{j} \frac{B_{2k+j(p-1)}}{2k + j(p - 1)} \equiv 0 \pmod{p^n}. \tag{3.9}$$

Concerning the divisibility properties of the numerators of the Bernoulli numbers, one of the earliest results is attributed to Adams (1878). In fact, he only proved a special case of a theorem previously stated by Sylvester (1861). The first proof of Sylvester's theorem is due to Glaisher (1900):

**(3B)** *If $p \nmid D_{2k}$, if $2k = p^t r$, with $t \geq 1$, and if $p \nmid r$, then $p^t \mid N_{2k}$.*

Letting $B_{2k}/2k = N'_{2k}/D'_{2k}$ with $D'_{2k} > 0$, $\gcd(N'_{2k}, D'_{2k}) = 1$, it follows:

**(3C)** *If $p$ is a prime, then $p \mid D_{2k}$ if and only if $p \mid D'_{2k}$.*

In 1845, von Staudt determined some factors of the numerator $N_{2k}$. Let

$$2k = k_1 k_2 \quad \text{with } \gcd(k_1, k_2) = 1 \tag{3.10}$$

such that $p \mid k_2$ if and only if $p \mid D_{2k}$. von Staudt proved:

**(3D)** *With above notation: $k_1 \mid N_{2k}$.*

This theorem however does not help to determine whether a prime $p$ divides $N_{2k}$ for $2 \leq 2k \leq p - 3$, since it only pinpoints factors $k_1 \leq k$.

The following irregularity criterion appears already in Vandiver's paper of 1932, though it is not stated explicitly as a lemma or theorem. It was found again by Montgomery in 1965; see also Carlitz, 1954.

**(3E)** *A prime p is irregular if and only if there exists an integer k such that* $p$ *divides* $N_{2k}/k_1$ *(where* $k_1$ *was defined in (3.10))*.

PROOF. As an illustration, I give this proof. If $p$ is irregular, then there exists $k$, $1 \leq k \leq (p-3)/2$ such that $p \mid N_{2k}$. But $p \nmid k_1$, so $p \mid (N_{2k}/k_1)$.

Conversely, since $p \mid N_{2k}$, then $p \nmid D_{2k}$ so $p - 1 \nmid 2k$. Let $2k \equiv 2h \pmod{p-1}$ with $2 \leq 2h \leq p - 3$. By Kummer's congruence $B_{2k}/2k \equiv B_{2h}/2h \pmod{p}$. So

$$2h\,\frac{N_{2k}}{k_1}\,D_{2h} \equiv k_2 N_{2h} D_{2k} \pmod{p}.$$

Since $p \nmid D_{2k}$, $p \nmid k_2$; by hypothesis, $p \mid N_{2h}$ and so by Kummer's regularity criterion, $p$ is an irregular prime.                                                                        □

## 4. The Abundance of Irregular Primes

In 1847 Kummer did not know whether every prime is regular. But in 1850 and 1851, he discovered that 37, 59, 67 are the only irregular primes less than 100. In 1874, he extended his computations up to 164. At that time and based on probabilistic arguments (which he himself regarded as doubtful), Kummer advanced the conjecture that asymptotically there should be as many regular as irregular primes.

Today, with extensive tables, it has been observed that the ratio

$$\frac{\text{number of irregular primes less than } N}{\text{number of primes less than } N}$$

is about 0.39 when $N$ is large. I'll return to this point at the final part of this lecture, in connection with a heuristic prediction of Siegel.

Despite the observed plurality of regular primes, it has not yet been shown that there exist infinitely many regular primes.

On the other hand, quite surprisingly, Jensen proved in 1915 that there are infinitely many irregular primes. I give below a proof which is due to Carlitz (1954):

**(4A)** *There exist infinitely many irregular primes.*

PROOF. Let $p_1, \ldots, p_m$ be irregular primes. Due to the growth of $|B_{2k}|$ with $k$, there is an index $k$ such that $2k$ is a multiple of $(p_1 - 1) \cdots (p_m - 1)$ and $|B_{2k}| > 2k$. Let $|B_{2k}|/2k = a/b$ with $a > b \geq 1$, $\gcd(a,b) = 1$.

Let $p$ be a prime dividing $a$. I show that $p \nmid D_{2k}$. Indeed, from

$$|N_{2k}|b = 2kD_{2k}a \quad \text{and} \quad p \mid a,\, p \nmid b$$

it follows that $p \mid N_{2k}$, so $p \nmid D_{2k}$.

Let $2k = k_1 k_2$ as in (3.10). Then by (3D)

$$\frac{|N_{2k}|}{k_1}\,b = k_2 D_{2k} a.$$

It follows that $p$ divides $N_{2k}/k_1$. By the Vandiver and Montgomery irregularity criterion, $p$ is irregular. Since $p - 1 \nmid 2k$ (by the theorem of von Staudt and Clausen) $p$ is distinct from $p_1, \ldots, p_m$. $\qquad\square$

As a matter of fact, Jensen had proved more, namely:

**(4B)** *There exists an infinite number of irregular primes $p$ such that $p \equiv 3$* (mod 4).

His ingenious proof, written in Danish, was made available in English by Vandiver in 1955.

Jensen's result was the object of several generalizations. In 1965, Montgomery proved that if $m > 2$, there exist infinitely many irregular primes $p$ such that $p \not\equiv 1 \pmod{m}$.

In 1976, Metsänkylä published the following result, which had been obtained independently in 1975 by Yokoi, for $m$ prime:

**(4C)** *Let $m > 2$, let $G$ be the group of invertible residue classes modulo $m$. If $H$ is any proper subgroup of $G$, there exist infinitely many irregular primes $p$ such that $p$ modulo $m$ is not in $H$.*

Up to now it is not known whether, given $m > 2$, there exist infinitely many irregular primes $p$ such that $p \equiv 1 \pmod{m}$. Montgomery conjectures that this is indeed true. But the only available result in this direction is due to Metsänkylä (1971):

**(4D)** *There exists an infinite number of irregular primes $p$ such that $p \equiv 1$* (mod 3) or $p \equiv 1$ (mod 4).

# 5. Computation of Irregular Primes

To determine whether a prime $p$ is irregular requires the computation of the residues modulo $p$ of the Bernoulli numbers $B_{2k}$ for $2 \leq 2k \leq p - 3$. Due to the size of the numbers involved, it is essential to derive congruences which reduce the amount of calculation.

A very general congruence for the sums $S_k(n)$ was discovered by Voronoï (1889). It is reproduced in the book of Uspensky and Heaslet (1939) and it was rediscovered by Grün in 1940.

**(5A)** *Let $m \geq 2$, $k \geq 1$, $a \geq 1$ and let $\gcd(a,m) = 1$. Then*

$$(a^{2k} - 1)S_{2k}(m - 1) \equiv 2kma^{2k-1} \sum_{j=1}^{m-1} j^{2k-1}\left[\frac{ja}{m}\right] \pmod{m^2} \qquad (5.1)$$

*and*

$$(a^{2k} - 1)N_{2k} \equiv D_{2k}2ka^{2k-1} \sum_{j=1}^{m-1} j^{2k-1}\left[\frac{ja}{m}\right] \pmod{m}. \qquad (5.2)$$

One of the various applications of Voronoï's congruences is to derive a simple proof of the following congruence of Vandiver (1917, and an easier proof in 1937):

**(5B)** *Let $p$ be an odd prime, $k \geq 1$, and $a \geq 1$, such that $a < p$ and $p - 1$ does not divide $2k$. Then*

$$(1 - a^{2k})B_{2k} \equiv 2ka^{2k-1} \sum_{j=1}^{a-1} S_{2k-1}\left(\left[\frac{jp}{a}\right]\right) \pmod{p}. \qquad (5.3)$$

I will now show how Stafford and Vandiver derived in 1930 some interesting congruences for $B_{2k}$. They are very useful in determining whether a given prime is regular.

**(5C)** *If $p$ is a prime, $p \geq 7$, $k \geq 1$, and $p - 1 \nmid 2k$, then*

$$\frac{4^{p-2k} + 3^{p-2k} - 6^{p-2k} - 1}{4k} B_{2k} \equiv \sum_{p/6 < j < p/4} j^{2k-1} \pmod{p}. \qquad (5.4)$$

Note that the sum in the right-hand side is only for $j$ between $p/6$ and $p/4$, which is a relatively small range.

In 1954, Lehmer, Lehmer, and Vandiver made use of (5.4) to determine whether a given prime $p$ is regular. By Kummer's regularity criterion, this is equivalent to $p \nmid B_{2k}$ for $2 \leq 2k \leq p - 3$. If $p$ does not divide

$$S_{2k} = \sum_{p/6 < j < p/4} j^{2k-1},$$

then also $p \nmid B_{2k}$. Again, if $p \mid S_{2k}$, but $p$ does not divide

$$6^{p-2k} - 4^{p-2k} - 3^{p-2k} + 1, \text{ then } p \mid B_{2k}.$$

No conclusion may be drawn if $p \mid S_{2k}$ and $p \mid 6^{p-2k} - 4^{p-2k} - 3^{p-2k} + 1$. In this situation, the same method may be applied (up to now with success) to one of the following congruences, which were also proved by Vandiver in 1937:

**(5D)** *If $p$ is a prime, $p \geq 5$, $k \geq 1$, and $p - 1 \nmid 2k$, then*

$$\frac{2^{p-2k} + 3^{p-2k} - 4^{p-2k} - 1}{4k} B_{2k} \equiv \sum_{p/4 < j < p/3} j^{2k-1} \pmod{p}. \qquad (5.5)$$

**(5E)** *If $p$ is a prime, $p \geq 11$, $k \geq 1$, and $p - 1 \nmid 2k$, then*

$$\frac{4^{p-2k} + 5^{p-2k} - 8^{p-2k} - 1}{4k} B_{2k} \equiv \sum_{p/8 < j < p/5} j^{2k-1} + \sum_{3p/8 < j < 2p/5} j^{2k-1} \pmod{p}.$$

$$(5.6)$$

The evaluation of the residues modulo $p$ of the sums in the right-hand side is laborious, but may still be carried out by computer for quite large values of $p$.

None of the above tests is theoretically assured to work.

E. Lehmer proved in 1938 the following congruence:

$$\sum_{j=1}^{(p-1)/2} (p - 2j)^{2k} \equiv 2^{2k-1} p B_{2k} \pmod{p^3} \tag{5.7}$$

which is valid when $2k \not\equiv 2 \pmod{p - 1}$.

Carrying out the computation of the sum modulo $p^2$, if it is congruent to 0, then $p \mid B_{2k}$ and conversely. The major inconvenience with the above sum is that $j$ runs from 1 to $(p - 1)/2$.

After the original computations of Kummer, the irregular primes up to 617 were determined with desk calculators by Stafford and Vandiver (1930) and Vandiver (1931, 1937).

With the advent of electronic computers this program was extended by Lehmer, Lehmer, and Vandiver (in 1954) up to 4001, and then by various authors (Selfridge, Nicol, Pollack, Kobelev, and Johnson) up to 30000. Finally in 1976 with the IBM 360/65 and 370, Wagstaff determined all the irregular primes less than 125000.

In the recent computations, especially those of Johnson and Wagstaff, much more numerical data has been accumulated, which I will now explain.

If $p$ is a prime, if $2 \le 2k \le p - 3$ the couple $(p,2k)$ is called an *irregular couple* if $p \mid B_{2k}$.

The *index of irregularity* of $p$, denoted ii$(p)$, is the number of irregular couples $(p,2k)$, where $2 \le 2k \le p - 3$. Thus $p$ is regular when ii$(p) = 0$.

For every $N > 1$ and $r \ge 1$, let $\pi_r(N) = \#\{p \le N \mid \text{ii}(p) = r\}$. Let also

$$\pi'(N) = \#\{p \le N \mid p \text{ is irregular}\}$$

and

$$\pi(N) = \#\{p \text{ prime}, p \le N\}.$$

I summarize the results obtained by Wagstaff, which he kindly communicated to me:

For $N = 125000$:

$$\pi(N) = 11734$$
$$\pi'(N) = 4605$$
$$\pi_1(N) = 3559$$
$$\pi_2(N) = 875$$
$$\pi_3(N) = 153$$
$$\pi_4(N) = 16$$
$$\pi_5(N) = 2$$
$$\pi_r(N) = 0 \quad \text{for } r \ge 6.$$

Moreover

$$\frac{\pi'(N)}{\pi(N)} = 0.39248.$$

All the irregular couples (for $p \le N$) have been completely determined.

Two pairs of "successive" irregular couples were found: (491, 336), (491, 338) and (587, 90), (587, 92).

If $h > 1$, no successive irregular couples $(p, 2k), (p, 2k + 2), \ldots, (p, 2k + 2h)$ were ever found.

Such occurrences of successive irregular couples turn out to be important in the light of other theorems which I will explain in a later lecture.

No prime $p$ was found such that $p^2$ divides some $B_{2k}$.

Among the open questions, I single out: Given $s \geq 1$, do there exist infinitely many primes $p$ such that ii$(p) = s$? Similarly, if $s \geq 2$ do there exist infinitely many primes $p$ such that ii$(p) \geq s$? This is only known to be true for $s = 1$ (Jensen's theorem).

According to a heuristic argument of Siegel (1964)

$$\lim_{N \to \infty} \frac{\pi'(N)}{\pi(N)} = 1 - \frac{1}{\sqrt{e}} \cong 0.39.$$

This is very close to the observed distribution of irregular primes. His argument may be extended and yields:

$$P_r = \lim_{N \to \infty} \frac{\pi_r(N)}{\pi(N)} = \frac{1}{2^r r! \sqrt{e}} \quad \text{(for } r \geq 0\text{)}.$$

This also agrees very closely with the observed values.

In a letter to Serre, Shanks wrote (3 December, 1975):

. . . one does have

$$1 = \sum_{k=0}^{\infty} P_k$$

as one should. But this implies, and I do not know whether anyone has noted it earlier, that

$$\frac{1}{2} = \sum_{k=0}^{\infty} k P_k. \tag{5.8}$$

So, among the $\pi(N) - 1$ odd primes $\leq N$ one should expect about

$$\tfrac{1}{2}(\pi(N) - 1)$$

pairs of irregular primes and the Bernoulli numbers they divide [=irregular couples]. Here is recent data of Wagstaff for $N = 10^5$, $\pi(N) - 1 = 9591$

| $k$ | primes of index $k$ | |
|---|---|---|
| 0 | 5802 | $0 \times 5802$ |
| 1 | 2928 | $1 \times 2928$ |
| 2 | 728 | $2 \times 728$ |
| 3 | 123 | $3 \times 123$ |
| 4 | 8 | $4 \times 8$ |
| 5 | 2 | $5 \times 2$ |
| | 9591 | 4795 |

$\tfrac{1}{2}(9591) = 4795.5$. Mama Mia!

Actually with $N = 125000$, the coincidence is not so accurate:

$$\tfrac{1}{2}(\pi(N) - 1) = \frac{11733}{2} = 5866.5$$

while $1 \times 3559 + 2 \times 875 + 3 \times 153 + 4 \times 16 + 5 \times 2 = 5842$.

Yet it would be worth while to prove (5.8) since "it would imply that F.L.T. is true for at least one-half of the primes."

# Bibliography

1713  Bernoulli, J.
*Ars Conjectandi* (opus posthumum), Bâle, 1713. Also in *Die Werke von Jakob Bernoulli*, vol. 3, Birkhäuser, Basel, 1975, 107–286.

1755  Euler, L.
*Institutiones Calculi Differentialis, Paris Posterioris* (Caput V), Imperial Acad. of Sciences, Petrograd, 1755. Also in *Opera Omnia*, ser. I, vol. X, Teubner, Leipzig and Berlin, 1913, 321–328.

1840  Clausen, T.
Auszug aus einem Schreiben an Herrn Schumacher. *Astronomische Nachr.*, **17**, 1840, 351–352.

1840  von Staudt, C.
Beweis eines Lehrsatzes, die Bernoullischen Zahlen betreffend. *J. reine u. angew. Math.*, **21**, 1840, 372–376.

1845  von Staudt, C.
"De Numeris Bernoullianis" and "De Numeris Bernoullianis Commentatio Altera". Erlangen, 1845.

1850  Kummer, E. E.*
Bestimmung der Anzahl nicht äquivalenter Classen fur die aus $\lambda$ten Wurzel der Einheit gebildeten complexen Zahlen und die Idealen Factoren derselben. *J. reine u. angew. Math.*, **40**, 1850, 93–116.

1850  Kummer, E. E.
Zwei besondere untersuchungen über die Classen-Anzahl und über die Einheiten der aus $\lambda$ten Wurzeln der Einheit gebildeten complexen Zahlen. *J. reine u. angew. Math.*, **40**, 1850, 117–129.

1851  Kummer, E. E.
Mémoire sur la théorie des nombres complexes composés de racines de l'unité et de nombres entiers. *J. Math. Pures et Appl.*, **16**, 1851, 377–498.

1861  Kummer, E. E.
Über die Klassenanzahl der aus $n$-ten Einheitswurzeln gebildeten complexen Zahlen. *Monatsber. Akad. d. Wiss., Berlin*, 1861, 1051–1053.

1861  Sylvester, J. J.
Sur une propriété de nombres premiers qui se rattache au théorème de Fermat. *C. R. Acad. Sci. Paris*, **52**, 1861, 161–163. Also in *Mathematical Papers*, vol. 2, Cambridge University Press, Cambridge, 229–231.

---

* See also *Collected Papers*, vol. I, edited by A. Weil, Springer-Verlag, Berlin, 1975.

1863   Kronecker, L.
Über die Klassenanzahl der aus Wurzeln der Einheit gebildeten komplexen
Zahlen. *Monatsber. Akad. d. Wiss.*, *Berlin*, 1863, 340–345. Reprinted in *Werke*,
I, Teubner, Leipzig, 1895, 125–131.

1863   Kummer, E.E.
Über die Klassenanzahl der aus zusammengesetzten Einheitswurzeln gebildeten
idealen complexen Zahlen. *Monatsber. Akad. d. Wiss, Berlin*, 1863, 21–28.

1874   Kummer, E. E.
Über diejenigen Primzahlen $\lambda$ fur welche die Klassenzahl der aus $\lambda$-ten Einheits-
wurzeln gebildeten complexen Zahlen durch $\lambda$ theilbar ist. *Monatsber. Akad. d.
Wiss.*, *Berlin*, 1874, 239–248.

1878   Adams, J. C.
Table of the values of the first 62 numbers of Bernoulli. *J. reine u. angew. Math.*,
**85**, 1878, 269–272.

1900   Glaisher, J. W. L.
Fundamental theorems relating to the Bernoullian numbers. *Messenger of Math.*,
**29**, 1900, 49–63.

1908   Hensel, K.
*Theorie der algebraischen Zahlen*, Teubner, Leipzig, 1908.

1915   Jensen, K. L.
Om talteoretiske Egenskaber ved de Bernoulliske tal. *Nyt Tidsskrift f. Math.*,
B, **26**, 1915, 73–83.

1917   Vandiver, H. S.
Symmetric functions formed by systems of elements of a finite algebra and their
connection with Fermat's quotient and Bernoulli numbers. *Annals of Math.*, **18**,
1917, 105–114.

1930   Stafford, E. and Vandiver, H. S.
Determination of some properly irregular cyclotomic fields. *Proc. Nat. Acad. Sci.
U.S.A.*, **16**, 1930, 139–150.

1932   Vandiver, H. S.
Note on the divisors of the numerators of Bernoulli's numbers. *Proc. Nat. Acad.
Sci. U.S.A.*, **18**, 1932, 594–597.

1934   Rado, R.
A new proof of a theorem of von Staudt. *J. London Math. Soc.*, **9**, 1934, 85–88.

1937   Vandiver, H. S.
On Bernoulli numbers and Fermat's last theorem. *Duke Math. J.*, **3**, 1937, 569–584.

1938   Lehmer, E.
On congruences involving Bernoulli numbers and the quotients of Fermat and
Wilson. *Annals of Math.*, **39**, 1938, 350–359.

1939   Uspensky, J. V. and Heaslet, M. A.
*Elementary Number Theory*, McGraw-Hill, New York, 1939.

1940   Grün, O.
Eine Kongruenz für Bernoullische Zahlen. *Jahresber. d. Deutschen Math. Verein.*,
**50**, 1940, 111–112.

1954   Carlitz, L.
Note on irregular primes, *Proc. Amer. Math. Soc.*, **5**, 1954, 329–331.

1954   Lehmer, D. H., Lehmer, E., and Vandiver, H. S.
An application of high speed computing to Fermat's last theorem. *Proc. Nat. Acad.
Sci. U.S.A.*, **40**, 1954, 25–33.

1955   Selfridge, J. L., Nicol, C. A., and Vandiver, H. S.
Proof of Fermat's last theorem for all exponents less than 4002. *Proc. Nat. Acad. Sci. U.S.A.*, **41**, 1955, 970–973.

1955   Vandiver, H. S.
Is there an infinity of regular primes? *Scripta Mathematica*, **21**, 1955, 306–309.

1964   Siegel, C. L.
Zu zwei Bemerkungen Kummers. *Nachr. Akad. Wiss. Göttingen. Math. Phys. Kl.* II, 1964, 51–57. Also in *Gesammelte Abhandlungen*, vol. III, Springer-Verlag, New York, 1966, 436–442.

1965   Montgomery, H. L.
Distribution of irregular primes. *Illinois J. Math.*, **9**, 1965, 553–558.

1967   Selfridge, J. L., and Pollack, B. W.
Fermat's last theorem is true for any exponent up to 25000. *Notices Amer. Math. Soc.*, **11**, 1967, 97, Abstract 608–138.

1970   Kobelev, V. V.
Proof of Fermat's last theorem for all prime exponents less than 5500. *Soviet Math. Dokl.*, **11**, 1970, 188–190.

1971   Metsänkylä, T.
Note on the distribution of irregular primes. *Annales Acad. Sci. Fennicae*, Ser. A, I, No. **492**, 1971, 7 pages.

1974   Johnson, W.
Irregular prime divisors of the Bernoulli numbers. *Math. Comp.*, **28**, 1974, 653–657.

1975   Shanks, D.
Letter to Serre (Dec. 3, 1975).

1975   Yokoi, H.
On the distribution of irregular primes. *J. Number Theory*, **7**, 1975, 71–76.

1976   Metsänkylä, T. Distribution of irregular prime numbers. *J. reine u. angew. Math.*, **282**, 1976, 126–130.

1977   Wagstaff, S. S.
The irregular primes to 125000. (Preprint, 1977). *Math. Comp.*, **32**, 1978, 583–591.

# LECTURE VII

# Kummer Exits

After having established Fermat's theorem for regular prime exponents, Kummer continued his work, considering the first case for arbitrary prime exponents. He was able to derive congruences, involving Bernoulli numbers, which must be satisfied by any would-be solution. From these congruences, he derived specific divisibility properties about Bernoulli numbers.

The results of Kummer were expanded and given another form by Mirimanoff, and more recently, extended considerably by Krasner. I will explain these more recent results in my next lecture.

As in his preceding work, Kummer began by breaking new ground in the theory of cyclotomic fields. This is what I will consider first.

## 1. The Periods of the Cyclotomic Equation

I will use the following notations:

$p = $ odd prime
$\zeta = $ primitive $p$th root of 1.
$\Phi_p(X) = X^{p-1} + X^{p-2} + \cdots + X + 1$ cyclotomic polynomial
$\lambda = 1 - \zeta$
$K = \mathbb{Q}(\zeta) = $ cyclotomic field, of degree $p - 1$.
$A = \mathbb{Z}[\zeta] = $ ring of cyclotomic integers
$g = $ primitive root modulo $p$
$G = \mathrm{Gal}(K \mid \mathbb{Q}) = $ Galois group of the cyclotomic extension
$\sigma : \zeta \to \zeta^g$ generator of $G$

Gauss introduced the periods of the cyclotomic equation. If $f$, $r$ are integers, $p - 1 = fr$, the $r$ periods of $f$ terms (relative to the primitive root

$g$ modulo $p$) are the cyclotomic integers $\eta_0, \eta_1, \ldots, \eta_{r-1}$ defined by

$$
\begin{aligned}
\eta_0 &= \zeta + \zeta^{g^r} + \zeta^{g^{2r}} + \cdots + \zeta^{g^{(f-1)r}}, \\
\eta_1 &= \zeta^g + \zeta^{g^{r+1}} + \zeta^{g^{2r+1}} + \cdots + \zeta^{g^{(f-1)r+1}}, \\
&\;\vdots \\
\eta_{r-1} &= \zeta^{g^{r-1}} + \zeta^{g^{2r-1}} + \zeta^{g^{3r-1}} + \cdots + \zeta^{g^{fr-1}}.
\end{aligned}
\tag{1.1}
$$

Note that

$$
\sum_{j=0}^{r-1} \eta_j = -1. \tag{1.2}
$$

For convenience, if $j \equiv j_0 \pmod{r}$, $0 \le j_0 \le r - 1$, I also define $\eta_j = \eta_{j_0}$. The periods are conjugate to each other:

$$
\sigma^i(\eta_j) = \eta_{i+j}. \tag{1.3}
$$

In particular, $\sigma^r(\eta_i) = \eta_i$ (for $i = 0,1,\ldots,r-1$). So the fields $\mathbb{Q}(\eta_0)$, $\mathbb{Q}(\eta_1), \ldots, \mathbb{Q}(\eta_{r-1})$ are conjugate over $\mathbb{Q}$. But $K \mid \mathbb{Q}$ is a cyclic extension, so $\mathbb{Q}(\eta_0) = \cdots = \mathbb{Q}(\eta_{r-1})$.

This field will be denoted by $K'$. It is the subfield left invariant by the subgroup of $G$ generated by $\sigma^r$, and $[K:K'] = f$, $[K':\mathbb{Q}] = r$.

Let $A'$ be the ring of integers of $K'$.

In his papers of 1846, 1847, and 1857, Kummer proved the following facts:

**(1A)** $\{\eta_0, \eta_1, \ldots, \eta_{r-1}\}$ is an integral basis of $A'$; in particular $A' = \mathbb{Z}[\eta_0, \eta_1, \ldots, \eta_{r-1}]$.

However, it is not true, in general, that $\mathbb{Z}[\eta_0] = \cdots = \mathbb{Z}[\eta_{r-1}] = A'$. Secondly, he showed also:

**(1B)** $A$ is a free $A'$-module with basis $\{1, \zeta, \ldots, \zeta^{f-1}\}$. That is, every algebraic integer $\alpha \in A$ is, in unique way, of the form

$$
\alpha = \sum_{j=0}^{f-1} \alpha'_j \zeta^j,
$$

where $\alpha'_j \in A'$ $(j = 0, 1, \ldots, f-1)$.

Let

$$
F(X) = \prod_{i=0}^{r-1} (X - \eta_i) \in \mathbb{Z}[X]. \tag{1.4}
$$

Kummer studied the congruence

$$
F(X) \equiv 0 \pmod{q}, \tag{1.5}
$$

where $q$ is any prime.

I should say, at this time, that in his paper of 1846, Kummer made a mistake in his study; however, in 1857, he gave a correct proof the theorem which follows (see comments by Weil, in volume I of Kummer's *Collected Works*). In 1975, Maury presented a corrected proof of Kummer's result in today's language.

**(1C)** *Let $q$ be any prime, let $fr = p - 1$. If $Q$ is any prime ideal of $A$ dividing $Aq$, then for every period $\eta_k$ ($0 \le k \le r - 1$) there exists a unique integer $u_k$, $0 \le u_k \le q - 1$, such that $\eta_k \equiv u_k \pmod{Q}$. In particular*

$$F(X) \equiv \prod_{k=0}^{r-1} (X - u_k) \pmod{q}. \tag{1.6}$$

I wish to stress that the numbers $u_0, u_1, \ldots, u_{r-1}$ need not be distinct, for example, they are certainly not when $q < r$.

If $q$ is a prime different from $p$, then it is unramified in $K$, that is, $Aq$ is the product of distinct prime ideals:

$$Aq = Q_0 Q_1 \cdots Q_{r-1}. \tag{1.7}$$

Let $fr = p - 1$, let $\eta_0, \eta_1, \ldots, \eta_{r-1}$ be the $r$ periods of $f$ terms and $K' = \mathbb{Q}(\eta_0) = \cdots = \mathbb{Q}(\eta_{r-1})$, $A' = \mathbb{Z}[\eta_0, \ldots, \eta_{r-1}]$. Then $A'q$ is also the product of $r$ distinct prime ideals in $A'$, each having degree 1:

$$\begin{aligned}
A'q &= Q'_0 Q'_1 \cdots Q'_{r-1}, \\
A'/Q_i &= \mathbb{F}_q \quad (i = 0,1,\ldots,r-1).
\end{aligned} \tag{1.8}$$

The numbering is such that

$$AQ'_i = Q_i \quad (\text{for } i = 0,1,\ldots,r-1).$$

# 2. The Jacobi Cyclotomic Function

Another tool in Kummer's research was the generalization by Jacobi of Gauss's and Lagrange's resolvents. These were quite essential in studying the question of solution by radicals.

Besides the notation already introduced, I shall fix the following:
Let $q$ be an odd prime, $q \ne p$, and

$h = $ primitive root modulo $q$,
$\rho = $ primitive $q$th root of 1,
$\theta = $ primitive $(q - 1)$th root of 1.

If $t$ is not a multiple of $q$, there exists a unique integer $s$, $0 \le s \le q - 2$, such that $t \equiv h^s \pmod{q}$. $s$ is called the *index* of $t$, relative to $h$, and denoted $s = \text{ind}_h(t)$, or simply, $s = \text{ind}(t)$. Thus $\text{ind}(1) = 0$, $\text{ind}(-1) = (q - 1)/2$. Clearly, if $t \equiv t' \pmod{q}$, then $\text{ind}(t) = \text{ind}(t')$. Also

$$\text{ind}(tt') \equiv \text{ind}(t) + \text{ind}(t') \pmod{q - 1} \text{ when } q \nmid tt'.$$

The Lagrange *resolvents* are sums of the form:

$$\sum_{i=0}^{q-2} \theta^i \rho^{h^i} = \sum_{t=1}^{q-1} \theta^{\text{ind}_h(t)} \rho^t. \tag{2.1}$$

Jacobi considered in 1837 (in a paper reprinted in 1846) the analogous sums when $q \equiv 1 \pmod{p}$. Let $q = pk + 1$, so $k$ is an even integer. If $m, n$ are integers, $p \nmid n, q \nmid m$, the Jacobi resolvents are

$$\langle \zeta^n, \rho^m \rangle = \sum_{t=1}^{q-1} \zeta^{n\,\text{ind}_h(t)} \rho^{mt} = \sum_{u=0}^{q-2} \zeta^{nu} \rho^{mh^u}. \tag{2.2}$$

Clearly $\langle \zeta^n, \rho^m \rangle$ belongs to the ring of integers of $\mathbb{Q}(\zeta, \rho)$.

The Jacobi cyclotomic functions are defined as follows: If $l, n$ are integers, $p \nmid n$, let

$$\psi_l(\zeta^n) = \sum_{t=1}^{q-2} \zeta^{n[\text{ind}_h(t) - (l+1)\text{ind}_h(t+1)]}. \tag{2.3}$$

Then $\psi_l(\zeta^n)$ is an integer of $\mathbb{Q}(\zeta)$.

Note that $\mathbb{Q}(\zeta, \rho)$ is a Galois extension of $\mathbb{Q}(\zeta)$. Let $\bar{\sigma}$ be the automorphism of $\mathbb{Q}(\zeta, \rho)$ such that $\bar{\sigma}(\zeta) = \zeta^g$, $\bar{\sigma}(\rho) = \rho$.

I'll now list various properties of these sums and of the Jacobi cyclotomic functions.

**(2A)** *If* $p \nmid n, q \nmid m$, *then*

a. $\langle \zeta^n, \rho^m \rangle = \langle \zeta^n, \rho \rangle \zeta^{-n\,\text{ind}(m)}$.
b. $\langle \zeta, \rho \rangle^p$ *and* $\bar{\sigma}\langle \zeta, \rho \rangle / \langle \zeta, \rho \rangle^g$ *belong to* $\mathbb{Q}(\zeta)$.
c. $\langle \zeta, \rho \rangle \langle \zeta^{-1}, \rho \rangle = q$. *In particular* $\langle \zeta, \rho \rangle \neq 0$.
d. *If* $l$ *is any integer such that* $p \nmid l + 1$, *then*

$$\psi_l(\zeta) = \frac{\langle \zeta, \rho \rangle \langle \zeta^l, \rho \rangle}{\langle \zeta^{l+1}, \rho \rangle}.$$

e. $\langle \zeta, \rho \rangle^p = q \psi_1(\zeta) \psi_2(\zeta) \cdots \psi_{p-2}(\zeta)$.
f. *If* $l = 1, 2, \ldots, p - 2$, *then* $\psi_l(\zeta) \psi_l(\zeta^{-1}) = q$.

The proof of these properties is of course somewhat laborious, but involves no essential difficulties. Most useful is the expression of $\langle \zeta, \rho \rangle^p$ given in (e).

The main result of Kummer concerns the decomposition into prime ideals of the ideal of $A = \mathbb{Z}[\zeta]$ generated by $\langle \zeta, \rho \rangle^p$. A preliminary result is the following:

**(2B)** *If* $q = kp + 1$ *and if* $Q = Aq + A(h^k - \zeta)$, *then* $Q$ *is a prime ideal with norm* $q$ *and* $Aq = \prod_{i=0}^{p-2} \sigma^i(Q)$.

Now let $\pi = (p - 1)/2$ and for every integer $k$ let $g_k$ be the unique integer such that $1 \leq g_k \leq p - 1$ and $g^k \equiv g_k \pmod{p}$. If $k < 0$ this is to be understood as $g_k g^{-k} \equiv 1 \pmod{p}$.

In 1857, Kummer proved:

**(2C)** *If* $1 \le d \le p - 2$, *let*

$$I_d = \{i \mid 0 \le i \le p - 2, g_{\pi - i} + g_{\pi - i + \text{ind}_g(d)} > p\}. \qquad (2.4)$$

*Then:*

$$A\psi_d(\zeta) = \prod_{i \in I_d} \sigma^i(Q) \qquad (2.5)$$

*and*

$$A\langle \zeta, \rho \rangle^p = \prod_{i=0}^{p-2} [\sigma^i(Q)]^{g_{\pi-i}}. \qquad (2.6)$$

This was the main basic result of Kummer concerning the Jacobi resolvents. It was generalized by Stickelberger in 1890, who allowed any natural number $m \ge 3$ in place of $p$, with $q$ prime, $q \equiv 1 \pmod{m}$.

As an immediate corollary:

**(2D)** *If* $Q_1$ *is a prime ideal of degree 1 of* $K = \mathbb{Q}(\zeta)$, *then*

$$\prod_{i=0}^{p-2} [\sigma^i(Q_1)]^{g_{p-1-i}}$$

*is a principal ideal.*

The virtue of the above result is to produce a principal ideal.

# 3. On the Generation of the Class Group of the Cyclotomic Field

In 1847, Kummer proved a theorem about the generation of the class group of the cyclotomic field. Later, with analytic methods, Dirichlet strengthened Kummer's result in a very substantial way, by proving the density theorem. From this result, he was able to prove the celebrated theorem on primes in arithmetic progressions.

Kummer's results were:

**(3A)** *The group of ideal classes of the cyclotomic field is generated by the classes of prime ideals of degree 1.*

As I said above, much more is true, but I'll comment later about it.

With this result and all those indicated before, Kummer proved the following result, which enabled him to produce principal ideals from any given ideal $J$:

**(3B)** *If* $1 \le d \le p - 2$, *for every ideal* $J$ *of* $A$, *the product* $\prod_{i \in I_d} \sigma^i(J)$ *is a principal ideal.*

To convey an idea how the preceding results unite to establish this one, I will briefly sketch the proof.

PROOF. By (3A), $J$ is in the same ideal class as a product of prime ideals $Q_j$ of degree 1:

$$J = A\alpha \prod_{j=1}^{s} Q_j.$$

Let $N(Q_j) = q_j$, so $q_j \equiv 1 \pmod{p}$ because the inertial degree of $Q_j$ is the order of $q_j$ modulo $p$. Let $q_j = k_j p + 1$ and let $Q_j^* = Aq_j + A(h_j^{k_j} - \zeta)$, where $h_j$ is a primitive root modulo $q_j$. By (2B)

$$Aq_j = \prod_{i=0}^{p-2} \sigma^i(Q_j^*)$$

so $Q_j$ is conjugate to $Q_j^*$, say $Q_j^* = \sigma^{j*}(Q_j)$, where $0 \le j^* \le p - 2$.
By (2.5)

$$\prod_{i \in I_d} \sigma^i(Q_j^*) = \sigma^{j*}\left(\prod_{i \in I_d} \sigma^i(Q_j)\right)$$

is a principal ideal of $A$. Hence

$$\prod_{i \in I_d} \sigma^i(J) = A\left(\prod_{i \in I_d} \sigma^i(\alpha)\right) \prod_{j=1}^{s} \left(\prod_{i \in I_d} \sigma^i(Q_j)\right)$$

is also a principal ideal of $A$.                                                        □

The strengthening of (3A) done by Dirichlet asserts that in *every* ideal class there is, not only one prime ideal of degree 1, but in fact, infinitely many such. Even more, these ideals have a positive density. But I'll not elaborate on this point.

## 4. Kummer's Congruences

Having proved Fermat's theorem for regular exponents, Kummer next considered arbitrary exponents. Initially, he dealt only with the first case.

A preliminary result in the line of Kummer's research was obtained by Cauchy in 1847. Let $p$ be an odd prime.

**(4A)** *If $p$ does not divide the sum:*

$$S_{p-4}\left(\frac{p-1}{2}\right) = 1^{p-4} + 2^{p-4} + \cdots + \left(\frac{p-1}{2}\right)^{p-4},$$

*then the first case of Fermat's theorem holds for the exponent $p$.*

As Genocchi pointed out in 1852 (and again in 1866), the above sum is related to a Bernoulli number:

$$-2B_{p-3} \equiv S_{p-4}\left(\frac{p-1}{2}\right) \pmod{p} \tag{4.1}$$

and Cauchy's result may be rephrased:

**(4B)** *If the first case of Fermat's theorem fails for the exponent $p$ then $p$ divides $B_{p-3}$.*

Kummer extended this criterion, as a consequence of his congruences. This is what I shall now explain.

Assume that $x$, $y$, $z$ are relatively prime integers such that $p \nmid z$ and

$$x^p + y^p + z^p = 0. \tag{4.2}$$

Then

$$(Az)^p = A(x^p + y^p) = \prod_{k=0}^{p-2} A(x + \zeta^{g^k}y).$$

As I have indicated in my fifth lecture, prior to proving Kummer's main theorem, there exist pairwise relatively prime ideals $L$, $J$ of $A$ such that

$$A(x + y) = L^p,$$
$$A(x + \zeta y) = J^p. \tag{4.3}$$

Then

$$A(x + \zeta^{g^k}y) = (\sigma^k(J))^p \quad \text{for } k = 0, 1, \dots, p - 2. \tag{4.4}$$

Taking norms in the extension $\mathbb{Q}(\zeta)|\mathbb{Q}$, $|x + y|^{p-1} = l^p$, where $N(L) = l$. Then $x + y$ has to be the $p$th power of an integer, $x + y = t^p$, and $L = At$. By (3B), for each $d$, $1 \le d \le p - 2$, the ideal $\prod_{i \in I_d} \sigma^i(J)$ is principal, say

$$A\alpha_0 = \prod_{i \in I_d} \sigma^i(J).$$

Hence

$$A\alpha_0^p = \prod_{i \in I_d} A(x + \zeta^{g^i}y)$$

so there is a unit $\varepsilon$ of $A$ such that

$$\varepsilon\alpha_0^p = \prod_{i \in I_d} (x + \zeta^{g^i}y). \tag{4.5}$$

Let $I_d' = \{i \mid 0 \le i \le p - 2, i \notin I_d\}$. It is easy to see that $i' \in I_d'$ if and only if $i' \equiv i + \pi \pmod{p}$, where $i \in I_d$ and $\pi = (p - 1)/2$. Hence taking complex-conjugates

$$\overline{\varepsilon}\overline{\alpha_0^p} = \prod_{i \in I_d'} (x + \zeta^{g^i}y). \tag{4.6}$$

Multiplying (4.5) and (4.6):

$$-\left(\frac{z}{t}\right)^p = N(x + \zeta y) = \prod_{i \in I_d} (x + \zeta^{g^i}y) \prod_{i \in I_d'} (x + \zeta^{g^i}y) = \varepsilon\overline{\varepsilon}(\alpha_0\overline{\alpha}_0)^p.$$

Hence $\varepsilon\overline{\varepsilon}$ is the $p$th power of a unit $\omega : \varepsilon\overline{\varepsilon} = \omega^p$.

Since $\varepsilon = \zeta^m \delta$, where $\delta$ is a real unit, $\bar{\varepsilon} = \zeta^{-m}\delta$ so $\omega^p = \delta^2$ and therefore $\delta = (\omega^r \delta^s)^p$, where $sp + 2r = 1$.

In conclusion,

$$\prod_{i \in I_d} (x + \zeta^{g^i} y) = \zeta^m \alpha^p \tag{4.7}$$

with $\alpha = \omega^r \delta^s \alpha_0 \in A$.

Since $\alpha \in A$ there exists a polynomial $F(X) \in \mathbb{Z}[X]$, of degree at most $p - 2$, such that $\alpha = F(\zeta)$. Hence the polynomial

$$\prod_{i \in I_d} (x + X^{g^i} y) - X^m (F(X))^p$$

vanishes at $\zeta$. So it is a multiple of the cyclotomic polynomial $\Phi_p(X)$ and I may write

$$\prod_{i \in I_d} (x + X^{g^i} y) = X^m (F(X))^p + \Phi_p(X) M(X), \tag{4.8}$$

where $M(X) \in \mathbb{Z}[X]$.

Look at the corresponding polynomial functions of the positive real variable $t$. Then

$$\prod_{i \in I_d} (x + t^{g^i} y) = t^m (F(t))^p + \Phi_p(t) M(t).$$

Since $t > 0$, it may be written as $t = e^v$, where $v$ an arbitrary real number. Hence

$$\prod_{i \in I_d} (x + e^{v g^i} y) = e^{mv} (F(e^v))^p + \Phi_p(e^v) M(e^v)$$

$$= e^{mv} (F(e^v))^p \left[ 1 + \frac{\Phi_p(e^v) M(e^v)}{e^{mv} (F(e^v))^p} \right]. \tag{4.9}$$

Note that $x$, $y$ and the coefficients of $M(X)$, $F(X)$ may well be negative integers. So, taking the logarithms of both sides of (4.9) it is necessary to consider the complex logarithmic function; this leads to an equality up to some multiple of $2\pi i$, which will be irrelevant, after taking derivatives:

$$\sum_{i \in I_d} \log(x + e^{v g^i} y) = mv + p \log F(e^v)$$

$$+ \log\left[ 1 + \frac{\Phi_p(e^v) M(e^v)}{e^{mv} (F(e^v))^p} \right] + 2k\pi i \quad (\text{where } k \in \mathbb{Z}). \tag{4.10}$$

The following notation will be used. If $G(v)$ is a differentiable function of $v$, let $D^n G(v)$ be the value of $d^n G(v)/dv^n$ at $v = 0$.

One might wonder, at this point, how Kummer could untangle himself and still reach some worthwhile conclusion from (4.10). But, with his mastery, nothing should astonish us anymore. What he succeeded in proving (1857) is:

**(4C)** *Assume that $p$ is an odd prime, $x$, $y$, $z$ are relatively prime integers, $p$ does not divide $z$ and $x^p + y^p + z^p = 0$. Then the following congruences are* *satisfied*:

$$[D^{p-2s}\log(x + e^v y)]B_{2s} \equiv 0 \;(\text{mod } p)$$

*for* $2s = 2, 4, \ldots, p - 3$.

Similar congruences hold when $x$ and $y$ are interchanged.

PROOF. Since the proof of this theorem is quite long, I have to restrict myself and only provide a sketch of its main points.

(a) If $n = 2, \ldots, p - 2$, then

$$\sum_{i \in I_d} D^n \log(x + e^{vg^i} y) \equiv 0 \;(\text{mod } p). \tag{4.11}$$

For this purpose, one computes the value at $v = 0$ of the $n$th derivative of the right-hand side of (4.10). For example

$$D^n(p \log F(e^v)) = \frac{p}{F(1)^n} H(F(1), F'(1), \ldots, F^{(n)}(1)),$$

where $H$ is a polynomial with coefficients in $\mathbb{Z}$ and $p$ does not divide $F(1)$, as may be seen without difficulty.

For the last term one proceeds in a somewhat similar manner.

(b) $[D^n \log(x + e^v y)](\sum_{i \in I_d} g^{ni}) \equiv 0 \;(\text{mod } p)$. This follows at once from (a). Let $S = \sum_{i \in I_d} g^{(p-2s)i}$. In order to evaluate the residue of $S$ mod $p$, various facts have to be proved.

(c)

$$\frac{1}{p}[g_{\pi-i} + g_{\pi+\text{ind}_g(d)-i} - g_{\pi+\text{ind}_g(d+1)-i}] = \begin{cases} 1 & \text{when } i \in I_d, \\ 0 & \text{when } i \notin I_d. \end{cases}$$

Indeed, the number in brackets is congruent to $g^{\pi-i}(1 + d - (d + 1)) \equiv 0 \;(\text{mod } p)$. From this, it is easy to establish the assertion. Hence

$$S = \sum_{i=0}^{p-2} \frac{1}{p}[g_{\pi-i} + g_{\pi-i+\text{ind}_g(d)} - g_{\pi-i+\text{ind}_g(d+1)}]g^{(p-2s)i}.$$

Since $g^p \equiv g \;(\text{mod } p)$,

$$pS \equiv S_1 + S_d - S_{d+1} \;(\text{mod } p^2), \tag{4.12}$$

where

$$S_j = \sum_{i=0}^{p-2} g_{\pi-i+\text{ind}_g(j)} g^{p(p-2s)i}. \tag{4.13}$$

(d)

$$S_j \equiv -j^{p(p-2s)} \sum_{i=0}^{p-2} g_{-i} g^{p(p-2s)i} \;(\text{mod } p^2).$$

This is easy to prove. It follows from (4.12) that

$$pS \equiv -[1 + d^{p(p-2s)} - (d + 1)^{p(p-2s)}]\left[\sum_{i=0}^{p-2} g_{-i} g^{p(p-2s)i}\right] \;(\text{mod } p^2). \tag{4.14}$$

Writing $g_{-i} = k$, the above congruence, becomes

$$pS \equiv -[1 + d^{p(p-2s)} - (d+1)^{p(p-2s)}]\left[\sum_{k=1}^{p-1} k^{1+p(2s-1)}\right]. \qquad (4.15)$$

Let $1 + p(2s - 1) = 2m$. Then, as I have indicated in Lecture VI,

$$\sum_{k=1}^{p-1} k^{1+p(2s-1)} = S_{2m}(p - 1)$$

$$= \frac{1}{2m+1}\left[p^{2m+1} + \binom{2m+1}{1}B_1 p^{2m} + \binom{2m+1}{2}B_2 p^{2m-1}\right.$$

$$\left. + \binom{2m+1}{4}B_4 p^{2m-3} + \cdots + \binom{2m+1}{2m}B_{2m}p\right]. \qquad (4.16)$$

From the theorem of von Staudt and Clausen, the denominator of a Bernoulli number is square-free. Hence

$$S_{2m}(p - 1) \equiv B_{2m}p \pmod{p^2} \qquad (4.17)$$

and

$$S \equiv -[1 + d^{p-2s} - (d+1)^{p-2s}]B_{2m} \pmod{p}. \qquad (4.18)$$

But $2m = 2s + (p - 1)(2s - 1)$. By Kummer's congruence (explained in Lecture VI, Formula (3.7))

$$\frac{B_{2m}}{2m} \equiv \frac{B_{2s}}{2s} \pmod{p}.$$

Hence

$$B_{2m} \equiv [1 + p(2s - 1)]\frac{B_{2s}}{2s} \equiv \frac{B_{2s}}{2s} \pmod{p}. \qquad (4.19)$$

From (b), written for $n = p - 2s$ $(2s = 2,4,\ldots,p - 3)$ and (4.18), (4.19), it follows that

$$[D^{p-2s}\log(x + e^v y)][1 + d^{p-2s} - (d+1)^{p-2s}]\frac{B_{2s}}{2s} \equiv 0 \pmod{p}. \qquad (4.20)$$

(e) *Conclusion.* The polynomial $1 + X^{p-2s} - (X + 1)^{p-2s}$ is not identically zero and has degree $p - 2s - 1 < p - 2$. So there exists $d$, $1 \leq d \leq p - 2$, not satisfying

$$1 + X^{p-2s} - (X + 1)^{p-2s} \equiv 0 \pmod{p}.$$

From (4.20), it follows that I may divide by $1 + d^{p-2s} - (d+1)^{p-2s}$ (for this choice of $d$), hence:

$$[D^{p-2s}\log(x + e^v y)]\frac{B_{2s}}{2s} \equiv 0 \pmod{p}. \qquad (4.21)$$

From this, the congruences of Kummer follow immediately.                                              □

If $x^p + y^p + z^p = 0$ and $p$ does not divide $x$, $y$, $z$, then the congruences of Kummer hold also for the pairs $(x,z)$, $(z,x)$, $(y,z)$, $(z,y)$.

By computing the derivative explicitly, it is easily seen that

$$D^j \log(x + e^v y) = \frac{R_j(x,y)}{(x + y)^j},\tag{4.22}$$

where $R_j(X,Y)$ is a homogeneous polynomial, with coefficients in $\mathbb{Z}$, of total degree $j$, which is a multiple of $Y$.

Moreover, it has the following properties. If $j > 1$, then $R_j(Y,X) = (-1)^j R_j(X,Y)$ and if $j$ is odd, $j > 1$, then $XY(X - Y)$ divides $R_j(X,Y)$.

Writing $Y = XT$ and $P_j(T) = R_j(1,T)$ gives $R_j(X,Y) = X^j P_j(T)$.

With these notations, Kummer's congruence may be easily rewritten as follows:

(4D) *If* $x^p + y^p + z^p = 0$, $p \nmid xyz$, *for* $2s = 2, 4, \ldots, p - 3$, *and*

$$t \in \left\{ \frac{x}{y}, \frac{y}{x}, \frac{x}{z}, \frac{z}{x}, \frac{y}{z}, \frac{z}{y} \right\}$$

*then*

$$P_{p-2s}(t)B_{2s} \equiv 0 \pmod p.$$

This is a more useful form of Kummer's criterion; instead of derivatives, it makes use of the values of certain polynomials.

In order to apply this criterion, it is important to compute these polynomials. For example, for small indices:

$P_1(T) = T,$
$P_2(T) = T,$
$P_3(T) = T(1 - T),$
$P_4(T) = T(1 - 4T + T^2),$
$P_5(T) = T(1 - T)(1 - 10T + T^2),$
$P_6(T) = T(1 - 26T + 66T^2 - 26T^3 + T^4).$

The first application by Kummer of his criterion was the extension of the results of Cauchy and Genocchi:

(4E) *If* $x^p + y^p + z^p = 0$, *with* $p \nmid xyz$, *then* $p$ *divides both* $B_{p-3}$ *and* $B_{p-5}$.

PROOF. Assume that $R_3(a,b) \equiv 0 \pmod p$ where $a$, $b$ are distinct, and $a, b \in \{x,y,z\}$. Then $P_3(t) \equiv 0 \pmod p$, where $t = a/b$ (note that $p \nmid xyz$). Hence $p \mid a - b$, that is, $a \equiv b \pmod p$. For the various choices of $a$, $b$, this yields $x \equiv y \equiv z \pmod p$. Hence

$$3x \equiv x + y + z \equiv x^p + y^p + z^p = 0 \pmod p$$

and because $p \neq 3$, it follows that $p \mid x$, against the hypothesis. Thus, for some $a$, $b$ necessarily $P_3(t) \not\equiv 0 \pmod p$. So, taking $2s = p - 3$, it follows from (4D) that $B_{p-3} \equiv 0 \pmod p$.

The proof that $B_{p-5} \equiv 0 \pmod p$ follows similar lines, and is just a little bit more involved.                                                               □

There have been various proofs of other criteria, analogous to those of Kummer's. See, for example, Vandiver (1919). In 1922, Fueter was able to again prove Kummer's congruences using only methods developed in Hilbert's *Zahlbericht*. Let me add that this didn't make matters any easier!

## 5. Kummer's Theorem for a Class of Irregular Primes

In his memoir of 1857, Kummer proved a theorem which establishes the truth of Fermat's last theorem for certain irregular exponents. To arrive at this theorem, Kummer investigated thoroughly the structure of the group of units of the cyclotomic field. He also used $\lambda$-adic logarithms. However, while Kummer's results were ultimately correct, there were several important gaps and mistakes in his proof. These were pointed out initially by Mertens in 1917. Vandiver clarified the doubtful points and provided rigorous proofs (1920, 1922, 1926, 1926).

I keep the same notation as before. If $\alpha \in A$, then $\alpha = a_0 + a_1\zeta + \cdots + a_{p-2}\zeta^{p-2}$ with $a_i \in \mathbb{Z}$.

Associated with $\alpha$, let $\alpha(x)$ be the function of the real variable $x$, defined by

$$\alpha(x) = a_0 + a_1 x + \cdots + a_{p-2}x^{p-2}.$$

Thus $\alpha(\zeta) = \alpha$, $\alpha(1) = a_0 + a_1 + \cdots + a_{p-2}$. Since $\lambda = 1 - \zeta$, $\alpha \equiv \alpha(1) \pmod{A\lambda}$. If $\alpha \notin A\lambda$, then $\alpha(1)$ is not a multiple of $p$.

Let $v_\lambda$ be the $\lambda$-adic valuation of $K$; it is the one associated to the prime ideal $A\lambda$. Then

$$v_\lambda\left(1 - \frac{\alpha}{\alpha(1)}\right) \geq 1.$$

Extending $v_\lambda$ to the completion $\hat{K}$ of $K$ (with respect to the topology defined by $v_\lambda$) Kummer considered the $\lambda$-adic logarithmic function (1852):

$$\log_\lambda\left(\frac{\alpha}{\alpha(1)}\right) = \sum_{n=1}^{\infty} \frac{(-1)^{n-1}}{n}\left(\frac{\alpha}{\alpha(1)} - 1\right)^n. \tag{5.1}$$

Eisenstein also had considered the $\lambda$-adic logarithmic function in 1850.

In 1851, Kummer also considered the function $l^{(s)}$ (for $s = 1, 2, \ldots, p-1$) defined as follows. If $t$ is a real variable, $\alpha \in A \backslash A\lambda$, then:

$$l^{(s)}(\alpha) = \left[\frac{d^s \log \alpha(e^t)}{dt^s}\right]_{t=0} \tag{5.2}$$

$l^{(s)}(\alpha)$ is a $p$-integral rational number.

The generalized concept of index was introduced by Kummer in 1852. Let $q$ be a prime, $q \neq p$, let $Q$ be a prime ideal of $A$ dividing $Aq$, with norm

$N(Q) = q^f$. Let $\alpha \in A\backslash Q$. Then it is clear that $\alpha^{q^f - 1} \equiv 1 \pmod{Q}$. Moreover, there exists a unique integer $c$, $0 \leq c \leq p - 1$, such that

$$\alpha^{(q^f - 1)/p} \equiv \zeta^c \pmod{Q}. \tag{5.3}$$

$c$ is called the *index* of $\alpha$ and denoted $\mathrm{ind}_Q(\alpha) = c$.

The $p$th *residue power character* modulo $Q$ is defined (for $\alpha \in A\backslash Q$) by:

$$\left\{ \frac{\alpha}{Q} \right\} = \zeta^{\mathrm{ind}_Q(\alpha)}. \tag{5.4}$$

So

$$\left\{ \frac{\alpha}{Q} \right\} \equiv \alpha^{(q^f - 1)/p} \pmod{Q}.$$

Among the properties, I note:

1. If $\alpha, \beta \in A\backslash Q$ and $\alpha \equiv \beta \pmod{Q}$, then $\{\alpha/Q\} = \{\beta/Q\}$.
2. If $\alpha, \beta \in A\backslash Q$, then $\{\alpha\beta/Q\} = \{\alpha/Q\}\{\beta/Q\}$.

Moreover $\{\alpha/Q\} = 1$ if and only if there exists $\beta \in A\backslash Q$ such that $\alpha \equiv \beta^p \pmod{Q}$.

In 1852 Kummer generalized the Jacobian cyclotomic functions. First, he noted that if $\zeta$ is a given primitive $p$th root of 1, there exists $\rho$, a primitive root modulo $Q$ (that is, the residue class of $\rho$ modulo $Q$ is a generator of the multiplicative group of the residue class field $A/Q$) such that

$$\zeta \equiv \rho^{(q^f - 1)/p} \pmod{Q}.$$

Let $q$ be an odd prime, $q \neq p$. For every $k = 1, 2, \ldots, p - 1$ and every integer $j$, let

$$\psi_j(\zeta^k) = \sum_{\substack{t=0 \\ t \neq (q^f - 1)/2}}^{q^f - 2} \zeta^{k[\mathrm{ind}_Q(\rho^t + 1) - (j+1)t]}. \tag{5.5}$$

If $q \equiv 1 \pmod{p}$, that is, if $f = 1$, these functions are essentially the Jacobi cyclotomic functions, as defined in this lecture.

Mitchell extended this definition in 1916 to also include the case where $q = 2$.

Among the properties of the Kummer cyclotomic functions, I mention:

$$\psi_j(\zeta^q) = \psi_j(\zeta). \tag{5.6}$$

If, $j \not\equiv 0, -1 \pmod{p}$, then

$$\psi_j(\zeta)\psi_j(\zeta^{-1}) = q^f. \tag{5.7}$$

If $f$ is even, then

$$\psi_j(\zeta) = q^{f/2}. \tag{5.8}$$

If $f$ is odd, then

$$\psi_j(\zeta) = \pm Q^{m_0}\sigma(Q)^{m_1} \cdots \sigma^{r-1}(Q)^{m_{r-1}}, \tag{5.9}$$

where $rf = p - 1$, $\sigma$ is the automorphism of $K$ such that $\sigma(\zeta) = \zeta^g$, $g$ a primitive root modulo $p$,

$$m_i = S_{\frac{1}{2}(p-1)-i} + S_{\frac{1}{2}(p-1)-i+\text{ind}_\varrho(j)} - S_{\frac{1}{2}(p-1)-i+\text{ind}_\varrho(j+1)}$$

$$S_k = \frac{1}{p}(g_k + g_{k+r} + \cdots + g_{k+(f-1)r}),$$

and

$$g_k \equiv g^k \ (\text{mod } p), \qquad 1 \le g_k \le p - 1.$$

Let

$$\delta = \sqrt{\frac{1 - \zeta^g}{1 - \zeta} \cdot \frac{1 - \zeta^{-g}}{1 - \zeta^{-1}}}$$

be one of the Kummer circular units ($=$ *Kreiseinheiten*). Let

$$\theta_s = \prod_{i=0}^{(p-3)/2} \sigma^i(\delta^{g^{-2is}}) \tag{5.10}$$

for $s = 1, 2, \ldots, (p-3)/2$.

In 1852, Kummer computed the index of $\theta_s$ in terms of the cyclotomic functions:

(5A) *If $s = 1, 2, \ldots, (p-3)/2$ and if $1 < j < p - 1$, then*

$$\text{ind}_\varrho(\theta_s) \equiv \frac{g^{2s} - 1}{2} \times \frac{1}{1 + j^{p-2s} - (j+1)^{p-2s}} \, l^{(p-2s)}\psi_j(\zeta) \ (\text{mod } p).$$

Also, if $f$ does not divide $p - 2s$, then $\text{ind}_\varrho(\theta_s) \equiv 0 \ (\text{mod } p)$.

Using this arsenal, Kummer studied Fermat's equation under certain working hypotheses.

Assume:

I*. The first factor $h^*$ of the class number of $K$ is divisible by $p$ but not by $p^2$.

The following result was assumed tacitly by Kummer; its first complete proof is due to Vandiver (1922):

(5B) *If Hypothesis (I*) is satisfied, then*:

I'. *There exists a unique index $2s$, $2 \le 2s \le p - 3$, such that $p | B_{2s}$; moreover $p^2 \nmid B_{2s}$.*

Next, Kummer assumed:

II. There exists an ideal $J$ of $A$ such that the unit $\theta_s$ [corresponding to the index $s$ of (I')] is not congruent modulo $J$ to the $p$th power of any element of $A$.

He proved:

(5C) *Let $q$ be an odd prime, $q \neq p$, let $Q$ be a prime ideal dividing $Aq$, and assume that $Q^h = A\alpha$ ($h$ is the class number of $K$). Under Hypothesis (I') and (II):*

$$\mathrm{ind}_Q(\theta_s) \equiv \frac{(-1)^{s+\frac{1}{2}(p-1)}(g^{2s} - 1)B_{2ps-p+1}}{2h} \times l^{(p-2s)}(\alpha) \ (\mathrm{mod}\ p),$$

*where $s$ is defined in Hypothesis (I').*

If $J = \prod_{i=1}^{m} Q_i$ (where each $Q_i$ is a prime ideal not dividing $Ap$), let $\mathrm{ind}_J(\alpha) = \sum_{i=1}^{m} \mathrm{ind}_{Q_i}(\alpha)$.
With this definition:

(5D) *If (I') and (II) are satisfied and if $J$ is any ideal whose ideal class has order (in the ideal class group) not a multiple of $p$, then $\mathrm{ind}_J(\theta_s) \equiv 0 \ (\mathrm{mod}\ p)$.*

(5E) *If (I') and (II) are satisfied and if $J$ is any ideal such that $J^p = A\alpha$, a principal ideal, then $J$ itself is a principal ideal if and only if*

$$l^{(p-2s)}(\alpha) \equiv 0 \ (\mathrm{mod}\ p),$$

[*where $s$ is defined in (I')*].

From all these results, Kummer concluded:

(5F) *If (I') and (II) are satisfied, then $p$ does not divide the second factor $h^+$ of the class number.*

(5G) *If (I') is satisfied and if $p$ divides $h^+$, then the unit $\theta_s$ [ for the index $s$ defined in (I')] is the $p$th power of a unit of $K$.*

Kummer's proof of the next theorem was incorrect. But, thanks to Vandiver's work, (1926), it is now rigorously established:

(5H) *If (I') is satisfied, if $\alpha \in A \backslash A\lambda$ and $\alpha$ is real, and if $l^{(s)}(\alpha) \equiv 0 \ (\mathrm{mod}\ p)$, then there exists a unit $\varepsilon$ such that $\varepsilon\alpha \equiv m \ (\mathrm{mod}\ p)$, where $m \in \mathbb{Z}$.*

Let

III. If $s$ is given by (I'), then $B_{2sp}$ is not a multiple of $p^3$.

The following result was again put on firm ground by Vandiver:

(5I) *If (I') and (III) are satisfied, if $p$ does not divide $h^+$, and if $\varepsilon$ is any unit of $K$ congruent modulo $Ap^2$ to an integer $m \in \mathbb{Z}$, then there exists a unit $\varepsilon_0$ such that $\varepsilon = \varepsilon_0^p$.*

With these preliminary results, Kummer finally proved:

**(5J)** *If the conditions* (I\*), (II), *and* (III) *are satisfied, then Fermat's theorem holds for the exponent p.*

This theorem applied to the irregular prime exponents $p = 37, 59$ and $67$. He verified by actual computation that the conditions (I\*), (II) and (III) were satisfied.

I should add that in 1893, Mirimanoff gave an independent proof of Fermat's theorem for $p = 37$. So, even discounting the inaccuracies of Kummer, Fermat's theorem was established without any doubt for the smallest irregular prime.

# 6. Computations of the Class Number

Kummer had made, all through his life, extensive computations of the class number of cyclotomic fields. In my sixth lecture, I have already briefly mentioned his findings about the irregular primes less than 164. As a matter of fact, not only did he determine whether a given prime $p$ would or would not divide some Bernoulli number $B_{2s}$ $(2 \le 2s \le p - 3)$, but he actually determined the first factor of the class number, its factorization into primes (for all $p < 164$). All this was done without the help of any machine. And amazingly, a recent check by Newmann (who extended Kummer's table) in 1970, uncovered only three mistakes: $p = 103, 139, 163$.

I think that it is instructive to reproduce Kummer's table (as amended by Newman), to give a better feel of the growth of the first factor.

| Prime $p$ | $h^*(p)$ into prime factors |
|:---:|:---|
| 3 | 1 |
| 5 | 1 |
| 7 | 1 |
| 11 | 1 |
| 13 | 1 |
| 17 | 1 |
| 19 | 1 |
| 23 | 3 |
| 29 | $2 \times 2 \times 2$ |
| 31 | $3 \times 3$ |
| 37 | 37 |
| 41 | $11 \times 11$ |
| 43 | 211 |
| 47 | $5 \times 139$ |
| 53 | 4889 |

| Prime $p$ | $h^*(p)$ into prime factors |
|---|---|
| 59 | $3 \times 59 \times 233$ |
| 61 | $41 \times 1861$ |
| 67 | $67 \times 12739$ |
| 71 | $7 \times 7 \times 79241$ |
| 73 | $89 \times 134353$ |
| 79 | $5 \times 53 \times 377911$ |
| 83 | $3 \times 279405653$ |
| 89 | $113 \times 118401449$ |
| 97 | $577 \times 3457 \times 206209$ |
| 101 | $5 \times 5 \times 5 \times 5 \times 5 \times 101 \times 601 \times 18701$ |
| 103 | $5 \times 103 \times 1021 \times 17247691$ |
| 107 | $3 \times 743 \times 9859 \times 2886593$ |
| 109 | $17 \times 1009 \times 9431866153$ |
| 113 | $2 \times 2 \times 2 \times 17 \times 11853470598257$ |
| 127 | $5 \times 13 \times 43 \times 547 \times 883 \times 3079 \times 626599$ |
| 131 | $3 \times 3 \times 3 \times 5 \times 5 \times 53 \times 131 \times 1301 \times 4673706701$ |
| 137 | $17 \times 17 \times 47737 \times 46890540621121$ |
| 139 | $3 \times 3 \times 47 \times 47 \times 277 \times 277 \times 967 \times 1188961909$ |
| 149 | $3 \times 3 \times 149 \times 512966338320040805461$ |
| 151 | $7 \times 11 \times 11 \times 281 \times 25951 \times 1207501 \times 312885301$ |
| 157 | $5 \times 13 \times 13 \times 157 \times 157 \times 1093 \times 1873 \times 418861 \times 3148601$ |
| 163 | $2 \times 2 \times 181 \times 23167 \times 365473 \times 441845817162679$ |

Other recent tables of $h^*(p)$ and their factors were compiled by Schruttka von Rechtenstamm (up to 257) and by Lehmer and Masley (up to 521).

I mentioned in Lecture VI the formula for the first factor of the class number, namely

$$h^*(p) = \frac{1}{(2p)^{(p-3)/2}} |G(\eta)G(\eta^2) \cdots G(\eta^{p-2})|, \tag{6.1}$$

where $\eta$ is a primitive $(p-1)$th root of 1, $G(X) = \sum_{j=0}^{p-2} g_j X^j$, $g$ is a primitive root modulo $p$, $1 \le g_j \le p - 1$, and $g_j \equiv g^j \pmod{p}$. Kummer has used this formula for his computations. However, for larger values of $p$, it is rather inefficient. Lehmer and Masley based the computations on the evaluation of the Maillet determinant.

If $p$ is an odd prime, and $r$ is not a multiple of $p$, let $R(r)$ denote the least positive residue of $r$ modulo $p$. For every integer $r$, $1 \le r \le p - 1$, let $r'$ be the unique integer such that $1 \le r' \le p - 1$ and $rr' \equiv 1 \pmod{p}$. The Maillet matrix for $p$ is

$$M_p = (R(rs'))_{r,s=1,\ldots,(p-1)/2}. \tag{6.2}$$

Its determinant will be denoted by $D_p$.

The following formula was discovered by Weil, who did not publish it. Carlitz and Olson discovered it independently (1955):

$$h^*(p) = \frac{1}{p^{(p-3)/2}} |D_p|. \tag{6.3}$$

With this formula, $h^*(p)$ was evaluated by Masley, in the range indicated.

Concerning the growth of $h^*(p)$, Kummer stated a very interesting conjecture which is as yet unproved:

$$h^*(p) \sim \gamma(p) = \frac{p^{(p+3)/4}}{2^{(p-3)/2}\pi^{(p-1)/2}} \tag{6.4}$$

(asymptotically as $p$ tends to infinity). This conjecture has stimulated a number of deep recent investigations. I would like to mention some of the outstanding results obtained thus far, in this vein.

In 1951, Ankeny and Chowla proved

$$\lim_{p \to \infty} \frac{\log \dfrac{h^*(p)}{\gamma(p)}}{\log p} = 0, \tag{6.5}$$

or equivalently

$$\log h^*(p) = \tfrac{1}{4}(p+3)\log p - \tfrac{1}{2}p \log 2\pi + o(\log p).$$

In 1964, Siegel proved a weaker result: $\log h^*(p) = \log \gamma(p) + o(p \log \log p)$. It follows that

$$\log h^*(p) \sim \frac{p}{4} \log p.$$

Computations by Pajunen in 1976 showed that for $5 < p \le 641$:

$$\frac{2}{3} < \frac{h^*(p)}{\gamma(p)} < \frac{3}{2}.$$

At the present, the best result is due to Montgomery and Masley (1976): If $p > 200$, then

$$(2\pi)^{-p/2} p^{(p-25)/4} \le h^*(p) \le (2\pi)^{-p/2} p^{(p+31)/4}. \tag{6.6}$$

Despite the growth of $h^*(p)$, it is nevertheless bounded by some quite nice functions. In Lepistö (1974) and Metsänkylä (1974) it is proved that

$$h^*(p) < 2p \left( \frac{p}{24} \right)^{(p-1)/4}. \tag{6.7}$$

In 1961, Carlitz gave the following other upper bound (which is ultimately weaker):

$$h^*(p) \le \begin{cases} \left( \dfrac{p-5}{4} \right)! & \text{when } p \equiv 1 \pmod 4, \\[2mm] \left( \dfrac{p-7}{4} \right)! \sqrt{\dfrac{p-3}{4}} & \text{when } p \equiv 3 \pmod 4, \, p > 3. \end{cases} \tag{6.8}$$

Ankeny and Chowla also proved in 1951 that there exists $p_0$ such that $h^*(p)$ is strictly increasing with $p$, for $p \geq p_0$. It has been conjectured by Lepistö that $p_0 = 19$. The latest computations of Lehmer and Masley confirm this conjecture up to 521.

A fact that was long suspected and has now been established, is the following theorem, obtained independently by Uchida and Montgomery in 1971.

*If the class number $h(p)$ of $\mathbb{Q}(\zeta_p)$ is 1, then $p \leq 19$.*

Even better, if $h^*(p) = 1$, then $h^+(p) = 1$, so that $h^*(p) = 1$ already implies $p \leq 19$.

I should say also that in his thesis (1972), Masley determined all the cyclotomic fields $\mathbb{Q}(\zeta_m)$ (where $m \not\equiv 2 \pmod 4$, without loss of generality) having class number 1 (see Masley and Montgomery, 1976). There are 29 distinct such fields, namely, for $m =$ 3, 4, 5, 7, 8, 9, 11, 12, 13, 15, 16, 17, 19, 20, 21, 24, 25, 27, 28, 32, 33, 35, 36, 40, 44, 45, 48, 60, 84.

The little that is known about the first factor of the class number of cyclotomic fields seems immense when compared to what has been established about the second factor.

It is again Kummer who proved, as I have stated in my sixth lecture: An odd prime $p$ divides $h(p) = h^*(p)h^+(p)$ if and only if $p$ divides $h^*(p)$. In other words, if $p \mid h^+(p)$, then $p \mid h^*(p)$.

As for the parity of the class-number factors, Kummer proved in 1870 that if $h^*(p)$ is odd, then so is $h^+(p)$.

In his computations up to 163, only $h^*(29)$, $h^*(113)$, $h^*(163)$ are even. However, he proved that $h^+(29) = 1$, $h^+(113)$ is also odd, while $h^+(163)$ is even. He also established that $h^+(257)$ is divisible by 3, and that $h^+(937)$ is even.

A method to produce cyclotomic fields with second factor greater than 1 was invented by Ankeny, Chowla, and Hasse in 1965. It was based on a lemma by Davenport and uses class field theory. This is the lemma:

*If $l, m$ are positive integers, if $m$ is not a square, and if the equation*

$$X^2 - (l^2 + 1)Y^2 = \pm m$$

*has solution in integers, then $m \geq 2l$.*

Next, it is shown that the class number $H(p)$ of the quadratic field $\mathbb{Q}(\sqrt{p})$, where $p = (2qn)^2 + 1$, $q$ a prime, $n > 1$, is greater than 1.

A comparison by means of class field theory of the field $\mathbb{Q}(\sqrt{p})$ and the real cyclotomic field $\mathbb{Q}(\zeta_p + \zeta_p^{-1})$ yields the following result:

*If $q$ is a prime, $n > 1$, and if $p = (2qn)^2 + 1$ is also a prime, then $h^+(p) > 2$.* In particular, for $p = $ 257, 401, 577, 1297, 1601, 2917, 3137, 4357, 7057, 8101, $h^+(p) > 2$.

A variant of the above result was presented by S.-D. Lang in 1977. He proved that if $q$ is a prime, $n \geq 1$, and if $p = [(2n + 1)q]^2 + 4$ is also a prime, then $h^+(p) > 2$. This gives $h^+(229) > 2$ and many more examples.

Does this method provide infinitely many examples? This question is at least of the level of difficulty of the problem of Sophie Germain on the problem of twin primes.

# Bibliography

1801   Gauss, C. F.
*Disquisitiones Arithmeticae*, G. Fleischer, Leipzig, 1801. Translated into English by Arthur A. Clarke. Yale University Press, New Haven, 1966.

1837   Jacobi, C. G. J.
Über die Kreistheilung und ihre Anwendung auf die Zahlentheorie. *Monatsber. Akad. d. Wiss.*, Berlin, 1837. Reprinted *J. reine u. angew. Math.*, **130**, 1846, 166–182.

1846   Kummer, E. E.*
Über die Divisoren gewisser Formen der Zahlen, welche aus der Theorie der Kreistheilung entstehen. *J. reine u. angew. Math.*, **30**, 1846, 107–116.

1847   Cauchy, A.
Mémoire sur diverses propositions relatives à la Théorie des Nombres. *C. R. Acad. Sci. Paris*, **25**, 1847, 177–183. Reprinted in *Oeuvres Complètes* (1), 10, Gauthier-Villars, Paris, 1897, 360–366.

1847   Kummer, E. E.
Zur Theorie der complexen Zahlen. *J. reine u. angew. Math.*, **35**, 1847, 319–326.

1847   Kummer, E. E.
Über die Zerlegung der aus Wurzeln der Einheit Gebildeten complexen Zahlen in ihren Primfactoren. *J. reine u. angew Math.*, **35**, 1847, 327–367.

1851   Kummer, E. E.
Mémoire sur la théorie des nombres complexes composés de racines de l'unité et de nombres entiers. *J. Math. Pures et Appl.*, **16**, 1851, 377–498.

1852   Genocchi, A.
Intorno all espressioni generali di numeri Bernoulliani. *Annali di scienze mat. e fisiche, compilati da Barnaba Tortolini*, **3**, 1852, 395–405.

1852   Kummer, E. E.
Über die Ergänzungssätze zu den allgemeinen Reciprocitätsgesetzen. *J. reine u. angew. Math.*, **44**, 1852, 93–146.

1857   Kummer, E. E.
Über die den Gaussischen Perioden der Kreistheilung entsprechenden Kongruenzwurzeln. *J. reine u. angew. Math.*, **53**, 1857, 142–148.

1857   Kummer, E. E.
Einige Sätze über die aus den Wurzeln der Gleichung $\alpha^\lambda = 1$ gebildeten complexen Zahlen, für den Fall dass die Klassenzahl durch $\lambda$ theilbar ist, nebst Anwendungen derselben auf einen weiteren Beweis des letztes Fermat'schen Lehrsatzes. Math. Abhandl. Akad. d. Wiss., Berlin, 1857, 41–74.

* See also *Collected Papers*, vol. I, edited by A. Weil, Springer-Verlag, Berlin, 1975.

1861   Kummer, E. E.
Über die Klassenanzahl der aus $n$-ten Einheitswurzeln gebildeten complexen Zahlen. Monatsber, Akad. d. Wiss., Berlin, 1861, 1051–1053.

1870   Kummer, E. E.
Über eine Eigenschaft der Einheiten der aus den Wurzeln der Gleichung $\alpha^\lambda = 1$ gebildeten complexen Zahlen, und über den zweiten Factor der Klassenzahl. Monatsber. Akad. d. Wiss., Berlin, 1870, 855–880.

1874   Kummer, E. E.
Über diejenigen Primzahlen $\lambda$ fur welche die Klassenzahl der aus $\lambda$-ten Einheitswurzeln gebildeten complexen Zahlen durch $\lambda$ theilbar ist. Monatsber. Akad. d. Wiss., Berlin, 1874, 239–248.

1886   Genocchi, A.
Sur les nombres de Bernoulli (extrait d'une lettre adressée à M. Kronecker). *J. reine u. angew. Math.*, **99**, 1886, 315–316.

1890   Stickelberger, L.
Über eine Verallgemeinerung der Kreistheilung. *Math. Annalen*, **37**, 1890, 321–367.

1893   Mirimanoff, D.
Sur l'équation $x^{37} + y^{37} + z^{37} = 0$. *J. reine u. angew. Math.* **111**, 1893, 26–36.

1897   Hilbert, D.
Die Theorie der algebraischen Zahlkörper. *Jahresber. d. Deutschen Math. Verein.*, **4**, 1897, 175–546. Also in *Gesammelte Abhandlungen*, vol. 1, Springer-Verlag, Berlin, 1932 (Reprinted by Chelsea Publ. Co., New York, 1965).

1913   Maillet, E.
Question 4269. *L'Interm. des Math.*, **20**, 1913, 218.

1914   Malo, E.
Réponse à la question 4269. *L'Interm. des Math.*, **21**, 1914, 173–176.

1916   Mitchell, H. H.
On the generalized Jacobi-Kummer cyclotomic function. *Trans. Amer. Math. Soc.*, **17**, 1916, 165–177.

1917   Mertens, F.
Über den Kummer'schen Logarithmus einer komplexen Zahl des Bereich einer primitiven $\lambda$-ten Einheitswurzeln in bezug auf den Modul $\lambda^{1+n}$, wo $\lambda$ eine ungerade Primzahl bezeichnet. *Sitzungsberichte d. Akad. der Wiss. zu Wien*, Abt. IIa, **126**, 1917, 1337–1343.

1919   Vandiver, H. S.
A property of cyclotomic integers and its relation to Fermat's last theorem. *Annals of Math.*, **21**, 1919, 73–80.

1920   Vandiver, H. S.
On Kummer's memoir of 1857 concerning Fermat's last theorem. *Proc. Nat. Acad. Sci. U.S.A.*, **6**, 1920, 266–269.

1922   Fueter, R.
Kummers Kriterien zum letzten Theorem von Fermat. *Math. Annalen*, **85**, 1922, 11–20.

1922   Vandiver, H. S.
On Kummer's memoir of 1857 concerning Fermat's last theorem. *Bull. Amer. Math. Soc.*, **28**, 1922, 400–407.

1926   Vandiver, H. S.
Summary of results and proofs concerning Fermat's last theorem. *Proc. Nat. Acad. Sci. U.S.A.*, **12**, 1926, 106–109.

1926   Vandiver, H. S.
Summary of results and proofs concerning Fermat's last theorem (second note).
*Proc. Nat. Acad. Sci. U.S.A.*, **12**, 1926, 767–772.

1928   Vandiver, H. S. and Wahlin, G. E.
*Algebraic Numbers*, II, Bull. Nat. Res. Council, **62**, 1928. Reprinted by Chelsea
Publ. Co., New York, 1967.

1951   Ankeny, N. C. and Chowla, S.
The class number of the cyclotomic field. *Can. J. Math.*, **3**, 1951, 486–494.

1955   Carlitz, L. and Olson, F. R.
Maillet's determinant. *Proc. Amer. Math. Soc.*, **6**, 1955, 265–269.

1961   Carlitz, L.
A generalization of Maillet's determinant and a bound for the first factor of the
class number. *Proc. Amer. Math. Soc.*, **12**, 1961, 256–261.

1964   Schrutka von Rechtenstamm, G.
Tabelle der Relativ-Klassenzahlen der Kreiskörper deren $\varphi$-Funktion des Wurzel
exponenten (Grad) nicht grösser als 256 ist. Abhandl. Deutsche Akad. Wiss., Berlin,
Kl. Math. Phys., No. 2, 1964, 1–64.

1965   Ankeny, N. C., Chowla, S., and Hasse, H.
On the class number of the maximal real subfield of a cyclotomic field. *J. reine u.
angew. Math.*, **217**, 1965, 217–220.

1966   Siegel, C. L.
Zu zwei Bemerkungen Kummers. Nachr. Akad. Wiss. Göttingen, Math. Phys.
Kl. II, 1964, 51–57. Reprinted in *Gesammelte Abhandlungen.*, III, Springer-Verlag,
New York, 1966, 436–442.

1970   Newman, M.
A table of the first factor for prime cyclotomic fields. *Math. Comp.*, **24**, 1970,
215–219.

1971   Uchida, K.
Class numbers of imaginary abelian number fields, III. *Tôhoku Math. J.*, **23**, 1971,
573–580.

1972   Masley, J. M.
On the Class Number of Cyclotomic Fields. Thesis, Princeton University, 1972,
51 pages.

1972   Metsänkylä, T.
On the growth of the first factor of the cyclotomic class number. *Ann. Univ. Turku*,
Ser. A, I, 1972, No. **155**, 12 pages.

1974   Lepistö, T.
On the growth of the first factor of the class number of the prime cyclotomic field.
*Ann. Acad. Sci. Fennicae*, Ser. A. I, 1974, No. **577**, 18 pages.

1974   Metsänkylä, T.
Class numbers and $\mu$-invariants of cyclotomic fields. *Proc. Amer. Math. Soc.*, **43**,
1974, 299–300.

1975   Maury, G.
Idéaux des Corps Cyclotomiques. Thèse, Univ. Paul Sabatier, Toulouse, 1975,
54 pages.

1978   Lehmer, D. H. and Masley, J. M.
Table of the cyclotomic class numbers $h^*(p)$ and their factors $200 < p < 512$.
*Math. Comp.*, **32**, 1978, 577–582.

1976   Masley, J. M. and Montgomery, H. L.
       Unique factorization in cyclotomic fields. *J. reine u. angew. Math.*, **286/7**, 1976,
       248–256.

1976   Pajunen, S.
       Computation of the growth of the first factor for prime cyclotomic fields. *BIT*, **16**,
       1976, 85–87.

1977   Lang, S.-D.
       Note on the class-number of the maximal real subfield of a cyclotomic field. *J. reine
       u. angew. Math.*, **290**, 1977, 70–72.

# After Kummer, a New Light

I'll report on the work of Mirimanoff, inspired by the last of Kummer's papers. With his great ability, he refined Kummer's treatment for the first case and obtained new congruences. On the other hand, due to the difficulty in achieving these comparatively meager improvements, it was obvious that no further progress would be forthcoming along these lines.

Then in 1909 came Wieferich. He discovered a criterion for the first case of an entirely different nature.

His first proof was an enigma. Few people were able to understand how Wieferich succeeded, like a magician, in unravelling from very complicated formulas, so simple and beautiful a criterion as:

*If the first case fails for the exponent p, then p must satisfy the congruence*

$$2^{p-1} \equiv 1 \pmod{p^2}.$$

Note that this condition is the first, in the history of Fermat's problem, which does not involve the would-be solutions $x$, $y$, $z$ of $X^p + Y^p + Z^p = 0$. Just the exponent, and as I'll show, it is a very stringent condition indeed.

With every breakthrough, come the followers. In this case a distinguished group including Mirimanoff, Frobenius, Vandiver, Pollaczek, and Morishima.

## 1. The Congruences of Mirimanoff

In 1905, Mirimanoff transformed Kummer's congruences. I recall that Kummer showed:

*If  x, y, z are relatively prime integers, not multiples of p, such that*

$x^p + y^p + z^p = 0$, *then*

$$[D^{p-2s}\log(x + e^v y)]B_{2s} \equiv 0 \;(\text{mod } p) \tag{1.1}$$

*for* $2s = 2, 4, \ldots, p - 3$. *Similar congruences hold by permuting* $x$, $y$, $z$.

Here $D^n f(v)$ denotes the $n$th derivative of $f(v)$ computed at $v = 0$. As I pointed out,

$$D^j \log(x + e^v y) = \frac{R_j(x, y)}{(x + y)^j},$$

where $R_j(X, Y)$ is a homogeneous polynomial with coefficients in $\mathbb{Z}$ and total degree $j$.

I recall the properties of these polynomials. Putting $Y = XT$, then $R_j(X, Y) = X^j P_j(T)$, where $P_j(T)$ has coefficients in $\mathbb{Z}$ and degree $j - 1$. For $j > 1$ and odd, $T(1 - T)$ divides $P_j(T)$, so $P_j(T) = T(1 - T)L_j(T)$, and $L_j(T)$ has degree $j - 3$.

The congruences of Kummer become

$$P_{p-2s}(t)B_{2s} \equiv 0 \;(\text{mod } p), \tag{1.2}$$

where $2s = 2, 4, \ldots, p - 3$ and

$$t \in G = \left\{ \frac{x}{y}, \frac{y}{x}, \frac{x}{z}, \frac{z}{x}, \frac{y}{z}, \frac{z}{y} \right\}.$$

Since $x^p + y^p + z^p = 0$, $x + y + z \equiv 0 \;(\text{mod } p)$.

If $t = x/y$, then the elements of $G$ are congruent modulo $p$, to those of the set

$$H = \left\{ t, \frac{1}{t}, -\frac{1}{1 + t}, -(1 + t), -\frac{t}{1 + t}, -\left(1 + \frac{1}{t}\right) \right\}. \tag{1.3}$$

In some special cases, $H$ degenerates: If $t \equiv 1$, or $-2$, or $-\frac{1}{2} \;(\text{mod } p)$, then $H = \{1, -2, -\frac{1}{2}\}$. If $t^2 + t + 1 \equiv 0 \;(\text{mod } p)$, then $p \equiv 1 \;(\text{mod } 6)$ and $H$ has only 2 distinct elements, In all other cases, $H$ has 6 distinct elements.

I note also that none of the elements in $G$ may be congruent to 0 or to $-1$ modulo $p$.

Mirimanoff's first result extended an earlier criterion of Kummer:

**(1A)** *If* $x^p + y^p + z^p = 0$ *and* $p \nmid xyz$, *then* $B_{p-7}$ *and* $B_{p-9}$ *are multiples of* $p$.

SKETCH OF THE PROOF. Since $[D^7 \log(x + e^v y)]B_{p-7} \equiv 0 \;(\text{mod } p)$ (and similarly, for any permutation of $\{x, y, z\}$), it suffices to show that $D^7 \log(x + e^v y)$ (or that any of the derivatives obtained by such a permutation) is not congruent to 0 modulo $p$.

Assume that this is false, so $P_7(t) \equiv 0 \;(\text{mod } p)$ for every $t \in G$. As I mentioned in my last lecture, $P_7(T) = T(1 - T)L_7(T)$, where $L_7(T)$ has coefficients in $\mathbb{Z}$ and degree 4.

If the elements of $G$ are pairwise incongruent, they cannot all be roots of the congruence $P_7(T) \equiv 0 \pmod{p}$ (note that the elements of $G$ are not congruent to $0 \pmod{p}$).

If the elements of $G$ are congruent to those of $\{1, -2, -\frac{1}{2}\}$, then $L_7(-2) \equiv 0 \pmod{p}$.

But

$$L_7(T) = 1 - 56T + 246T^2 - 56T^3 + T^4.$$

Hence $p$ divides $7 \times 223$, so $p = 223$ (since 7 is excluded).

Also, from the naïve approach mentioned in Lecture IV,

$$(t + 1)^p \equiv t^p + 1 \pmod{p^2}$$

for any $t \in G$.

Hence, with $t \equiv 1 \pmod{p}$, it follows that

$$2^p \equiv 2 \pmod{p^2},$$

where $p = 223$. However, an easy calculation shows that this is not possible.

There remains the case where $t^2 + t + 1 \equiv 0 \pmod{p}$. Then $p \equiv 1 \pmod 6$ and also $t \not\equiv 1 \pmod p$, hence $t$ is a root of $L_7(T) \equiv 0 \pmod{p}$. But

$$L_7(T) = (T^2 + T + 1)(T^2 - 57T + 302) + (-301T - 301)$$

hence $-301t - 301 \equiv 0 \pmod p$ and $p$ divides $7 \times 43$ since $t \not\equiv -1 \pmod p$. Thus $p = 43$. But, $4p + 1 = 173$ is also a prime. Hence from the Legendre and Sophie Germain criterion, the first case holds for $p = 43$, and $p = 43$ is also excluded.

A similar proof may be repeated for the 9th derivatives $D^9 \log(x + e^v y)$ and those obtained by permutation of $x$, $y$, $z$. It still works, but requires a more careful analysis. ☐

This method cannot be pushed further, without great pain since it leads to polynomials $L_j(T)$ of degree greater than 6, when $j \geq 11$.

However, Morishima extended this result in 1932 to guarantee that if the first case fails for the exponent $p$, then $B_{p-11}$ and $B_{p-13}$ must also be divisible by $p$.

Later I shall describe a very powerful theorem of Krasner, along the same lines, but obtained through a totally different method.

To derive Mirimanoff's congruences, it is necessary to study in more detail the polynomials $P_j(T)$. The first property to note is

$$P_j(T) = (-1)^j T^j P_j\left(\frac{1}{T}\right), \quad \text{for } j > 1. \tag{1.4}$$

So, writing

$$P_j(T) = a_{j,1}T - a_{j,2}T^2 + \cdots + (-1)^{k+1}a_{j,k}T^k + \cdots$$
$$+ (-1)^j a_{j,j-1}T^{j-1}, \tag{1.5}$$

it follows that

$$a_{j,k} = a_{j,j-k} \quad \text{for } 1 < j, 1 \leq k \leq j - 1. \tag{1.6}$$

The computation of the coefficients $a_{j,k}$ ($1 \leq k \leq j - 1$) is done recursively. The following lemma, due to Euler, is needed:

**Lemma 1.1.** *Let $t \neq 1$ be any real number, let $g(v) = 1/(1 + e^v t)$. Then the Taylor development of $g$ around $v = 0$ is*

$$g(v) = C_0(t) + C_1(t)v + C_2(t)\frac{v^2}{2!} + \cdots + C_n(t)\frac{v^n}{n!} + \cdots,$$

*where*

$$C_0(t) = \frac{1}{1 + t}$$

$$C_1(t) = -\frac{t}{(1 + t)^2}$$

$$C_2(t) = -\frac{t - t^2}{(1 + t)^3}$$

$$\vdots$$

$$C_n(t) = -\frac{c_{n,1}t - c_{n,2}t^2 + \cdots + (-1)^{n-1}c_{n,n}t^n}{(1 + t)^{n+1}}$$

*and*

$$c_{n,h} = h^n - \binom{n+1}{1}(h-1)^n + \binom{n+1}{2}(h-2)^n - \binom{n+1}{3}(h-3)^n + \cdots$$

$$+ (-1)^{h-1}\binom{n+1}{h-1}$$

*for $h = 1, 2, \ldots, n$.*

With this lemma, the following may be shown, without much difficulty:

*The coefficients of $P_j(T)$ are given by the formula:*

$$a_{j,k} = k^{j-1} - \binom{j}{1}(k-1)^{j-1} + \binom{j}{2}(k-2)^{j-1} + \cdots + (-1)^{k-1}\binom{j}{k-1}$$

*for $1 < j, 1 \leq k \leq j - 1$.* $\tag{1.7}$

So the coefficients $a_{j,k}$ are the sums of the first $k$ summands, computed for $X = k$, of

$$G_j^{(j-1)}(X) = X^{j-1} - \binom{j}{1}(X-1)^{j-1} + \binom{j}{2}(X-2)^{j-1} - \cdot\cdot$$

$$+ (-1)^j(X - j + 1)^{j-1}. \tag{1.8}$$

Another fact about these polynomials is that $G_j^{(j-1)}(X) = 0$. This is not difficult to show; the easiest proof makes use of the theory of finite differences.

A somewhat involved algebraic manipulation leads to the following:

If $1 < j \le p - 1$, then

$$(1 + T)^{p-j}P_j(T) \equiv T - 2^{j-1}T^2 + 3^{j-1}T^3 - \cdots - (p-1)^{j-1}T^{p-1} \pmod{p}.$$

$$(1.9)$$

Rewriting this as

$$(1 - T)^{p-j}P_j(-T) \equiv T + 2^{j-1}T^2 + 3^{j-1}T^3 + \cdots + (p-1)^{j-1}T^{p-1} \pmod{p},$$

$$(1.10)$$

the Mirimanoff polynomials are brought to light:

$$\varphi_1(T) = T + T^2 + T^3 + \cdots + T^{p-1},$$
$$\varphi_2(T) = T + 2T^2 + 3T^3 + \cdots + (p-1)T^{p-1},$$
$$\vdots$$
$$\varphi_j(T) = 1^{j-1}T + 2^{j-1}T^2 + 3^{j-1}T^3 + \cdots + (p-1)^{j-1}T^{p-1}, \quad (1.11)$$
$$\vdots$$
$$\varphi_{p-1}(T) = 1^{p-2}T + 2^{p-2}T^2 + 3^{p-2}T^3 + \cdots + (p-1)^{p-2}T^{p-1},$$
$$\varphi_p(T) = 1^{p-1}T + 2^{p-1}T^2 + 3^{p-1}T^3 + \cdots + (p-1)^{p-1}T^{p-1}.$$

So (1.10) can be rewritten as

$$\varphi_j(T) \equiv -(1 - T)^{p-j}P_j(-T) \pmod{p} \qquad (1.12)$$

for $j = 2, \ldots, p - 1$.

Due to their nice form, the Mirimanoff polynomials display many interesting properties. First, some obvious facts:

$$\varphi_1(T) = \frac{T(1 - T^{p-1})}{1 - T}, \qquad (1.13)$$

$$\varphi_1(T) \equiv \varphi_p(T) \pmod{p}, \qquad (1.14)$$

$$\varphi_1(T) \equiv -P_p(-T) \pmod{p}. \qquad (1.15)$$

Let $\Gamma$ be the following operator on polynomials of $\mathbb{Q}[T]$:

$$\Gamma(G) = T\frac{dG}{dT}. \qquad (1.16)$$

Let $\Gamma^k$ be the $k$th iterated operator of $\Gamma$ (for $k \ge 1$). Then:

$$\Gamma(\varphi_j) = \varphi_{j+1} \quad (\text{for } j = 1, \ldots, p - 1), \qquad (1.17)$$

$$\Gamma(\varphi_{p-1}) \equiv \varphi_1 \pmod{p}. \qquad (1.18)$$

Therefore

$$\Gamma^{p-1}(\varphi_j) \equiv \varphi_j \pmod{p} \quad \text{for } j = 1, 2, \ldots, p - 1.$$

Mirimanoff also considered the polynomials

$$\psi_j(T) = \varphi_j(1 - T) \quad \text{for } j = 1, 2, \ldots, p. \tag{1.19}$$

$\psi_j(T)$ has degree $p - 1$ and is a multiple of $1 - T$. Moreover: $T^{p-j}$ divides $\psi_j(T)$ modulo $p$ (for $j = 2, \ldots, p - 1$) and $\psi_1(T) \equiv T^{p-1} - 1 \pmod{p}$.

Let $\Lambda$ be the following operator on polynomials of $\mathbb{Q}[T]$:

$$\Lambda(G) = (T - 1)\frac{dG}{dT}. \tag{1.20}$$

It follows at once that

$$\Lambda(\psi_j) = \psi_{j+1} \quad (\text{for } j = 1, \ldots, p - 2), \tag{1.21}$$

$$\Lambda(\psi_{p-1}) \equiv \psi_1(T) \pmod{p}. \tag{1.22}$$

Further relations between the polynomials $\varphi_j(T)$ and $\psi_j(T)$ are the following:

$$\varphi_{p-1}(T) \equiv \psi_{p-1}(T) \equiv \frac{1 - T^p - (1 - T)^p}{p} \pmod{p}. \tag{1.23}$$

Taking $T = -1$, then

$$\sum_{j=1}^{p-1} \frac{(-1)^{j-1}}{j} \equiv \sum_{j=1}^{p-1} (-1)^{j-1} j^{p-2} \equiv -\varphi_{p-1}(-1) \equiv \frac{2^p - 2}{p} \pmod{p}. \tag{1.24}$$

The explicit computation of $\psi_{p-2}(T)$ gives:

$$-\psi_{p-2}(T) \equiv \frac{T^2}{2} + \frac{T^3}{3}\left(1 + \frac{1}{2}\right) + \frac{T^4}{4}\left(1 + \frac{1}{2} + \frac{1}{3}\right)$$

$$+ \cdots + \frac{T^{p-1}}{p-1}\left(1 + \frac{1}{2} + \cdots + \frac{1}{p-2}\right) \pmod{p}. \tag{1.25}$$

This leads, with some further work, to the congruence

$$\varphi_{p-2}(T) \equiv T^p \psi_{p-2}\left(\frac{1}{T}\right) + \psi_{p-2}(T) \pmod{p}. \tag{1.26}$$

At this point, Mirimanoff introduced logarithms. He made the happy observation, which may be established by induction on $j$: If $j = 1, 2, \ldots, p - 2$, then

$$[\log(1 - T)]^j \equiv (-1)^j j! \psi_{p-j}(T) \pmod{T^p, p}, \tag{1.27}$$

where the congruence means that the coefficients of $T^k$ (for $k \le p - 1$) in both sides are congruent modulo $p$. This served as starting point to prove the following congruences:

$$[\varphi_{p-1}(T)]^2 \equiv -2\left[\psi_{p-2}(T) + T^{2p}\psi_{p-2}\left(\frac{1}{T}\right)\right] \pmod{p}. \tag{1.28}$$

As a corollary

$$-\frac{1}{2}[\varphi_{p-1}(T)]^2 \equiv \varphi_{p-2}(T) + (T-1)^{2p}\varphi_{p-2}\left(\frac{T}{T-1}\right) \pmod{p}. \quad (1.29)$$

With this groundwork, Mirimanoff was ready to prove:

**(1B)** *If* $x$, $y$, $z$ *are pairwise relatively prime integers, not divisible by the prime* $p > 2$, *and such that* $x^p + y^p + z^p = 0$, *and if* $-t \in G = \{x/y, y/x, x/z, z/x, y/z, z/y\}$, *then the following congruences are satisfied:*

$$\varphi_{p-1}(t) \equiv 0 \pmod{p},$$
$$\varphi_{p-2}(t)\varphi_2(t) \equiv 0 \pmod{p},$$
$$\varphi_{p-3}(t)\varphi_3(t) \equiv 0 \pmod{p},$$
$$\vdots$$
$$\varphi_{(p+1)/2}(t)\varphi_{(p-1)/2}(t) \equiv 0 \pmod{p}.$$

PROOF. Despite all the preparation, the proof is still long. So, I'll sketch its main points, trying to bring the main idea into view. To begin, the Kummer congruences

$$P_{p-2s}(-t)B_{2s} \equiv 0 \pmod{p}$$

are satisfied for $2s = 2, 4, \ldots, p-3$.
    Since $\varphi_{p-2s}(t) \equiv -(1-t)^{p-2s}P_{p-2s}(-t) \pmod{p}$,

$$\varphi_{p-2s}(t)B_{2s} \equiv 0 \pmod{p}.$$

Setting $j = 3, 5, \ldots, p-2$ yields

$$\varphi_j(t)B_{p-j} \equiv 0 \pmod{p}. \quad (1.30)$$

(a) $\varphi_{p-1}(t) \equiv 0 \pmod{p}$. For this, use (1.29):

$$-\frac{1}{2}[\varphi_{p-1}(t)]^2 \equiv \varphi_{p-2}(t) + (t-1)^{2p}\varphi_{p-2}\left(\frac{t}{1-t}\right) \pmod{p}.$$

Next, $\varphi_{p-2}(t) \equiv 0 \pmod{p}$ and $t/(1-t) \in G$ so $\varphi_{p-2}(-t/(1-t)) \equiv 0 \pmod{p}$. This establishes (a).
    (b) For $j = 1, 2, \ldots, p-2$,

$$\Gamma^{p-2-j}\left(\frac{\varphi_{j+1}(T)}{1-T}\right) \equiv T + 2^{p-2-j}(1^j + 2^j)T^2 + 3^{p-2-j}(1^j + 2^j + 3^j)T^3$$
$$+ \cdots + (p-1)^{p-2-j}(1^j + 2^j + \cdots$$
$$+ (p-1)^j)T^{p-1} \pmod{p}.$$

Start with the defining expression for $\varphi_{j+1}(T)$, which is rewritten as

$$\varphi_{j+1}(T) = (1-T)[T + (1^j + 2^j)T^2 + (1^j + 2^j + 3^j)T^3$$
$$+ \cdots + (1^j + 2^j + \cdots + (p-1)^j)T^{p-1}]$$
$$+ (1^j + 2^j + \cdots + (p-1)^j)T^p. \quad (1.31)$$

But

$$(j + 1)[1^j + 2^j + \cdots + (p - 1)^j]$$
$$= p^{j+1} + \binom{j+1}{1} B_1 p^j + \binom{j+1}{2} B_2 p^{j-1} + \cdots + \binom{j+1}{j} B_j p$$

(as seen in Lecture VI). Since $j \leq p - 2$, by von Staudt and Clausen's theorem $p$ does not divide the denominator of each $B_i$ ($1 \leq i \leq j$), so the above sum is congruent to 0 modulo $p$ and (1.31) becomes the congruence

$$\varphi_{j+1}(T) \equiv (1 - T)[T + (1^j + 2^j)T^2$$
$$+ \cdots + (1^j + 2^j + \cdots + (p - 1)^j)T^{p-1}] \pmod{p}. \quad (1.32)$$

Applying $\Gamma$ to $\varphi_{j+1}(T)/(1 - T)$ and iterating this process leads to the congruence (b).

(c) For $j = 1, 2, \ldots, p - 2$,

$$\Gamma^{p-2-j}\left(\frac{\varphi_{j+1}(T)}{1 - T}\right) \equiv \frac{1}{j+1} \varphi_1(T) + \frac{1}{2} \varphi_{p-1}(T) + \binom{j}{1} \frac{B_2}{2} \varphi_{p-2}(T)$$
$$+ \binom{j}{3} \frac{B_4}{4} \varphi_{p-4}(T) + \binom{j}{5} \frac{B_6}{6} \varphi_{p-6}(T) + \cdots \pmod{p}.$$

$$(1.33)$$

To arrive at this congruence it is enough to recall that

$$1^j + 2^j + \cdots + n^j = \frac{n^{j+1}}{j+1} + \frac{n^j}{2} + \binom{j}{1} \frac{B_2}{2} n^{j-1}$$
$$+ \binom{j}{3} \frac{B_4}{4} n^{j-3} + \binom{j}{5} \frac{B_6}{6} n^{j-5} + \cdots.$$

(d) For $j = 1, \ldots, p - 2$,

$$\Gamma^{p-2-j}[\varphi_{j+1}(T)(1 - T)^{p-1}] \equiv \varphi_{p-1}(T) + \Gamma^{p-2-j}[\varphi_{j+1}(T)\varphi_1(T)] \pmod{p}.$$
$$(1.34)$$

Indeed, by repeated application of $\Gamma$:

$$\Gamma^{p-2-j}[\varphi_{j+1}(T)(1 - T)^{p-1}] = \Gamma^{p-2-j}\left[\frac{\varphi_{j+1}(T)}{1 - T} \times (1 - T)^p\right]$$
$$\equiv \Gamma^{p-2-j}\left(\frac{\varphi_{j+1}(T)}{1 - T}\right)(1 - T)^p$$
$$\equiv \Gamma^{p-2-j}(\varphi_{j+1}(T))$$
$$+ \Gamma^{p-2-j}[\varphi_{j+1}(T)\varphi_1(T)] \pmod{p} \quad (1.35)$$

because $\Gamma[(1 - T)^p] \equiv 0 \pmod{p}$ and $(1 - T)^{p-1} \equiv 1 + \varphi_1(T) \pmod{p}$. From (1.17), $\Gamma^{p-2-j}(\varphi_{j+1}(T)) \equiv \varphi_{p-1}(T) \pmod{p}$, showing that (d) is true.

(e) Conclusion:

$$\Gamma^{p-2-j}(\varphi_{j+1}(T)\varphi_1(T)) = \varphi_{p-1}(T)\varphi_1(T) + \binom{p-2-j}{1}\varphi_{p-2}(T)\varphi_2(T)$$

$$+ \binom{p-2-j}{2}\varphi_{p-3}(T)\varphi_3(T)$$

$$+ \cdots + \binom{p-2-j}{1}\varphi_{j+2}(T)\varphi_{p-2-j}(T)$$

$$+ \varphi_{j+1}(T)\varphi_{p-1-j}(T). \tag{1.36}$$

Let $-t \in G$, and compute the residue of $\Gamma^{p-2-j}(\varphi_{j+1}(t)/(1-t))$. Since $t \not\equiv 1 \pmod{p}$, by (1.13) $\varphi_1(t) \equiv 0 \pmod{p}$. By (a), (1.30) and (1.33):

$$\Gamma^{p-2-j}\left(\frac{\varphi_{j+1}(t)}{1-t}\right) \equiv 0 \pmod{p}. \tag{1.37}$$

So, by (1.35) and (1.37)

$$\Gamma^{p-2-j}[\varphi_{j+1}(t)(1-t)^{p-1}] \equiv 0 \pmod{p}. \tag{1.38}$$

On the other hand, by (1.34), (a), and (1.36),

$$\Gamma^{p-2-j}[\varphi_{j+1}(t)(1-t)^{p-1}] \equiv \binom{p-2-j}{1}\varphi_{p-2}(t)\varphi_2(t)$$

$$+ \binom{p-2-j}{2}\varphi_{p-3}(t)\varphi_3(t)$$

$$+ \cdots + \binom{p-2-j}{1}\varphi_{j+2}(t)\varphi_{p-2-j}(t)$$

$$+ \varphi_{j+1}(t)\varphi_{p-1-j}(t). \tag{1.39}$$

By (1.38) the last sum is congruent to 0 modulo $p$, for $j = 1, 2, \ldots, p - 2$. Letting $j = p - 3, p - 4, \ldots$, successively, gives

$$\varphi_{p-2}(t)\varphi_2(t) \equiv 0 \pmod{p},$$
$$\varphi_{p-3}(t)\varphi_3(t) \equiv 0 \pmod{p},$$
$$\vdots$$
$$\varphi_{(p+1)/2}(t)\varphi_{(p-1)/2}(t) \equiv 0 \pmod{p}. \qquad \square$$

In 1967, Le Lidec gave another form to Mirimanoff congruences, using other polynomials. I use the following notation: If $p \nmid s$, let $\bar{s}$ be the unique integer such that $s \equiv \bar{s} \pmod{p}$ and $1 \leq \bar{s} \leq p - 1$. Let $s'$ be any integer such that $s's \equiv 1 \pmod{p}$. For each $n = 1, 2, \ldots, p - 2$, let $E_n$ be the set of all integers $s$, $1 \leq s \leq p - 2$, such that $\overline{(n+1)'ns} < s$.

LeLidec considered the following polynomials:

$$\Lambda_n(T) = \sum_{s \in E_n} \bar{s}^r T^{p-s}. \tag{1.40}$$

The basic relationship with Mirimanoff polynomials is the following:

**(1C)** *If* $x \not\equiv 0, 1 \pmod{p}$, *then* $x$ *is a common zero of the congruences*

$$\varphi_{p-1}(T) \equiv 0 \pmod{p}$$
$$\Lambda_2(T) \equiv 0 \pmod{p}$$
$$\vdots \tag{1.41}$$
$$\Lambda_{p-2}(T) \equiv 0 \pmod{p}$$

*if and only if it is a common zero of the congruences*

$$\varphi_{p-1}(T) \equiv 0 \pmod{p}$$
$$\varphi_{p-2}(T)\varphi_2(T) \equiv 0 \pmod{p}$$
$$\vdots$$
$$\varphi_{(p+1)/2}(T)\varphi_{(p-1)/2}(T) \equiv 0 \pmod{p}.$$

From this result, Le Lidec proved:

**(1D)** *If* $x$, $y$, $z$ *are pairwise relatively prime integers, not multiples of the odd prime* $p$, *if* $x^p + y^p + z^p = 0$, *and if* $-t \in G = \{x/y, y/x, x/z, z/x, y/z, z/y\}$, *then* $t$ *is a root of the congruences* (1.41).

## 2. The Theorem of Krasner

As I have said, the theorems of Cauchy and Genocchi, of Kummer, Mirimanoff, and Morishima established that if the first case fails for the exponent $p$, then the Bernoulli numbers $B_{p-3}$, $B_{p-5}$, $B_{p-7}$, $B_{p-9}$, $B_{p-11}$, $B_{p-13}$ are congruent to 0 modulo $p$.

The idea behind these theorems is that this is an unlikely event. At any rate it can be checked in a finite number of steps.

So it is desirable to obtain stronger restrictions, say, that a longer sequence of successive Bernoulli numbers be congruent to 0 modulo $p$.

In 1934, Krasner proved a most striking result along this line. Yet, in its formulation there is a condition which makes the theorem unfit for any practical application: it holds for primes $p$ larger than $n_0 = (45!)^{88} \cong 7.0379 \times 10^{4934}$.

Krasner's theorem again relies on Kummer's congruences. So it is necessary to investigate carefully the quantities $D^i \log(x + e^v y)$.

I begin with an expression already known since Herschel (1816). An analogous formula is found in Hilbert's *Zahlbericht* in §132. Let $\Delta^k 0^i$ denote the first term of the sequence of $k$th iterated differences obtained from

$$\{0,1^i,2^i,3^i,\ldots\}.$$

As easily seen

$$\Delta^k 0^i = \sum_{j=0}^k (-1)^{k-j} \binom{k}{j} j^i. \tag{2.1}$$

Herschel's lemma is the following:

**Lemma 2.1.** *If* $x, y$ *are real numbers,* $x \neq -y$, *if* $t > 0$ *is a real variable and* $t = e^v$, *then for every* $i \geq 1$

$$\left[ \frac{d^i \log(x + e^v y)}{dv^i} \right]_{v=0} = \sum_{k=1}^i \frac{\Delta^k 0^i}{k!} \left[ \frac{d^k \log(x + ty)}{dt^k} \right]_{t=1}. \tag{2.2}$$

Putting $\theta = y/(x + y)$ yields

$$\left[ \frac{d^k \log(x + ty)}{dt^k} \right]_{t=1} = (-1)^{k-1}(k-1)! \theta^k$$

so (2.2) may be rewritten as

$$\left[ \frac{d^i \log(x + e^v y)}{dv^i} \right]_{v=0} = -\sum_{k=1}^i (-1)^k \frac{\Delta^k 0^i}{k} \theta^k.$$

Let $T$ be an indeterminate and consider the polynomials

$$M_i(T) = \sum_{k=1}^i (-1)^k \frac{\Delta^k 0^i}{k} T^k. \tag{2.3}$$

Among the relevant properties of these polynomials, I mention:

$$M_1(T) = -T,$$

and $M_{i+1}(T) = T(1 - T)M_i'(T)$ for $i \geq 1$. The leading coefficient of $M_i(T)$ is $(-1)^i(i - 1)!$ and the coefficient of $T$ is $-1$. The roots of $M_i(T)$ (for $i \geq 2$) are real, simple, and belong to the closed unit interval $[0,1]$.

Let $N_i(T) = M_i(T)/T(1 - T) = M_{i-1}'(T)$ and $P_i(T) = T^i M_i(1/T)(1 - T)$ (for $i \geq 2$). These polynomials have degree $i - 2$, no roots in common. Their resultant $R_i$ is not zero and satisfies the inequality

$$|R_i| < [(i - 1)!]^{2(i-2)}. \tag{2.4}$$

With these preparations, Krasner proved:

**(2A)** *Let* $n_0 = (45!)^{88}$. *If* $p$ *is a prime,* $p > n_0$, *if*

$$k(p) = [\sqrt[3]{\log p}],$$

*and if the first case of Fermat's theorem fails for the exponent* $p$, *then the* $k(p)$ *Bernoulli numbers* $B_{p-1-2i}$ [*for* $i = 1, \ldots, k(p)$] *are congruent to* $0$ *modulo* $p$.

PROOF. I'll omit the details of the proof and only give the main idea. Consider the sequence of polynomials $M_1(T)$, $M_2(T)$, .... Let $j_0 > 0$ be the smallest integer such that $j > j_0$ implies

$$[(j-1)!]^{2(j-2)} < e^{(\frac{1}{4}(j-1))^3}. \tag{2.5}$$

It may be seen that $j_0 = 46$.

For every $j = 1, 2, \ldots, j_0$, let $R_j$ be the resultant of $N_j(T)$ and $P_j(T)$. By (2.4) $\max|R_j| \leq (45!)^{88} = n_0$.

If $p > n_0$ and there exist $x$, $y$, $z$, pairwise relatively prime integers, not multiples of $p$, such that $x^p + y^p + z^p = 0$, it must be shown that $B_{p-i} \equiv 0 \pmod{p}$ for $i = 3, 5, \ldots, 2k(p) + 1$. Assume the contrary for some odd index $j$, $3 \leq j \leq 2k(p) + 1$. By Kummer's congruence,

$$\left(\frac{d^j \log(x + e^v y)}{dv^j}\right)_{v=0} \equiv 0 \pmod{p}.$$

By (2.3) $M_j(\theta) \equiv 0 \pmod{p}$, where $\theta = y/(x + y)$. Since $\theta \not\equiv 0$, $1 \pmod{p}$, $N_j(\theta) \equiv 0 \pmod{p}$. Again, by Kummer's criterion

$$\left(\frac{d^j \log(x + e^v z)}{dv^j}\right)_{v=0} \equiv 0 \pmod{p}$$

and similarly $M_j(1/\theta) \equiv 0 \pmod{p}$. Therefore $P_j(\theta) \equiv 0 \pmod{p}$, so $R_j \equiv 0 \pmod{p}$. But $R_j \neq 0$, hence $p \leq |R_j|$.

If $j \leq j_0$, then $|R_j| \leq n_0 < p$. If $j_0 < j$, then

$$|R_j| < [(j-1)!]^{2(j-2)} < e^{(\frac{1}{4}(j-1))^3} \leq e^{k(p)^3} \leq e^{\log p} = p.$$

In both cases, there is a contradiction.                                                      □

Because of his various results, Vandiver in his papers of 1946, 1953 stated his belief that the first case is true. Krasner wrote in 1953 to Vandiver about this:

> Concerning your discussion of the truth of Fermat's theorem, I think as you do, that in Case I it is certainly true. Your argument is in order. Even the numbers $p$ not satisfying my criterion given in Theorem VIII of your article must be very exceptional. But I think that my preceding result given in the *Comptes Rendus* of 1934 furnishes even much stronger arguments in this sense. It seems quite unlikely that all the $[\sqrt[3]{\log p}]$ consecutive Bernoullian numbers $B_{p-1-2i}$, $1 \leq i \leq [\sqrt[3]{\log p}]$, are divisible by $p$, if $p$ is not too small. If you admit that the probability for an unknown Bernoullian number to be divisible by $p$ is $1/p$ such a divisibility has only the probability
>
> $$\left(\frac{1}{p}\right)^{[\sqrt[3]{\log p}]} \approx \frac{1}{p^{\sqrt[3]{\log p}}}$$
>
> and a simple calculation shows that the mathematical expectation of the number of primes $p \geq n$ satisfying this condition does not exceed $2/n^{\sqrt[3]{\log n}}$.

When we use a little different and very likely hypothesis that, if $f_p$ is the frequency of the Bernoullian numbers divisible by $p$,

$$\frac{\sum_{p \le x} pf_p}{\pi(x)} \to 1,$$

the result is not very different, with only maybe, a greater dispersion.

# 3. The Theorems of Wieferich and Mirimanoff

As I mentioned already, Wieferich's proof of his famous theorem is quite difficult. Mirimanoff gave a simpler proof in 1909, soon after the original. He made use of his polynomials. His manipulations required a formula due to Euler, for the alternate sum of odd powers of consecutive integers:

$$\sum_{j=1}^{n} (-1)^{j-1} j^{2k+1} = (-1)^{n+1} \left\{ \frac{1}{2} n^{2k+1} + \binom{2k+1}{1} \frac{2^2 - 1}{2} B_2 n^{2k} \right.$$

$$+ \binom{2k+1}{3} \frac{2^4 - 1}{4} B_4 n^{2k-2}$$

$$+ \cdots + \binom{2k+1}{2k-1} \frac{2^{2k} - 1}{2k} B_{2k} n^2$$

$$\left. + \frac{2^{2k+2} - 1}{2k+2} B_{2k+2} \right\} + \frac{2^{2k+2} - 1}{2k+2} B_{2k+2}. \qquad (3.1)$$

With this formula, Mirimanoff proved ($p > 3$ and $T$ is an indeterminate):

$$-[1^{p-2} - 2^{p-2} + 3^{p-2} - \cdots - (p-1)^{p-2}]T^p$$

$$= \frac{1}{2}(T+1)\varphi_{p-1}(-T) + (T-1)\left\{ \binom{p-2}{1} \frac{2^2 - 1}{2} B_2 \varphi_{p-2}(-T) \right.$$

$$+ \binom{p-2}{3} \frac{2^4 - 1}{4} B_4 \varphi_{p-4}(-T) + \cdots + \binom{p-2}{p-4} \frac{2^{p-3} - 1}{p-3} B_{p-3} \varphi_3(-T) \right\}$$

$$- \frac{2^p - 2}{p-1} B_{p-1} \frac{T^p - T}{T+1}. \qquad (3.2)$$

I restate the theorem of Wieferich (1909) and give Mirimanoff's proof:

**(3A)** *If the first case of Fermat's last theorem fails for the exponent p, then*

$$2^{p-1} \equiv 1 \pmod{p^2}.$$

PROOF. Assume there exist integers $x$, $y$, $z$, not multiples of $p$, such that $x^p + y^p + z^p = 0$. Let $G = \{x/y, y/x, x/z, z/x, y/z, z/y\}$. By Mirimanoff congruences, if $t \in G$, then

$$\varphi_{p-2s}(-t)B_{2s} \equiv 0 \pmod{p}$$

for $s = 1, 2, \ldots, (p-3)/2$, and also

$$\varphi_{p-1}(-t) \equiv 0 \pmod{p}.$$

From the theorem of von Staudt and Clausen, it follows that $(2^p - 2)B_{p-1}$ is $p$-integral. Hence, by (3.2),

$$1^{p-2} - 2^{p-2} + 3^{p-2} - \cdots - (p-1)^{p-2} \equiv 0 \pmod{p}. \qquad (3.3)$$

By (1.24),

$$1^{p-2} - 2^{p-2} + 3^{p-2} - \cdots - (p-1)^{p-2} \equiv \frac{2^p - 2}{p} \pmod{p}.$$

Hence $2^{p-1} \equiv 1 \pmod{p^2}$.                                        □

The quantity $q_p(2) = (2^{p-1} - 1)/p$ is an integer, by Fermat's little theorem. It is called the *Fermat quotient* of $p$, with base 2. To say that $2^{p-1} \equiv 1 \pmod{p^2}$ is equivalent to saying $q_p(2) \equiv 0 \pmod{p}$.

From the time of its discovery, it was quite apparent that Wieferich's theorem represented a noteworthy advance over all previous results.

The search for primes $p$ satisfying the condition began immediately. It was not until 1913 that Meissner discovered after long calculations the first example:

$$2^{1092} \equiv 1 \pmod{1093^2}.$$

A second example was later encountered by Beeger (1922):

$$2^{3510} \equiv 1 \pmod{3511^2}.$$

These congruences may also be proved in a rather short, but perhaps artificial way (see Landau, 1927, Guy 1967).

As I said in my first lecture, no other example exists with $p < 6 \times 10^9$. Calculations have been performed by various researchers recently with computers, and brought to the above limit by Brillhart, Tonascia, and Weinberger in 1971. So, for every prime $p \neq 1093, 3511, p < 3 \times 10^9$, the theorem of Wieferich and the computations above, guarantee that the first case holds for $p$.

Having understood the reasons behind Wieferich's theorem, Mirimanoff proved an analogous criterion, this time for the base 3:

**(3B)** *If the first case of Fermat's last theorem fails for the exponent p, then*

$$3^{p-1} \equiv 1 \pmod{p^2}.$$

In other notation, the Fermat quotient with base 3, $q_p(3) = (3^{p-1} - 1)/p \equiv 0 \pmod{p}$.

The proof of this theorem is substantially more difficult. Let it be only said that it makes much use of properties of the Mirimanoff polynomials and also of logarithms.

A computation showed that for $p = 1093$ and $3511$ the Fermat quotient $q_p(3) \not\equiv 0 \pmod{p}$. This guarantees that the first case also holds for these two exponents, which were not covered by Wieferich's theorem.

In 1910, Frobenius gave a proof of the theorems of Wieferich and Mirimanoff. His proof was algebraic, without the use of Kummer's congruences.

Soon after, in 1912, Furtwängler proved a very general theorem, using class field theory, or more precisely, Eisenstein's reciprocity law. As a corollary, he derived both theorems of Wieferich and Mirimanoff. These theorems will be considered in my next lecture.

In the literature there is a paper by Linkovski (1968), in which he claims: If the first case of Fermat's theorems fails for $p$, then $2^{p-1} \equiv 1 \pmod{p^3}$. This would represent an outstanding strengthening of Wieferich's theorem. However, Linkovski's proof is not correct, since it is based on the following assertion published by Grebeniuk in 1956:

If $x$, $y$, $z$ are integers, not multiples of $p$, such that $x^p + y^p + z^p = 0$, if $l$ divides $x + y + z$, and $\gcd(l, xyz) = 1$, then $l$ divides $2^{p-1} - 1$. In 1975, Gandhi and Stuff analyzed Grebeniuk's proof and found a mistaken deduction, so the statements of Grebeniuk and Linkovski are now questionable.

Of course, this is only one of so many mistakes made about Fermat's theorem, by outstanding mathematicians (like Kummer himself), as well as by good and not-so-good mathematicians.

I'm only giving these facts, in order to tell a story. Terjanian doubted the veracity of the congruence obtained by Linkovski. He wrote to Peschl, since that congruence is quoted in Klösgen's monograph—but not used in it—and Klösgen was Peschl's student. Peschl asked Hasse, during his visit to Bonn, for the Colloquium in Krull's memory. Hasse then wrote to Dr. Kotov, of Minsk, with whom he exchanges stamps, asking to translate Linkovski's paper. In reply, Kotov wrote:

> In the paper of Grebeniuk, which is quoted by Linkovski, there is a mistake. This was established by my colleague V. I. Bernik (also a student of Sprindžuk). The following happened. A mathematician Jepimaschko of Vitebsk produced in 1970 a proof of Fermat's last theorem. This proof was based on the paper of Grebeniuk, quoted by Linkovski. Bernik and I (Kotov) have checked all the arguments of Jepimaschko without finding any mistake. Later, Bernik checked the paper of Grebeniuk, on which Jepimaschko was based, and there he found the mistake. We knew the work of Linkovski, but we have not called his attention to the mistake of Grebeniuk. We had already had so much trouble to understand Jepimaschko's that we didn't wish to enter again into unhealthy discussions about Fermat's problem.

Despite the best of intentions, and the cleverest of our manipulations, beware of a proof of Fermat's theorem. A mistake may be carefully hidden.

# 4. Fermat's Theorem and the Mersenne Primes

There are various nice consequences of the theorems of Wieferich and Mirimanoff. I show how it is possible to deduce that the first case of Fermat's theorem holds for a class of primes which includes the Mersenne primes.

I begin with this easy lemma:

**Lemma 4.1.** *Let $p$ be an odd prime, $k$ not a multiple of $p$. Then $kp$ cannot be written in the form*

$$kp = \pm m \pm n,$$

*where $m^{p-1} \equiv 1 \pmod{p^2}$ and $n^{p-1} \equiv 1 \pmod{p^2}$.*

Mirimanoff (1910), Landau (1913), and Vandiver (1914) proved repeatedly:

**(4A)** *If $p = 2^a 3^b \pm 1$ or $p = \pm 2^a \pm 3^b$, where $a \geq 0$, $b \geq 0$, then the first case of Fermat's last theorem holds for the exponent $p$.*

PROOF. If the theorem fails for $p$, then $2^{p-1} \equiv 1 \pmod{p^2}$ and $3^{p-1} \equiv 1 \pmod{p^2}$. By the lemma (with $k = 1$), $p$ cannot be of the form indicated.  □

In particular, if $p = 2^a \pm 1$ is a prime, the first case holds for $p$.

It is quite easy to see that if $p = 2^a + 1$ is a prime, then $a$ itself is a power of 2, and so $p = 2^{2^n} + 1$. The number $F_n = 2^{2^n} + 1$ is called a *Fermat number*. The only known prime Fermat numbers are $F_0, F_1, F_2, F_3, F_4$. Euler discovered that $F_5$ is a multiple of 641. It has been conjectured that there are only finitely many prime Fermat numbers, and perhaps only the ones already known. This is a very difficult question.

On the other hand, if $p = 2^a - 1$ is a prime, then $a$ is necessarily equal to a prime, $a = q$. Then $p = 2^q - 1 = M_q$ is a *Mersenne number*. At the present there are 24 known Mersenne numbers which are primes. The largest one is $M_q$, with $q = 19937$, discovered by Tuckerman in 1971; it has 6002 digits.[1]

In view of (4A) and the likelihood that at any given moment the largest known prime will be a Mersenne prime, I may state:

*The first case of Fermat's last theorem holds for the largest prime known today and this is likely to be true at any future time!*

Let me add that Schinzel has conjectured that there exist infinitely many square-free Mersenne numbers. Rotkiewicz proved in 1965 that if this conjecture is true, then there exist an infinite number of primes $p$ such that $2^{p-1} \not\equiv 1 \pmod{p^2}$. And therefore, by Wieferich's theorem, there are an infinite number of primes $p$ for which the first case of Fermat's theorem holds.

---

[1] As indicated in the footnote of page 24, newer larger Mersenne primes have been discovered Noll & Nickel, Noll, Nelson and Slowinski (1978, 1979, 1979).

# 5. Summation Criteria

The earliest criterion for the first case involving summations was discovered by Cauchy in 1847:
  If the first case fails for $p$, then

$$\sum_{j=1}^{(p-1)/2} j^{p-4} \equiv 0 \pmod{p}. \tag{5.1}$$

In view of the expression of summations, like the above, in terms of Bernoulli numbers, it was not unexpected to arrive at criteria for the first case involving summations. This idea was exploited by E. Lehmer and Vandiver, among others. Quite recently, I have also contributed some new criteria of the same kind.

The first group of congruences involves sums of powers $p-2$ of terms in arithmetic progression, Fermat quotients $q_p(2)$, $q_p(3)$ and the Wilson quotient $W(p)$, which I now define.

Wilson's theorem says that

$$(p-1)! \equiv -1 \pmod{p}$$

so

$$W(p) = \frac{(p-1)! + 1}{p} \tag{5.2}$$

is an integer, called *Wilson quotient* of $p$.

In 1938, Emma Lehmer proved the following congruences:

$$\sum_{j=1}^{(p-1)/2} (p-2j)^{p-2} \equiv q_p(2)[1 - pW(p)] \pmod{p^2}, \tag{5.3}$$

$$\sum_{j=1}^{[p/3]} (p-3j)^{p-2} \equiv q_p(3)\frac{1 - pW(p)}{2} \pmod{p^2} \quad \text{when } p > 3, \tag{5.4}$$

$$\sum_{j=1}^{[p/4]} (p-4j)^{p-2} \equiv \frac{3q_p(2)[1 - pW(p)] + p[q_p(2)]^2}{4} \pmod{p^2} \quad \text{when } p > 3, \tag{5.5}$$

$$\sum_{j=1}^{[p/6]} (p-6j)^{p-2} \equiv \frac{[4q_p(2) + 3q_p(3)][1 - pW(p)] + pq_p(2)q_p(3)}{12} \pmod{p^2}$$

$$\text{when } p > 5, \tag{5.6}$$

$$\sum_{j=1}^{(p-1)/2} j^{p-2} \equiv -2q_p(2)[1 - pW(p)] + 2p[q_p(2)]^2 \pmod{p^2}. \tag{5.7}$$

From these congruences, Emma Lehmer derived the ones below: (Let me add that already in 1901, Glaisher had established (5.8) and (5.10), while in

1905, Lerch proved (5.8) and (5.9) for the modulus $p$.)

$$\sum_{j=1}^{(p-1)/2} \frac{1}{p-2j} \equiv q_p(2) - \frac{1}{2} p[q_p(2)]^2 \pmod{p^2}, \tag{5.8}$$

$$\sum_{j=1}^{[p/3]} \frac{1}{p-3j} \equiv \frac{1}{2} q_p(3) - \frac{1}{4} p[q_p(3)]^2 \pmod{p^2} \quad \text{when } p > 3, \tag{5.9}$$

$$\sum_{j=1}^{[p/4]} \frac{1}{p-4j} \equiv \frac{3}{4} q_p(2) - \frac{3}{8} p[q_p(2)]^2 \pmod{p^2} \quad \text{when } p > 3, \tag{5.10}$$

$$\sum_{j=1}^{[p/2]} \frac{1}{p-6j} \equiv \frac{1}{4} q_p(3) + \frac{1}{3} q_p(2) - p\left\{ \frac{1}{8}[q_p(3)]^p + \frac{1}{6}[q_p(2)]^2 \right\} \pmod{p^2}$$

$$\text{when } p > 5, \tag{5.11}$$

$$\sum_{j=1}^{(p-1)/2} \frac{1}{j} \equiv -2q_p(2) + p[q_p(2)]^2 \pmod{p^2}. \tag{5.12}$$

Combining the above congruences with the theorems of Wieferich and Mirimanoff, Emma Lehmer obtained various criteria given in (5A) below. As a matter of fact, the case $n = 2$ had been proved by Sylvester in 1861, while for $n = 3$ it was discovered by Lerch in 1905. Subsequently, Yamada proved the result for $n = 3, 6$ (see his second paper of 1941, correcting errors in the one of 1939).

(5A) *If the first case of Fermat's theorem fails for the exponent $p$ and if $n = 2, 3, 4, 6$, then*

$$\sum_{j=1}^{[p/n]} \frac{1}{j} \equiv 0 \pmod{p}.$$

PROOF. Just use the theorems of Wieferich, Mirimanoff and the congruences, modulo $p$, (5.8), (5.9), (5.10), (5.11).  □

The next result, due to Vandiver, is much more difficult to derive. It is based on a formula due to Frobenius (1914) connecting Bernoulli numbers and the Mirimanoff polynomials.

Let $p > 2$ be a prime, let $m \geq 2$ and let $\xi_0 = 1, \xi_1, \ldots, \xi_{m-1}$ be the $m$th roots of 1. Let $\varphi_n(X)$ denote the $n$th Mirimanoff polynomial.

**Lemma 5.1.** *Let $n \geq 1$ and $0 \leq l \leq m - 1$. Then*

$$(mB + l)^n - B^n \equiv (-1)^n n \sum_{i=1}^{m-1} \frac{\xi_i^l \varphi_n(\xi_i)}{\xi_i^p - 1} \pmod{p}.$$

I have used in the left-hand side the symbolic notation: $B$ is an indeterminate, the left-hand side is computed as a polynomial in $B$ and each power $B^k$ is to be interpreted as being the Bernoulli number $B_k$.

A special case is

$$(mB + 1)^{p-2} \equiv 2 \sum_{i=1}^{m-1} \frac{\xi_i^l \varphi_{p-2}(\xi_i)}{\xi_i^p - 1} \pmod{p}. \tag{5.13}$$

Setting $l = 0$ yields a congruence obtained explicitly by Vandiver (1917):

$$\frac{m^n - 1}{n} B_n \equiv (-1)^n \sum_{i=1}^{m-1} \frac{\varphi_n(\xi_i)}{\xi_i^p - 1} \pmod{p}. \tag{5.14}$$

Moreover, if $n = p - 1$ and $p \nmid m$, then

$$q_p(m) = \sum_{i=1}^{m-1} \frac{\varphi_{p-1}(\xi_i)}{\xi_i^p - 1} \pmod{p}. \tag{5.15}$$

With these formulas, Vandiver proved in 1925:

**(5B)** *If the first case of Fermat's theorem fails for the exponent $p$, then*

$$\sum_{j=1}^{[p/3]} \frac{1}{j^2} \equiv 0 \pmod{p}.$$

I totally omit his ingenious but long proof.

In 1933, Schwindt was able to transform Vandiver's criterion. He proved:

**Lemma 5.2.** *If $p > 3$, then*

$$5 \sum_{j=1}^{[p/3]} \frac{1}{j^2} \equiv \sum_{j=1}^{[p/6]} \frac{1}{j^2} \pmod{p}.$$

From this lemma, it follows:

**(5C)** *If the first case of Fermat's theorem fails for the exponent $p$, then*

$$\sum_{j=1}^{[p/6]} \frac{1}{j^2} \equiv 0 \pmod{p}.$$

I will conclude the report on summation criteria for the first case, by giving my own results. The statements of these make use of the Bernoulli polynomials

$$B_n(X) = \sum_{i=0}^{n} \binom{n}{i} B_i X^{n-i}. \tag{5.16}$$

Their constant term is $B_n$, the corresponding Bernoulli number. It is not my intention here to say more than strictly necessary about these polynomials.

One striking fact is that for $a = \frac{1}{2}, \frac{1}{3}, \frac{1}{4}, \frac{1}{6}, \frac{2}{3}, \frac{3}{4}, \frac{5}{6}$, there are formulas giving $B_{2n}(a)$, in terms of $n$ and the Bernoulli numbers. Explicitly:

$$B_{2n}\left(\frac{1}{2}\right) = -\left(1 - \frac{1}{2^{2n-1}}\right)B_{2n}, \tag{5.17}$$

$$B_{2n}\left(\frac{1}{3}\right) = B_{2n}\left(\frac{2}{3}\right) = -\frac{1}{2}\left(1 - \frac{1}{3^{2n-1}}\right)B_{2n}, \tag{5.18}$$

$$B_{2n}\left(\frac{1}{4}\right) = B_{2n}\left(\frac{3}{4}\right) = -\frac{1}{2^{2n}}\left(1 - \frac{1}{2^{2n-1}}\right)B_{2n}, \tag{5.19}$$

$$B_{2n}\left(\frac{1}{6}\right) = B_{2n}\left(\frac{5}{6}\right) = \frac{1}{2}\left(1 - \frac{1}{2^{2n-1}}\right)\left(1 - \frac{1}{3^{2n-1}}\right)B_{2n}. \tag{5.20}$$

However, no corresponding formulas are known for the values of $B_{2n+1}(X)$, say at $a = \frac{1}{3}, \frac{1}{4}, \frac{1}{6}, \frac{2}{3}, \frac{5}{6}$, only in terms of Bernoulli numbers.

I also need to use certain other congruences proved by Emma Lehmer (with $p > 3$):

$$\sum_{j=1}^{[p/3]} (p - 3j)^{p-3} \equiv \frac{3^{p-3}}{p-2}\left\{\frac{p-2}{3}pB_{p-3} - B_{p-2}\left(\frac{s}{3}\right)\right\} \pmod{p^2}, \tag{5.21}$$

where $p \equiv s \pmod 3$ so $s = 1$ or $2$, and

$$\sum_{j=1}^{[p/6]} (p - 6j)^{p-3} \equiv \frac{6^{p-3}}{p-2}\left\{\frac{p-2}{6}pB_{p-3} - B_{p-2}\left(\frac{t}{6}\right)\right\} \pmod{p^2}, \tag{5.22}$$

where $p \equiv t \pmod 6$, so $t = 1$ or $5$.

Using these congruences, I establish:

**Lemma 5.3.** *If $p > 3$, then*

1. $5B_{p-2}(\frac{1}{3}) \equiv B_{p-2}(\frac{1}{6}) \pmod p$.
2. $\sum_{j=0}^{p-2} (-1)^j(j + 1)3^j(2^{j+1} - 5)B_j \equiv 0 \pmod p$.

**Lemma 5.4.** *If $B_{p-(2n+1)} \equiv 0 \pmod p$, with $1 \le n \le (p - 3)/2$, then*

$$\sum_{j=1}^{[p/3]} \frac{1}{j^{2n+1}} \equiv 0 \pmod p$$

*and*

$$\sum_{j=1}^{[p/6]} \frac{1}{j^{2n+1}} \equiv 0 \pmod p.$$

And from these lemmas, I conclude:

**(5D)** *If the first case of Fermat's theorem fails for the exponent $p$, then*

1. $B_{p-2}(\frac{1}{3}) \equiv 0 \pmod{p}$, $B_{p-2}(\frac{1}{6}) \equiv 0 \pmod{p}$.
2. $2 \sum_{j=1}^{p/3} j^{p-3} \equiv B_{p-2}(s/3) \pmod{p^2}$ and $2 \sum_{j=1}^{p/6} j^{p-3} \equiv B_{p-2}(t/6) \pmod{p^2}$, where $s = 1$ or $2$, $t = 1$ or $5$, $p \equiv s \pmod 3$, $p \equiv t \pmod 6$.

Using the same method, I proved:

**(5E)** *If the first case of Fermat's theorem fails for the exponent* $p$, *then*

$$\sum_{j=1}^{[p/3]} \frac{1}{j^r} \equiv 0 \pmod{p}$$

*and*

$$\sum_{j=1}^{[p/6]} \frac{1}{j^r} \equiv 0 \pmod{p}$$

*for* $r = 1, 3, 5, 7, 9, 11, 13$.

Note that the case $r = 1$ was already stated in (5A). Note also that using Krasner's theorem (2A), if $p$ is sufficiently large the last congruences must also hold for $r$ odd, $1 \le r \le 2\sqrt[3]{\log p}$.

# 6. Fermat Quotient Criteria

The important theorems of Wieferich and Mirimanoff were extended by various authors to Fermat quotients with other bases: $q_p(m) = (m^{p-1} - 1)/p$.
In 1914, Vandiver proved:

**(6A)** *If the first case of Fermat's theorem fails for the exponent* $p$, *then*

$$q_p(5) \equiv 0 \pmod{p}$$

*and also*

$$1 + \frac{1}{2} + \frac{1}{3} + \cdots + \frac{1}{[p/5]} \equiv 0 \pmod{p}$$

In the same year, Frobenius proved:

**(6B)** *If the first case of Fermat's theorem fails for the exponent* $p$, *then*

$$q_p(11) \equiv 0 \pmod{p} \quad and \quad q_p(17) \equiv 0 \pmod{p}.$$

*Also, if* $p \equiv -1 \pmod 6$, *then* $q_p(7) \equiv 0 \pmod{p}$, $q_p(13) \equiv 0 \pmod{p}$, *and* $q_p(19) \equiv 0 \pmod{p}$.

New advances were made by Pollaczek in 1917 when he proved:

**(6C)** *If the first case of Fermat's theorem fails for the exponent* $p$ *and if* $m$ *is any prime,* $m \le 31$, *then with the exception of at most finitely many primes* $p$,

$$q_p(m) \equiv 0 \pmod{p}.$$

**(6D)** *If x, y, z are integers, not multiples of p, such that $x^p + y^p + z^p = 0$, then $x^2 + xy + y^2 \not\equiv 0 \pmod{p}$.*

These theorems are difficult to prove.
In 1931, Morishima completed Pollaczek's result (6C), showing:

**(6E)** *If the first case of Fermat's theorem fails for the exponent p, then, without any exception, $q_p(m) \equiv 0 \pmod{p}$ for all primes $m \leq 31$.*

As a matter of fact, Morishima proved also that $q_p(m) \equiv 0 \pmod{p}$ for $m = 37, 41, 43$, with finitely many exceptional primes $p$. Rosser wrote two papers in 1940 and 1941, in which he eliminated the exceptional primes in Morishima's result. However, in his doctoral thesis in 1948, at Rosser's suggestion, Gunderson thoroughly examined the paper of Morishima and found large gaps in the proofs. He proceeded to correct whatever he could and so he succeeded in establishing on firm ground (6E) above. But Agoh and Yamaguchi, who are presently visiting Queen's University, and have worked with Morishima, assured me that Morishima's proof really has no gaps.

Later, in my eleventh lecture, I'll indicate how the Lehmers, Rosser, and Gunderson used these criteria involving Fermat's quotient to assure the validity of the first case of Fermat's theorem for quite an extended range of the exponent $p$.

Now I want to indicate some other applications of these criteria involving Fermat's quotients. In a way analogous to the proof of (4A), Spunar proved the following result in 1931; it was rediscovered twice, by Gottschalk in 1938 and Ferentinou-Nicolacopoulou in 1963.

**(6F)** *Let p be an odd prime, and assume that there exists k, not a multiple of p, such that $kp = \pm m \pm n$, where each prime factor of m and of n is at most equal to 43. Then the first case of Fermat's theorem holds for p.*

PROOF. Otherwise, by the above criteria $m^{p-1} \equiv 1 \pmod{p^2}$, $n^{p-1} \equiv 1 \pmod{p^2}$ and this contradicts Lemma 4.1.                               $\square$

For example, this gives a direct way of seeing that the first case holds for $p = 1093$ and $p = 3511$ since

$$1093 = 2^2 + 3^2 \times 11^2$$

and

$$3511 = 11 + 2^2 \times 5^3 \times 7.$$

In fact many numbers may be written in the form indicated in (6F). But, is this set infinite? This is an open question. Just to comment on this problem, I want to compare it with a very useful classical theorem of Bang (1886), Zsigmondy (1892), Birkhoff and Vandiver (1904). This theorem says the following:

**(6G)** *If* $a > b \geq 1$ *and* $\gcd(a,b) = 1$, *then for every* $n \geq 2$, $a^n \pm n^n$ *has a prime factor* $p_n$ *which does not divide any of the numbers* $a^m \pm b^m$ *(with corresponding signs) and* $1 \leq m < n$. *The only exceptions to this statement are trivial, namely* $2^3 + 1$, $2^6 - 1$, $(2^{e-1} + h)^2 - (2^{e-1} - h)^2$ *for* $h$ *odd,* $2^{e-1} > h$.

So, taking $a \leq 43$, there are an infinite number of distinct primes $p_n$, with some $k_n$ such that $k_n p_n = a^n - b^n$ (for example). However, there is no guarantee that $p_n$ does not divide $k_n$. So these primes are not necessarily of the kind indicated in (6F).

In 1968, Puccioni examined what would follow if the set $\mathscr{P}_{43}$ of all primes of the form in (6F), is assumed to be finite.

For each prime $p$ let

$$\mathscr{M}_p = \{l \text{ prime} \,|\, p^{l-1} \equiv 1 \pmod{l^3}\}.$$

Numerical computations show that the sets $\mathscr{M}_p$ are quite small. In fact very few examples of such numbers are known.

Puccioni's theorem says:

**(6H)** *If* $\mathscr{P}_{43}$ *is a finite set, then for every prime* $p \leq 43$, *such that* $p \not\equiv \pm 1 \pmod 8$ *(i.e.,* $p = 2, 3, 5, 11, 13, 19, 29, 37, 43$*) the set* $\mathscr{M}_p$ *is infinite.*

This seems quite unlikely, however nothing has as yet been proved to the contrary. The offshoot of this discussion is, once more, that it is not yet known, even for the first case, whether Fermat's theorem holds for infinitely many prime exponents.

To conclude this section, I should mention that there are interesting papers of Lerch (1905) and Johnson (1977) about the properties of the Fermat quotient and the vanishing of these quotients modulo $p$.

# Bibliography

1816   Herschel, J. F. W.
    On the development of exponential functions, together with several new theorems relating to finite differences. *Philosophical Transactions London*, **106**, 1816, 25–45.

1847   Cauchy, A.
    Mémoire sur diverses propositions relatives à la théorie des nombres (4 parts). *C.R. Acad. Sci. Paris*, **24**, 1847, 996–999. *C.R. Acad. Sci. Paris*, **25**, 1847, 132–138, 177–183, 242–245. Reprinted in *Oeuvres Complètes* (1), 10, Gauthier-Villars, Paris, 296–299, 354–359, 360–365, 366–368.

1861   Sylvester, J. J.
    Sur une propriété des nombres premiers qui se rattache au théorème de Fermat. *C. R. Acad. Sci. Paris*, **52**, 1861, 161–163. Also in *Mathematical Papers*, vol. 2, 229–231, Cambridge, Cambridge University Press, 1908.

1861   Sylvester, J. J.
       Note relative aux communications faites dans les séances de 28 Janvier et 4 Février
       1861. *C. R. Acad. Sci. Paris*, **52**, 1861, 307–308. Also in *Mathematical Papers*,
       vol. 2, 234–235, Cambridge, Cambridge Univ. Press, 1908.

1886   Bang, A. S.
       Taltheoretiske Undersøgelser. *Tidskrift for Math.*, series 5, 4, 1886, 70–80,
       130–137.

1892   Zsigmondy, K.
       Zur Theorie der Potenzreste. *Monatsh f. Math.* **3**, 1892, 265–284.

1897   Hilbert, D.
       Die Theorie der algebraischen Zahlkörper. *Jahresbericht d. Deutschen Math.
       Vereinigung* **4**, 1897, 175–546. Reprinted in *Gesammelte Abhandlungen*, vol. I,
       Chelsea Publ. Co., New York, 1965, 63–363.

1901   Glaisher, J. W. L.
       On the residues of $r^{p-1}$ to modulus $p^2$, $p^3$, etc. . . . *Quart. J. Pure and Applied
       Math.*, **32**, 1901, 1–27.

1904   Birkhoff, G. D. and Vandiver, H. S.
       On the integral divisors of $a^n - b^n$. *Annals of Math.*, (2), **5**, 1904, 173–180.

1905   Lerch, M.
       Zur Theorie des Fermatschen Quotienten $(a^{p-1} - 1)/p = q(a)$. *Math. Ann.*, **60**,
       1905, 471–490.

1905   Mirimanoff, D.
       L'équation indéterminée $x^l + y^l + z^l = 0$ et le critérium de Kummer. *J. reine u.
       angew. Math.*, **128**, 1905, 45–68.

1909   Mirimanoff, D.
       Sur le dernier théorème de Fermat et le critérium de M. A. Wieferich. *Enseigne-
       ment Math.*, **11**, 1909, 455–459.

1909   Wieferich, A.
       Zum letzten Fermat'schen Theorem. *J. reine u. angew. Math.* **136**, 1909, 293–302.

1910   Frobenius, G.
       Über den Fermatschen Satz. *J. reine u. angew. Math.*, **137**, 1910, 314–316. Also in
       *Gesammelte Abhandlungen*, vol. 3, Springer-Verlag, Berlin, 1968, 428–430.

1910   Frobenius, G.
       Über den Fermatschen Satz II. Sitzungsber. Akad. d. Wiss. zu Berlin, 1910,
       200–208. Also in *Gesammelte Abhandlungen*, vol. 3, Springer-Verlag, Berlin, 1968,
       431–439.

1910   Frobenius, G.
       Über die Bernoullischen Zahlen und die Eulerschen Polynome. Sitzungsber. Akad.
       d. Wiss. zu Berlin, 1910, 809–847. Also in *Gesammelte Abhandlungen*, vol. 3,
       Springer-Verlag, Berlin, 1968, 440–478.

1910   Mirimanoff, D.
       Sur le dernier théorème de Fermat. *C. R. Acad. Sci. Paris*, **150**, 1910, 204–206.

1911   Mirimanoff, D.
       Sur le dernier théorème de Fermat. *J. reine u. angew. Math.*, **139**, 1911, 309–324.

1913   Landau, E.
       Réponse à une question de E. Dubouis. Existence d'une infinité de nombres
       premiers $2^a 3^b \pm 1$ et $\pm 2^a \pm 3^b$. *L'Interm. des Math.*, **20**, 1913, 180.

1913   Meissner, W.
       Über die Teilbarkeit von $2^p - 2$ durch das Quadrat der Primzahl $p = 1093$. Sit-
       zungsber. Akad. d. Wiss. zu Berlin, 1913, 663–667.

1914   Frobenius, G.
Über den Fermatschen Satz, III. Sitzungsber. Akad. d. Wiss. zu Berlin, 1914,
653–681. Also in *Gesammelte Abhandlungen*, vol. 3, Springer-Verlag, Berlin 1968,
648–676.

1914   Vandiver, H. S.
Extension of the criteria of Wieferich and Mirimanoff in connection with Fermat's
last theorem. *J. reine u. angew. Math.*, **144**, 1914, 314–318.

1917   Pollaczek, F.
Über den grossen Fermat'schen Satz. Sitzungsber. Akad. d. Wiss. Wien, Abt. IIa,
**126**, 1917, 45–59.

1917   Vandiver, H. S.
Symmetric functions formed by systems of elements of a finite algebra and their
connection with Fermat's quotient and Bernoulli numbers. *Annals of Math.*, **18**,
1917, 105–114.

1922   Beeger, N. G. W. H.
On a new case of the congruence $2^{p-1} \equiv 1(p^2)$. *Messenger of Math.*, **51**, 1922,
149–150.

1925   Vandiver, H. S.
A new type of criteria for the first case of Fermat's last theorem. *Annals of Math.*,
**26**, 1925, 88–94.

1927   Landau, E.
*Vorlesungen über Zahlentheorie*, vol. III. S. Hirzel, Leipzig, 1927. Reprinted by
Chelsea Publ. Co., New York, 1969.

1928   Morishima, T.
Über die Fermatsche Vermutung. *Proc. Imp. Acad. Japan*, **4**, 1928, 590–592.

1931   Morishima, T.
Über den Fermatschen Quotienten. *Jpn. J. Math.*, **8**, 1931, 159–173.

1931   Spunar, V. M.
On Fermat's last theorem, III. *J. Wash. Acad. Sci.*, **21**, 1931, 21–23.

1932   Morishima, T.
Über die Fermatsche Vermutung, VII. *Proc. Imp. Acad. Japan*, **8**, 1932, 63–66.

1933–1934   Schwindt, H.
Eine Bemerkung zu einem Kriterium von H. S. Vandiver. *Jahresber. d. Deutschen
Math. Verein.* **43**, 1933–1934, 229–232.

1934   Krasner, M.
Sur le premier cas du théorème de Fermat. *C. R. Acad. Sci. Paris*, **199**, 1934,
256–258.

1938   Gottschalk, E.
Zum Fermatschen Problem. *Math. Ann.*, **115**, 1938, 157–158.

1938   Lehmer, E.
On congruences involving Bernoulli numbers and the quotients of Fermat and
Wilson. *Annals of Math.*, **39**, 1938, 350–359.

1939   Yamada, K.
Ein Bemerkung zum Fermatschen Problem. *Tôhoku Math. J.*, **45**, 1939, 249–251.

1940   Rosser, J. B.
A new lower bound for the exponent in the first case of Fermat's last theorem. *Bull.
Amer. Math. Soc.*, **46**, 1940, 299–304.

1941   Rosser, J. B.
An additional criterion for the first case of Fermat's last theorem. *Bull. Amer.
Math. Soc.*, **47**, 1941, 109–110.

1941   Yamada, K.
Berichtigung zu der Note: Eine Bemerkung zum Fermatschen Problem. *Tôhoku Math. J.*, **48**, 1941, 193–198.

1946   Vandiver, H. S.
Fermat's last theorem; its history and the nature of the results concerning it. *Amer. Math. Monthly*, **53**, 1946, 555–578.

1948   Gunderson, N. G.
Derivation of Criteria for the First Case of Fermat's Last Theorem and the Combination of these Criteria to Produce a New Lower Bound for the Exponent. Thesis, Cornell University, 1948.

1953   Vandiver, H. S.
A supplementary note to a 1946 article on Fermat's last theorem. *Amer. Math. Monthly*, **60**, 1953, 164–167.

1956   Grebeniuk, D. G.
Obobtshenie Teoremii Wieferich (Generalization of Wieferich's theorem). *Doklady A. N. Uzbekskoï SSR*, **8**, 1956, 9–11.

1963   Ferentinou Nicolacopoulou I.
Une propriété des diviseurs du nombre $r^m + 1$. Applications au dernier théorème de Fermat. *Bull. Soc. Math. Grèce*, (N.S.), **4**, 1963, 121–126.

1965   Rotkiewicz, A.
Sur les nombres de Mersenne dépourvus de diviseurs carrés et sur les nombres naturels $n$ tels que $n^2 | 2^n - 2$. *Matematicky Vesnik*, 2, **17**, 1965, 78–80.

1967   Guy, R. K.
The primes 1093 and 3511. *Math. Student*, **35**, 1967, 204–206.

1967   LeLidec, P.
Sur une forme nouvelle des congruences de Kummer-Mirimanoff. *C. R. Acad. Sci. Paris*, **265**, 1967, 89–90.

1968   Linkovski, J.
Sharpening of a theorem of Wieferich. *Math. Nachr.*, **36**, 1968, 141.

1968   Puccioni, S.
Un teorema per una resoluzioni parziali del famoso problema de Fermat. *Archimede*, **20**, 1968, 219–220.

1969   LeLidec, P.
Nouvelle forme des congruences de Kummer-Mirimanoff pour le premier cas du théorème de Fermat. *Bull. Soc. Math. France*, **97**, 1969, 321–328.

1971   Brillhart, J., Tonascia, J., and Weinberger, P.
On the Fermat quotient, in *Computers in Number Theory*, Academic Press, New York, 1971, 213–222.

1971   Tuckerman, B.
The 24th Mersenne prime. *Proc. Nat. Acad. Sci. U.S.A.*, **68**, 1971, 2319–2310.

1975   Gandhi, J. M. and Stuff, M.
On the first case of Fermat's last theorem and the congruence $2^{p-1} \equiv 1 \pmod{p^3}$. *Notices Amer. Math. Soc.*, **22**, 1975, A-453.

1977   Johnson, W.
On the non-vanishing of Fermat quotients. *J. reine u. angew. Math.*, **292**, 1977, 196–200.

1978   Nickel, L. and Noll, L. C.
*Los Angeles Times*, November 16, 1978, part II, page 1. See also: Le dernier premier, *Gazette Sci. Math. Québec.* 3, 1979, 27.

# The Power of Class Field Theory

In 1912, Furtwängler used class field theory to derive two important criteria about the first case of Fermat's last theorem. As corollaries, he then gave new proofs of the theorems of Wieferich and Mirimanoff. In this way, the methods of class field theory entered into the game.

It is my intention in this lecture to give a succinct overview of those parts of the theory which are relevant to Fermat's problem. In no way will I attempt a systematic treatment of the theory. The advantage of this approach is that it brings directly into focus the tools that are most useful. Usually these lie hidden behind more refined statements.

## 1. The Power Residue Symbol

Let $p$ be an odd prime, let $\zeta$ be a primitive $p$th root of 1, $K = \mathbb{Q}(\zeta)$, $A$ the ring of integers of $K$, and $\lambda = 1 - \zeta$.

The first fact, which is the cornerstone of the theory, is quite simple and I have already mentioned it in Lecture VII, §5.

**(1A)** *If $Q$ is a prime ideal of $A$, $Q \neq A\lambda$, and if $\alpha \in A \backslash Q$, then there exists a unique integer $a$, $0 \leq a \leq p - 1$, such that*

$$\alpha^{(N(Q) - 1)/p} \equiv \zeta^a \pmod{Q}.$$

$N(Q)$ denotes the absolute norm of $Q$, and I already noted before that $N(Q) = q^f$, where $f$ is the order of $q$ modulo $p$ ($q$ being the only prime such that $Q$ divides $Aq$).

In view of the above result, the following definition makes sense:

$$\left\{\frac{\alpha}{Q}\right\} = \zeta^a \tag{1.1}$$

when $Q \neq A\lambda$, $\alpha \in A\backslash Q$.

The symbol $\{—\}$ has the following properties:

**(1B)** *If* $\alpha, \beta \in A\backslash Q$, $Q \neq A\lambda$, *then*

a. $\alpha \equiv \beta \pmod{Q}$ *implies* $\{\alpha/Q\} = \{\beta/Q\}$.
b. $\{1/Q\} = 1$.
c. $\{\alpha/Q\}\{\beta/Q\} = \{\alpha\beta/Q\}$.
d. $\{\alpha^p/Q\} = 1$.

This justifies calling the mapping $\alpha \in A\backslash Q \to \{\alpha/Q\}$, the *p*th *power residue character* or *symbol* defined by the prime ideal $Q \neq A\lambda$.

By multiplicativity, the symbol may be defined for all ideals $J$ of $A$ such that $A\lambda \nmid J$. If $J = \prod_{i=1}^{m} Q_i^{r_i}$, and if $\alpha \in A$, $\alpha \notin \bigcup_{i=1}^{m} Q_i$ (or in other words, if $\gcd(A\alpha, J) = A$), define

$$\left\{\frac{\alpha}{J}\right\} = \prod_{i=1}^{m} \left\{\frac{\alpha}{Q_i}\right\}^{r_i}. \tag{1.2}$$

In particular

$$\left\{\frac{\alpha}{A}\right\} = 1.$$

For the extended symbol, the properties are analogous:

**(1C)** *If the symbols below make sense, then*

a. $\alpha \equiv \beta \pmod{J}$ *implies* $\{\alpha/J\} = \{\beta/J\}$.
b. $\{1/J\} = 1$.
c. $\{\alpha/J\}\{\beta/J\} = \{\alpha\beta/J\}$.
d. $\{\alpha^p/J\} = 1$.
e. $\{\alpha/JJ'\} = \{\alpha/J\}\{\alpha/J'\}$.

If $\alpha, \beta \in A\backslash A\lambda$, and if $\gcd(A\alpha, A\beta) = A$, then the following notation is used:

$$\left\{\frac{\alpha}{A\beta}\right\} = \left\{\frac{\alpha}{\beta}\right\}.$$

I recall the notion of a semi-primary element from Lecture V: Any element $\alpha \notin A\lambda$ such that there exists $m \in \mathbb{Z}$ for which $\alpha \equiv m \pmod{A\lambda^2}$. I now list in a lemma the main properties of semi-primary integers, of which the first two were quoted before:

**Lemma 1.1.**

1. *If* $\alpha \in A\backslash A\lambda$, *there exists a root of unity* $\zeta^j$ *such that* $\zeta^j\alpha$ *is semi-primary.*
2. *If* $\alpha, \beta \in A\backslash A\lambda$ *are semi-primary, then so are* $\alpha\beta$ *and* $\alpha/\beta$.

3. *If $\alpha \in A \backslash A\lambda$, then $\alpha^p$ is semi-primary.*
4. *If $\alpha \in A \backslash A\lambda$ and $\alpha$ is a real number, then $\alpha$ is semi-primary.*
5. *If $\alpha$ is a unit and $\alpha$ is semi-primary, then $\alpha$ is a real number.*

The main theorem for the Legendre and Jacobi symbol is Gauss's quadratic reciprocity law. This is because it allows the computation of the symbol. Gauss also proved the reciprocity law for the biquadratic symbol and later (1844) Eisenstein discovered the reciprocity law for the cubic residue symbol. A more elementary proof was published recently by Kaplan (1969) for the biquadratic case and by Hayashi (1974) for the cubic case.

The more general reciprocity law for the power residue symbol was the object of deep studies by Eisenstein (1850) and later by Kummer (from 1850 to 1887).

Using the Jacobi cyclotomic functions, Eisenstein proved the following special reciprocity law for the power-residue symbol:

**(1D)** *If $m$ is an integer, $p \nmid m$, if $\alpha \in A$ is semi-primary, and $\gcd(A\alpha, Am) = A$, then*

$$\left\{\frac{m}{\alpha}\right\} = \left\{\frac{\alpha}{m}\right\}.$$

A proof, in modern notation, may be found in volume III of Landau's book (1927).

The following corollaries are required in the applications to Fermat's theorem:

**(1E)** *Let $m$ be an integer, $p \nmid m$, let $\alpha \in A$ be semi-primary, $\gcd(A\alpha, Am) = A$. If $A\alpha$ is the $p$th power of some ideal of $A$, $A\alpha = J^p$, then $\{\alpha/m\} = 1$.*

Similarly,

**(1F)** *Let $m$ be an integer, $p \nmid m$, let $\alpha \in A \backslash A\lambda$, and $\gcd(A\alpha, Am) = A$. If $\alpha$ is a real number, then $\{\alpha/m\} = 1$.*

**(1G)** *If $q$ is a prime, $q \neq p$, and if $\{\zeta/q\} = 1$, then $q^{p-1} \equiv 1 \pmod{p^2}$.*

The last result is used directly in studying Fermat's theorem. The proofs of these corollaries are also in Landau's book.

# 2. Kummer Extensions

In the study of the cyclotomic field $K = \mathbb{Q}(\zeta)$ ($\zeta$ a primitive $p$th root of 1, $p$ an odd prime) there comes a time when it is necessary to study the cyclic extensions of $K$ having degree $p$. Kummer considered such extensions and he showed that they are of the form $L = K(\sqrt[p]{\alpha})$, where $\alpha \in A$ is not the $p$th power of any element of $K$.

Let $B$ be the ring of integers of $L$. Let $\sigma$ be the generator of the Galois group of $L|K$ such that $\sigma(\sqrt[p]{\alpha}) = \zeta\sqrt[p]{\alpha}$.

The problem to investigate was the decomposition of (nonzero) prime ideals $P$ of $A$ in the extension $L|K$. Writing $BP = \bar{P}_1^e \bar{P}_2^e \cdots \bar{P}_g^e$, where $\bar{P}_1, \ldots, \bar{P}_g$ are distinct prime ideals of $B$ and the exponents are equal, $e \geq 1$, the problem is the determination of $e$, $g$ and $f$, where the relative norm of $\bar{P}_i$ is $N_{L|K}(\bar{P}_i) = P^f$. Since $efg = [L:K] = p$, the only possibilities are as follows:

$$\begin{array}{lll} e = p, & f = g = 1 & P \text{ is ramified,} \\ f = p, & e = g = 1 & P \text{ is inert,} \\ g = p, & e = f = 1 & P \text{ is split.} \end{array}$$

The situation is completely known.

*Case I. $P \mid A\alpha$.*
Writing $A\alpha = P^t J$, where $P \nmid J$, then:

a. If $p \nmid t$, then $BP = \bar{P}^p$, where $\bar{P}$ is a prime ideal of $B$, so $P$ is ramified.
b. If $p \mid t$, then there exists $\alpha' \in A$ such that $L = K(\sqrt[p]{\alpha'})$ and $P \nmid A\alpha'$.

*Case II. $P \nmid A\alpha$ and $P \neq A\lambda$.*

a. If there exists $\beta \in A$ such that $\alpha \equiv \beta^p \pmod{P}$, then $P$ is split.
b. Otherwise, $P$ is inert.

*Case III. $P = A\lambda \nmid A\alpha$.*

a. If there exists $\beta \in A$ such that $\alpha \equiv \beta^p \pmod{A\lambda^{p+1}}$, then $A\lambda$ splits.
b. If there exists $\beta \in A$ such that $\alpha \equiv \beta^p \pmod{A\lambda^p}$, but there does not exist $\gamma \in A$ such that $\alpha \equiv \gamma^p \pmod{A\lambda^{p+1}}$, then $A\lambda$ is inert.
c. If for every $\beta \in A$, $\alpha \not\equiv \beta \pmod{A\lambda^p}$, then $A\lambda$ is ramified.

The above results may easily be rephrased in terms of the power residue symbol. Assume that $P \neq A\lambda$ is a prime ideal unramified in $L|K$, where $\alpha \notin P$. Then $\{\alpha/P\} = 1$ exactly when $P$ splits in $L|K$, and, more generally, the order of $\{\alpha/P\}$ in the cyclic group of roots of 1 is equal to the inertial degree $f$ of $P$ in the extension $L|K$.

# 3. The Main Theorems of Furtwängler

These are just two theorems which brought class field theory as a method of studying Fermat's theorem.

Furtwängler first showed (1912):

**(3A)** *If $p$ is an odd prime and $x$, $y$, $z$ are relatively prime integers such that* $x^p + y^p + z^p = 0$, *if $r$ is any natural number, $p \nmid r$, $p \nmid z$, $\gcd(r,z) = 1$, then*

$$\left\{ \frac{\zeta^y x + \zeta^{-x} y}{r} \right\} = 1.$$

His first main theorem is the following:

**(3B)** *If $p$ is an odd prime, $x$, $y$, $z$ are relatively prime integers such that* $x^p + y^p + z^p = 0$, *and $r$ is a natural number such that $r \mid x$, $p \nmid x$, then $r^{p-1} \equiv 1$* (mod $p^2$).

PROOF. It may be assumed, for example, that $p \nmid z$. Since $p \nmid r$ and $r$ is relatively prime to $y$ and to $z$, then by (3A)

$$\left\{ \frac{\zeta^y x + \zeta^{-x} y}{r} \right\} = 1.$$

But $\zeta^y x + \zeta^{-x} y \equiv \zeta^{-x} y$ (mod $Ar$), hence by (1C)

$$1 = \left\{ \frac{\zeta^{-x} y}{r} \right\} = \left\{ \frac{\zeta^{-x}}{r} \right\} \left\{ \frac{y}{r} \right\}.$$

By (1F), $\{y/r\} = 1$, hence $\{\zeta/r\}^{-x} = 1$. Since $p \nmid x$, $\{\zeta/r\} = 1$, hence by (1G) $r^{p-1} \equiv 1$ (mod $p^2$). ☐

The proof of the second main theorem of Furtwängler is not much more difficult:

**(3C)** *If $p$ is an odd prime, $x$, $y$, $z$ are relatively prime integers such that* $x^p + y^p + z^p = 0$, *and $r$ is a natural number, such that $r \mid x \pm y$, $p \nmid x^2 - y^2$, then $r^{p-1} \equiv 1$* (mod $p^2$).

With these theorems, Furtwängler obtained both theorems of Wieferich and Mirimanoff in a very natural way. Here is the shortest proof of Wieferich criterion (backed however by considerable theory):

PROOF. If $p \neq 2$ and $x^p + y^p + z^p = 0$, $p \nmid xyz$, $\gcd(x,y,z) = 1$, then at least one of the integers is even, say $2 \mid x$. By (3B), $2^{p-1} \equiv 1$ (mod $p^2$). ☐

And now, a very short proof of Mirimanoff's theorem.

PROOF. If $3 \mid xyz$, then by the first theorem of Furtwängler $3^{p-1} \equiv 1$ (mod $p^2$).

If $3 \nmid xyz$, then $x^p \equiv x$ (mod 3), $y^p \equiv y$ (mod 3), and $z^p \equiv z$ (mod 3). So $0 = x^p + y^p + z^p \equiv x + y + z \equiv \pm 1 \pm 1 \pm 1$ (mod 3). Therefore $x \equiv y \equiv z$ (mod 3).

Note that $p$ does not divide all three numbers $x - y$, $y - z$, $z - x$, otherwise $0 = x^p + y^p + z^p \equiv x + y + z \equiv 3x$ (mod $p$). Since $p \neq 3$, $p \mid x$, against the hypothesis. Without loss of generality $p \nmid x - y$. Since $-z \equiv x + y \pmod{p}$ and $p \nmid z$, then $p \nmid x + y$. But $3 \mid x - y$, so by the second theorem of Furtwängler, $3^{p-1} \equiv 1 \pmod{p^2}$. $\qquad\qquad\qquad\qquad\qquad\qquad\qquad\qquad\qquad\qquad\qquad\square$

In view of their importance, the theorems of Furtwängler were proved again, extended and generalized.

Noteworthy are the two theorems proved by McDonnell in 1930:

**(3D)** *Let $p$ be an odd prime, let $x$, $y$, $z$ be relatively prime integers such that $x^p + y^p + z^p = 0$.*

1. *If $p \nmid xy + yz + zx$ and $r \mid x^2 - yz$, then $r^{p-1} \equiv 1 \pmod{p^2}$.*
2. *If $p \nmid x(y - z)(x^2 + yz)$ and $r \mid x^2 + yz$, then $r^{p-1} \equiv 1 \pmod{p^2}$.*

These theorems are not easily applicable, because the assumptions cannot be verified in practice.

In 1919 Vandiver gave the following application of Furtwängler's theorem. The same result, assuming that the first case fails, was indicated in Lecture IV, (3A).

**(3E)** *If $p \neq 2$ and if $x$, $y$, $z$ are relatively prime integers such that $x^p + y^p + z^p = 0$, then*

$$x^p \equiv x \pmod{p^3}$$
$$y^p \equiv y \pmod{p^3}$$
$$z^p \equiv z \pmod{p^3}$$

*and $x + y + z \equiv 0 \pmod{p^3}$. Moreover, if $p \mid z$, then $p^3 \mid z$.*

Furtwängler also proved the following result which parallels the theorem of Sophie Germain:

**(3F)** *Let $p \neq 2$ and $q = kp + 1$ be primes. Assume:*

1. *The first case of Fermat's theorem holds for every prime exponent dividing $k$.*
2. *The congruence $X^p + Y^p + Z^p \equiv 0 \pmod{q}$ has only the trivial solution.*

*Then the first case of Fermat's theorem holds for $p$.*

## 4. The Method of Singular Integers

While writing his important papers in class field theory, Furtwängler introduced the singular integers. These were later used by Takagi, Fueter, and Inkeri to give new proofs of the two main theorems of Furtwängler. These proofs ironically avoided class field theory, or more precisely, the use of Eisenstein's reciprocity law.

I keep the same notations: $p \neq 2$, $\zeta$, $K$, $A$, $\lambda$. An element $\alpha \in A$ is a *singular integer* if there exists an ideal $J$ such that $A\alpha = J^p$ (it is possible that $J$ be a principal ideal). More generally, I shall consider elements $\alpha$ such that $A\alpha = J^a$, where $J$ is some ideal and $a$ is odd, $a \geq 1$.

Let $g$ be a primitive root modulo $p$, $\sigma: \zeta \mapsto \zeta^g$ the generator of the Galois group of $K \mid \mathbb{Q}$. For every $i \geq 0$ let $g_i$ be the unique integer such that $1 \leq g_i \leq p - 1$ and $g_i \equiv g^i \pmod{p}$. Similarly, for every $i < 0$ let $g_i$, $1 \leq g_i \leq p - 1$, be defined by $g_i g^{-i} \equiv 1 \pmod{p}$. Let $h_i = (gg_i - g_{i+1})/p$ for every $i \in \mathbb{Z}$. So $h_i \in \mathbb{Z}$.

Consider the polynomials in $\sigma$:

$$G(\sigma) = \sum_{i=0}^{p-2} g_{-i}\sigma^i, \tag{4.1}$$

$$H(\sigma) = \sum_{i=0}^{p-2} h_{-i}\sigma^i. \tag{4.2}$$

Following Inkeri (1948), I will describe the action of these polynomials on the generalized singular integers:

**(4A)** *Let $\alpha \in A$ and assume that $A\alpha = J^a$, where $J$ is some ideal and $a \geq 1$ is odd. Then:*

1. $\alpha^{G(\sigma)} = \zeta^u \beta^a$, *where* $0 \leq u \leq p - 1$, $\beta \in A$.
2. $\alpha^{H(\sigma)} = \zeta^v \gamma^a$, *where* $0 \leq v \leq p - 1$, $\gamma \in A$.

Actually, these results had been obtained for $a = p$ by Fueter (1922) and Takagi (1922), but their proofs, contrary to Inkeri's, used some facts from class field theory.

Inkeri (1946) also proved a more precise result, under stronger conditions:

**(4B)** *If $\alpha \in A$ is semi-primary, and $A\alpha = J^a$, where $a$ is odd and $p \mid a$, then:*

1. $\alpha^{G(\sigma)} = \beta^a$ *with* $\beta \in A$.
2. $\alpha^{H(\sigma)} = \gamma^a$ *with* $\gamma \in A$.

With this method, Fueter in 1922 gave a proof of Kummer's congruences (explained in Lecture VII). As he stated in his paper, he wanted to show: "... how all the necessary theorems to derive Kummer's congruences are found already in the *Zahlbericht* [of Hilbert] and how, with a few strokes, Kummer's criterion arises. Yes, and still more! Also Mirimanoff's form and the criterion of Wieferich-Furtwängler may be immediately obtained, and therefore for the latter, it is required much less than Eisenstein's reciprocity law".

Inkeri also gave in 1948 another similar proof of Kummer's congruences for the first case.

In 1857, Kummer obtained a formula, which was extended by Fueter (1922).

**(4C)** *Let $\alpha \in A$ be such that $A\alpha = J^p$ ($\alpha$ is a singular integer). Then for every* $a = 1, 2, \ldots, p - 2$, $\alpha^{I_a(\sigma)} = \zeta^{u_a}\gamma_a^p$ *where* $0 \leq u_a \leq p - 1$, $\gamma_a \in A$ *and*

$$I_a(\sigma) = \sum_{j=0}^{p-2} \left( \left[ \frac{(a+1)g_j}{p} \right] - \left[ \frac{ag_j}{p} \right] \right) \sigma^{-j}.$$

In 1948, Inkeri gave new proofs of Furtwängler's theorems, using the method of singular integers.

In 1933, Moriya extended the theorems of Furtwängler for Fermat's equation with exponents $p^n$, $n \geq 1$. In 1946, using the method of singular integers, Inkeri gave a new proof of Moriya's theorem, and extended the theorems of McDonnell for the prime-power exponents.

These are Moriya's theorems:

**(4D)** *Let $n \geq 1$, let $p \neq 2$ be a prime, and assume that there exist relatively prime integers $x$, $y$, $z$ such that $x^{p^n} + y^{p^n} + z^{p^n} = 0$. Let $r$ be a natural number, satisfying any one of the following conditions:*

1. $r \mid x$, $p \nmid x$
2. $r \mid x - y$, $p \nmid x^2 - y^2$

*Then $r^{p-1} \equiv 1 \pmod{p^{n+1}}$.*

Again, with the same method, Inkeri (1946) gave the following generalization of Vandiver's theorem (3E):

**(4E)** *If $p \neq 2$, $n \geq 1$, if $x$, $y$, $z$ are relatively prime nonzero integers, such that $x^{p^n} + y^{p^n} + z^{p^n} = 0$, then*

$$x^p \equiv x \pmod{p^{2n+1}},$$
$$y^p \equiv y \pmod{p^{2n+1}},$$
$$z^p \equiv z \pmod{p^{2n+1}}.$$

# 5. Hasse

I have already described the success of Furtwängler using Eisenstein's reciprocity law for the power residue symbol. This is nothing more than a very special case of the general reciprocity law for this symbol, as it was developed by Kummer (1850, 1852, 1858, 1859, 1859, 1887), Furtwängler (1909, 1912), and later subsumed under Takagi's (1922) and Artin's (1927) reciprocity laws for arbitrary abelian extensions. These matters are explained with great care by Hasse in his *Bericht* (1927, 1930).

As often happens, it is not obvious how to deduce explicit consequences from a theorem which is very general. Hasse's great contribution was to give a very convenient form to the general reciprocity law for the power residue symbol. And from this, he could prove, in a more systematic manner, many of the theorems previously discovered by Kummer, Furtwängler and others.

I will not state here the general forms of any of the reciprocity laws. Instead, I'll select and present only those explicit reciprocity formulas used by Hasse.

**(5A)** *Let* $\alpha$, $\beta \in K = \mathbb{Q}(\zeta)$, $\alpha \neq 0$, $\beta \neq 0$. *Let* $F_\alpha$ *and* $F_\beta$ *be the conductors of the extensions* $K(\sqrt[p]{\alpha})|K$ *and* $K(\sqrt[p]{\beta})|K$. *If* $\gcd(A\alpha, F_\beta) = \gcd(A\beta, F_\alpha) = \gcd(F_\alpha, F_\beta) = A$, *then* $\{\alpha/\beta\} = \{\beta/\alpha\}$.

**(5B)** *Let* $\alpha$, $\beta \in K$, $\alpha \neq 0$, $\beta \neq 0$. *If* $\gcd(A\alpha, A\beta) = A$, *and if* $A\lambda$ *is unramified and each real place of* $K$ *splits completely in* $K(\sqrt[p]{\alpha})|K$ *(or in* $K(\sqrt[p]{\beta})|K$*), then* $\{\alpha/\beta\} = \{\beta/\alpha\}$.

Another form of the reciprocity law proved by Hasse, in volume II, page 77, is the following:

**(5C)** *Let* $\alpha$, $\beta \in A$ *be such that* $\gcd(A\alpha, A\beta) = A$. *Assume also that* $\alpha \equiv 1 \pmod{A\lambda^{p-1}}$, $\beta \equiv 1 \pmod{A\lambda}$ *and that* $\alpha$ *is totally positive. Then*

$$\left\{\frac{\alpha}{\beta}\right\}\left\{\frac{\beta}{\alpha}\right\}^{-1} = \zeta^{\operatorname{Tr}_{K|\mathbb{Q}}\left(\frac{\alpha-1}{p}\cdot\frac{\beta-1}{\lambda}\right)}.$$

If $p = 2$, and the symbol is the Jacobi symbol $(-)$, the above becomes Jacobi's reciprocity law:

If $a > 0$, if $a$, $b$ are odd, relatively prime integers, then

$$\left(\frac{a}{b}\right)\left(\frac{b}{a}\right)^{-1} = (-1)^{\frac{a-1}{2}\cdot\frac{b-1}{2}}.$$

As a corollary of (5C), it is easy to show:

**(5D)** *Let* $a \in \mathbb{Z}$, $\beta \in A$, *and assume that* $\gcd(Aa, A\beta) = A$. *Assume also that* $p$ *does not divide* $a$ *and* $\beta \equiv 1 \pmod{A\lambda}$. *Then:*

$$\left\{\frac{a}{\beta}\right\}\left\{\frac{\beta}{a}\right\}^{-1} = \zeta^{-\frac{a^{p-1}-1}{p}\operatorname{Tr}_{K|\mathbb{Q}}\left(\frac{\beta-1}{\lambda}\right)}.$$

As the conditions become more restrictive, the formulas are more special:

**(5E)** *Let* $a \in \mathbb{Z}$, $\beta \in A\backslash A\lambda$, *and* $\gcd(Aa, A\beta) = A$. *Assume that* $p \nmid a$ *and let* $b \in \mathbb{Z}$ *be such that* $\zeta^b\beta \equiv 1 \pmod{A\lambda^2}$ *(see Lemma 1.1). Then*

$$\left\{\frac{a}{\beta}\right\}\left\{\frac{\beta}{a}\right\} = \zeta^{\frac{a^{p-1}-1}{p}b}.$$

I note that Eisenstein's reciprocity law is nothing but a particular case of the above result. Indeed, if $\beta$ is semi-primary, say $\beta \equiv m \pmod{A\lambda^2}$, with

$m \in \mathbb{Z}$, then $\beta^{p-1} \equiv m^{p-1} \equiv 1 \pmod{A\lambda^2}$. Applying (5E) to $a$ and $\beta^{p-1}$ yields

$$\left\{\frac{a}{\beta^{p-1}}\right\}\left\{\frac{\beta^{p-1}}{a}\right\} = \zeta^{\frac{a^{p-1}-1}{p} \times 0} = 1.$$

So

$$\left\{\frac{a}{\beta}\right\}\left\{\frac{\beta}{a}\right\}^{-1} = \zeta^c \quad \text{where } c(p-1) \equiv 0 \pmod p$$

hence $c = 0$ and $\{a/\beta\} = \{\beta/a\}$.

With the above reciprocity laws, Hasse computed the values of various symbols.

**(5F)** If $\alpha \in A$ is totally positive and $\alpha \equiv 1 \pmod{A\lambda^{p-1}}$, then

$$\left\{\frac{\zeta}{\alpha}\right\} = \zeta^{\mathrm{Tr}_{K|\mathbb{Q}}\left(\frac{\alpha-1}{p}\right)}.$$

**(5G)** If $\alpha \in A$, $\alpha \equiv 1 \pmod{A\lambda^p}$, then

$$\left\{\frac{p}{\alpha}\right\} = \zeta^{\mathrm{Tr}_{K|\mathbb{Q}}\left(\frac{\alpha-1}{p\lambda}\right)} \quad \text{and} \quad \left\{\frac{\zeta}{\alpha}\right\} = \zeta^{-\mathrm{Tr}_{K|\mathbb{Q}}\left(\frac{\alpha-1}{p\lambda}\right)}.$$

The next expressions involve the $\lambda$-adic logarithms of Kummer, which I have already considered in Lecture VII, §5. As a reminder, if $\alpha \in A$ and $\alpha \equiv 1 \pmod{A\lambda}$, then the $\lambda$-adic logarithm is defined by

$$\log_\lambda \alpha = \sum_{n=1}^{\infty} \frac{(-1)^{n-1}}{n}(\alpha-1)^n,$$

the series being convergent in the $\lambda$-adic topology; so $\log_\lambda \alpha$ is an element of the completion $\hat{K}$ of $K$, relative to the valuation $v_\lambda$ belonging to the prime ideal $A\lambda$.

Artin and Hasse proved in 1925:

**(5H)** If $\alpha \in A$ and $\alpha \equiv 1 \pmod{A\lambda}$, then

$$\left\{\frac{\zeta}{\alpha}\right\} = \zeta^{\frac{1}{p}\mathrm{Tr}_\lambda(\log_\lambda \alpha)} \quad \text{and} \quad \left\{\frac{p}{\alpha}\right\} = \zeta^{-\mathrm{Tr}_\lambda\left(\frac{\log_\lambda \alpha}{\lambda^p}\right)},$$

where $\mathrm{Tr}_\lambda$ denotes the trace in the extension $\hat{K}|\hat{\mathbb{Q}}_p$.

Also in 1925, Hasse proved:

**(5I)** If $\alpha \in A$ is totally positive, $\gcd(A\alpha, A\beta) = A$, and $\alpha \equiv \beta \pmod{A\lambda^2}$, then $\{\alpha/\beta\}\{\beta/\alpha\}^{-1} = \zeta^c$, where

$$c = \sum_{i=1}^{p-1} i \frac{\mathrm{Tr}_\lambda(\zeta^{-i}\log_\lambda \alpha)}{p} \times \frac{\mathrm{Tr}_\lambda(\zeta^i \log_\lambda \alpha)}{p}.$$

I note that $c$ is a $\lambda$-adic integer, hence it is of the form $c = c_0 + c_1\lambda + c_2\lambda^2 + \cdots$ with $c_i \in \mathbb{Z}$. The power $\zeta^c$ is defined to be equal to $\zeta^{c_0}$.

Another way of expressing the product in (5I) was discovered by Takagi in 1927. It generalizes a previous formula by Kummer (1852) for the case of a regular prime $p$. To explain Takagi's formula, let $g$ be a primitive root modulo $p$, let $\sigma : \zeta \mapsto \zeta^g$ be the corresponding generator of the Galois group of $K|\mathbb{Q}$. For every $j = 1, \ldots, p$, let

$$\kappa_j = (1 - \lambda^j)^{-g^j \frac{\sigma^{p-1}-1}{\sigma - g^j}}.$$

In particular, $\kappa_1 = \zeta$. Then, it may be observed that

$$\kappa_j^{\sigma - g^j} \equiv 1 \pmod{A\lambda^{p+1}}$$
$$\kappa_j \equiv 1 - \lambda^j \pmod{A\lambda^{j+1}}$$

for $j = 1, \ldots, p$.

Moreover, every $\alpha \in A$, $\alpha \equiv 1 \pmod{A\lambda}$ may be written in a unique way in the form

$$\alpha \equiv k_1^{t_1(\alpha)} \cdots \kappa_{p-1}^{t_{p-1}(\alpha)} \kappa_p^{t_p(\alpha)} \pmod{A\lambda^{p+1}},$$

where each $t_j(\alpha)$ is an integer. Takagi's formula is the following:

**(5J)** *If* $\alpha, \beta \in A$, $\gcd(A\alpha, A\beta) = A$, *and* $\alpha \equiv 1 \pmod{A\lambda}$, $\beta \equiv 1$, $\pmod{A\lambda}$, *then*

$$\left\{\frac{\alpha}{\beta}\right\}\left\{\frac{\beta}{\alpha}\right\}^{-1} = \zeta^{-u}$$

*where*

$$u = \sum_{j=1}^{p-1} j t_j(\alpha) t_{p-j}(\beta).$$

Kummer's formula is in terms of the logarithmic differential quotient, which I have introduced in Lecture VII, §5. Namely, if $u$ is a real variable, $\alpha \in A \backslash A\lambda$ and $j = 1, 2, \ldots, p - 1$, then

$$l^{(j)}(\alpha) = \left[\frac{d^j \log \alpha(e^u)}{du^j}\right]_{u=0}.$$

These are $p$-integral rational numbers and they are related as follows to the above exponents $t_j(\alpha)$:

$$t_j(\alpha) \equiv \frac{(-1)^{j-1}}{j!} l^{(j)}(\alpha) \pmod{p}.$$

This gives Kummer's formula:

**(5K)** *If* $\alpha, \beta \in A$, $\gcd(A\alpha, A\beta) = A$, *and* $\alpha \equiv 1 \pmod{A\lambda}$, $\beta \equiv 1 \pmod{A\lambda}$ *then*

$$\left\{\frac{\alpha}{\beta}\right\}\left\{\frac{\beta}{\alpha}\right\}^{-1} = \zeta^w,$$

*where*

$$w = \sum_{j=1}^{p-1} (-1)^j l^{(j)}(\alpha) l^{(j)}(\beta).$$

As a corollary:

**(5L)** *If* $\alpha \in A$, $\alpha \equiv 1 \pmod{A\lambda}$, *then*

$$\left\{ \frac{\zeta}{\alpha} \right\} = \zeta^{-t_{p-1}(\alpha)} = \zeta^{-l^{(p-1)}(\alpha)}.$$

It follows from the definition

$$\left\{ \frac{\zeta}{\alpha} \right\} = \zeta^{\frac{N(\alpha) - 1}{p}}$$

that

$$t_{p-1}(\alpha) \equiv l^{(p-1)}(\alpha) \equiv -\frac{N(\alpha) - 1}{p} \pmod{p}.$$

Similarly, the following formula is due to Kummer (1858) and Takagi (1922):

**(5M)** *If* $\alpha \in A$, $\alpha \equiv 1 \pmod{A\lambda}$, *then*

$$\left\{ \frac{p}{\alpha} \right\} = \zeta^{-t_p(\alpha)}$$

*and moreover* $t_p(\alpha) \equiv \mathrm{Tr}_\lambda(\lambda^{-p} \log_\lambda \alpha) \pmod{p}$.

Also:

**(5N)** *If* $\alpha \in A$, $\alpha \equiv 1 \pmod{A\lambda}$, *then* $\{\lambda/\alpha\} = \zeta^{t_p(\alpha) + b}$, *where*

$$b = \sum_{j=1}^{p-2} \frac{(-1)^j B_j}{j!} t_{p-j}(\alpha) \equiv \sum_{j=1}^{p-3} \frac{(-1)^j B_j}{j} l^{(p-j)}(\alpha) \pmod{p}.$$

In particular, if $\varepsilon = -\lambda^{p-1}/p$, then $\{\varepsilon/\alpha\} = \zeta^{-b}$.

After this review of the main formulas for the power residue symbol, as they may be found in volume II of Hasse's *Bericht*, I will now describe their applications, which may be obtained by computing the power residue symbol for appropriate elements.

Assume that $x, y, z$ are nonzero integers such that $x^p + y^p + z^p = 0$. So

$$(-z)^p = \prod_{i=0}^{p-1} (x + \zeta^i y).$$

Let $\alpha = (x + \zeta y)/(x + y)$, so

$$\alpha = 1 - (y/(x + y))\lambda \equiv 1 \pmod{A\lambda}.$$

I write $t = y/(x + y) \equiv -y/z \pmod{p}$. Since $A\alpha = J^p$ for some ideal $J$ of $A$, then for every $\beta \in A$, $\beta \neq 0$, it follows that $\{\beta/\alpha\} = 1$. By considering various elements $\beta$, different conditions may be obtained.

Taking $\beta = \zeta$ leads to the criterion:

(5O) *With above notations,*

$$\sum_{i=1}^{p-1} \frac{t^i}{i} \equiv 0 \ (\mathrm{mod}\ p).$$

Taking $\beta = p$ gives the following:

(5P) *With above notations,*

$$\sum_{i=1}^{p} \frac{t^i}{i} \mathrm{Tr}_{K|Q}\left(\frac{1}{\lambda^{p-i}}\right) \equiv 0 \ (\mathrm{mod}\ p).$$

From $\beta = \lambda$, it follows:

(5Q) *With above notations,*

$$t^{p-1} \equiv 1 \ (\mathrm{mod}\ p^2).$$

From $\beta = \varepsilon = -\lambda^{p-1}/p$, the following may be deduced:

(5R) *With above notations,*

$$\sum_{j=1}^{(p-3)/2} \frac{B_{2j}}{2j} l^{(p-2j)}(\alpha) \equiv 0 \ (\mathrm{mod}\ p).$$

For the first case, the logarithmic differential quotients are expressible in terms of Mirimanoff polynomials:

$$l^{(j)}(\alpha) \equiv \frac{1}{(u-1)^p}\, \varphi_j(u) \ (\mathrm{mod}\ p)$$

for $j = 2, \ldots, p-2$, where $u = t/(t-1) = -y/x$ and

$$\varphi_j(X) = \sum_{i=1}^{p-1} i^{j-1} X^i.$$

The above criterion takes a form similar to Kummer's congruences (for the first case):

$$\sum_{j=1}^{(p-3)/2} \frac{B_{2j}}{2j} \varphi_{p-2j}(u) \equiv 0 \ (\mathrm{mod}\ p).$$

To obtain Kummer's congruences, put $\beta = j(1-\zeta)/(1-\zeta^j)$ (for $j = 1, \ldots, p-1$). This gives, after some calculations, Kummer's congruences for the first case:

$$B_{2j} l^{(p-2j)}(\alpha) \equiv 0 \ (\mathrm{mod}\ p)$$

or equivalently, $B_{2j}\varphi_{p-2j}(u) \equiv 0 \ (\mathrm{mod}\ p)$ (for $j = 1, 2, \ldots, (p-3)/2$).

Taking $\beta = \alpha^{\sigma^i}$ (for $i = 0,1,\ldots,p-2$), gives Mirimanoff's congruences for the first case:

$$l^{(j)}(\alpha)l^{(p-j)}(\alpha) \equiv 0 \,(\mathrm{mod}\,p)$$

or equivalently,

$$\varphi_j(u)\varphi_{p-j}(u) \equiv 0 \,(\mathrm{mod}\,p)$$

(for $j = 1,2,\ldots,(p-1)/2$).

The first theorem of Furtwängler is obtained by considering $\beta = l$, where $l$ is a prime, $l \neq p$, and $l\,|\,xyz$. Then $l^{p-1} \equiv 1 \,(\mathrm{mod}\,p^2)$.

The second theorem of Furtwängler is obtained similarly, taking $\beta = l$, where $l \neq 2$, $l\,|\,x - y$, $l \neq p$.

As I have already stated, the theorems of Wieferich and Mirimanoff follow at once from the above theorems of Furtwängler and therefore may be deduced by means of Hasse's theory.

In 1971, Gandhi performed some computations along the same lines; he proved:

(5S) *If* $x$, $y$, $z$ *are relatively prime integers,* $p \nmid xyz$, $x^p + y^p + z^p = 0$, *if* $l$ *is any prime such that* $l \nmid xyz$, $l \not\equiv 1 \,(\mathrm{mod}\,p)$, *and if* $t \equiv y/x \,(\mathrm{mod}\,l)$, $u \equiv z/y$ $(\mathrm{mod}\,l)$, $v \equiv y/z \,(\mathrm{mod}\,l)$, *then*

$$\left\{ \frac{x + \zeta^i y}{l} \right\} = \zeta^{-iq_p(l)v},$$

$$\left\{ \frac{x + \zeta^i z}{l} \right\} = \zeta^{-iq_p(l)u},$$

$$\left\{ \frac{z + \zeta^i y}{l} \right\} = \zeta^{-iq_p(l)t},$$

*for* $i = 1, 2, \ldots, p - 1$, *where* $q_p(l) = (l^{p-1} - 1)/p$ *is the Fermat quotient.*

Gandhi also found some analogous expressions if the second case fails for the exponent $p$. However, he gives an erroneous deduction of Vandiver's condition that if the first case fails for $p$, then $5^{p-1} \equiv 1 \,(\mathrm{mod}\,p^2)$.

# 6. The $p$-Rank of the Class Group of the Cyclotomic Field

In these lectures, I have often arrived at important conclusions from the study of the class number of the cyclotomic field. A more refined approach consists of studying the class group and examining its structure. More specifically, considering the $p$-primary component of the class group.

I introduce or recall some notation. Let $h = p^b t$, $b \geq 0$, $p \nmid t$, be the class number of $K$. If $J \neq 0$ is any fractional ideal of $K$, let $[J]$ denote the class of $J$, that is, the set of all fractional ideals of the form $A\alpha.J$, for some $\alpha \neq 0$. I write $J \sim J'$ when $[J] = [J']$. Let $\mathscr{C}_p = \mathscr{C}\ell_p(K)$ be the maximal *p*-subgroup of $\mathscr{C} = \mathscr{C}\ell(K)$; it has order $p^b$ and it is equal to $\{[J]^t \mid [J] \in \mathscr{C}\}$.

Since $\mathscr{C}_p$ is abelian and finite, $\mathscr{C}_p \cong \mathscr{Z}_1 \times \cdots \times \mathscr{Z}_m$, this is the unique decomposition as a product of cyclic *p*-groups. Let $\mathscr{Z}_i$ have order $p^{b_i}$, so $b_1 + \cdots + b_m = b$. The integer $m$ is called the *p*-rank of $\mathscr{C}$.

I choose an integral ideal $J_i$ such that $[J_i]$ is a generator of $\mathscr{Z}_i$. Thus $[J_i^{p^{b_i}}] = [A]$, but $[J_i^{p^{b_i-1}}] \neq [A]$. Hence, every ideal class $[J] \in \mathscr{C}_p$ may be written in unique way in the form

$$[J] = [J_1]^{x_1} \cdots [J_m]^{x_m} \tag{6.1}$$

with $0 \leq x_i \leq p^{b_i} - 1$. The set $\{[J_1], \ldots, [J_m]\}$ is a basis of $\mathscr{C}_p$.

Then there exists $I$ such that $[J] = [I]^p$ if and only if $p \mid x_1, \ldots, p \mid x_m$.

I denote by $\mathscr{C}_p^p$ the subgroup of $\mathscr{C}_p$ generated by $[J_1^p], \ldots, [J_m^p]$. Then $\mathscr{C}_p/\mathscr{C}_p^p$ is a vector space over the field $\mathbb{F}_p$ having dimension $m$.

For each $i = 1, \ldots, m$, let $\mathscr{G}_i$ be the subgroup of all $[J] \in \mathscr{C}_p$ such that in the representation (6.1), $p \mid x_i$. Then $\#(\mathscr{G}_i) = p^{b-1}$.

If $K'$ is a subfield of $K$, I use similar notations: $h' = p^a t'$, $a \geq 0$, $p \nmid t'$; $\mathscr{C}'$, $\mathscr{C}_p'$ stand for the ideal class group and its *p*-primary subgroup.

The following group homomorphisms arise naturally:

1. $\iota : \mathscr{C}' \to \mathscr{C}$ defined by $\iota([J']) = [AJ']$; observe that $\iota(\mathscr{C}_p') \subset \mathscr{C}_p$.
2. $N_{K|K'} : \mathscr{C} \to \mathscr{C}'$ defined by $N_{K|K'}([J]) = [N_{K|K'}(J)]$, where $J \neq 0$ is a fractional ideal of $K$ and $N_{K|K'}$ is the relative norm; I note that $N_{K|K'}(\mathscr{C}_p) \subset \mathscr{C}_p'$.

Let $\mathscr{N} = \{[J] \in \mathscr{C}_p \mid N_{K|K'}[J] = [A']\}$. Thus $\mathscr{N} = \mathrm{Ker}(N_{K|K'})$. $\mathscr{N}$ is also a finite *p*-group and has a basis.

I assemble in a lemma the following almost obvious facts:

**Lemma 6.1.** *If $p \nmid [K:K']$, then*

1. $\iota$ *is injective.*
2. $N_{K|K'} : \mathscr{C}_p \to \mathscr{C}_p'$ *is surjective.*
3. $\mathscr{C}_p \cong \mathscr{N} \times \mathscr{C}_p'$ *(direct product of groups).*
4. $\mathrm{rank}(\mathscr{C}_p) = \mathrm{rank}(\mathscr{N}) + \mathrm{rank}(\mathscr{C}_p')$.
5. *If $\{[J_1'], \ldots, [J_{m'}']\}$ is a basis of $\mathscr{C}_p'$, then there exist ideal classes $[J_{m'+1}], \ldots, [J_m]$ in $\mathscr{N}$ such that $\{[AJ_1'], \ldots, [AJ_{m'}'], [J_{m'+1}], \ldots, [J_m]\}$ is a basis of $\mathscr{C}_p$.*

After these generalities, let $K = \mathbb{Q}(\zeta)$, where $\zeta$ is a primitive *p*th root of 1, $p > 2$. Let $K^+ = \mathbb{Q}(\zeta + \zeta^{-1})$ be the real cyclotomic field. The notations are the following, and they are self-explanatory: $\mathscr{C}$, $\mathscr{C}^+$, $h = h^* h^+$, $h^* = p^{a_1} t_1$, $a_1 \geq 0$, $p \nmid t_1$; $h^+ = p^a t$, $a \geq 0$, $p \nmid t$; $\mathscr{C}_p$ has order $p^{a+a_1}$, $\mathscr{C}_p^+$ has order $p^a$.

$\mathcal{N}$ is the subgroup of all $[J] \in \mathscr{C}_p$ such that $N_{K|K^+}([J]) = [A^+]$; it has order $p^{a_1}$, $\mathscr{C}_p \cong \mathscr{C}_p^+ \times \mathcal{N}$. Let $e = \text{rank}(\mathscr{C}_p^+)$, $e_1 = \text{rank}(\mathcal{N})$, so $\text{rank}(\mathscr{C}_p) = e + e_1$.

Hecke proved a very interesting relation between these ranks $e$, $e_1$. His proof was based on the fundamental work of Furtwängler in class field theory, more specifically its connection with the general existence theorem for class fields. Whereas the full results of Furtwängler were not required by Hecke, nevertheless he needed the explicit description of the unramified cyclic extensions of degree $p$ of $K$. Of course this in turn may be easily obtained from the existence theorem of class field theory.

I describe the results obtained by Furtwängler in his papers of 1904 and 1907. Let $\alpha_1, \ldots, \alpha_m \in A$. They are $p$-independent when the following condition is satisfied: if $\alpha_1^{r_1} \cdots \alpha_m^{r_m} = \beta^p$, with $\beta \in A$ and $0 \le r_i \le p - 1$ (for $i = 1, \ldots, m$), then $r_1 = \cdots = r_m = 0$.

In 1859, while studying the reciprocity law for the power residue symbol, Kummer proved the following important theorem, which may be also found in Hilbert's *Zahlbericht* §135, Theorem 152:

**(6A)** *Let* $\alpha_1, \ldots, \alpha_m \in A$ *be* $p$-*independent. Let* $j_1, \ldots, j_m$ *be integers,* $0 \le j_1, \ldots, j_m \le p - 1$. *For every* $n = 1, \ldots, p - 1$ *let* $\mathcal{Q}_n$ *be the set of all prime ideals* $Q$, $Q \nmid A\alpha_i$ $(i = 1, \ldots, m)$, $Q \ne A\lambda$, *such that*

$$\left\{\frac{\alpha_1}{Q}\right\}^n = \zeta^{j_1}, \ldots, \left\{\frac{\alpha_m}{Q}\right\}^n = \zeta^{j_m}.$$

*Then*

1. *If* $s > 1$ *is a real variable, then*

$$p^m \sum_{n=1}^{p-1} \sum_{Q \in \mathcal{Q}_n} \frac{1}{N(Q)^s} = (p - 1)\log\frac{1}{s - 1} + f(s),$$

*where* $f(s)$ *remains bounded in the neighborhood of* $s = 1$, *and* $N(Q)$ *is the absolute norm of* $Q$.
2. *The set* $\bigcup_{n=1}^{p-1} \mathcal{Q}_n$ *is infinite.*

Actually, Čebotarev proved in 1923 the density theorem that asserts that each set $\mathcal{Q}_n$ is infinite. This was further extended by Schinzel in 1977 (see his Theorem 4).

Now, let $f = e + e_1 = \text{rank}(\mathscr{C}_p)$, let $\{[J_1], \ldots, [J_f]\}$ be a basis of $\mathscr{C}_p$. Without loss of generality, each $J_i$ may be taken to be an integral ideal and $A\lambda \nmid J_i$.

Let $p^{b_i}$ be the order of $[J_i]$, and let $\rho_i \in A$ be such that

$$[J_i]^{p^{b_i}} = A\rho_i.$$

I write $\pi = \frac{1}{2}(p - 1)$. Let $\{\varepsilon_1, \ldots, \varepsilon_{\pi-1}\}$ be a fundamental system of units, let $\varepsilon_\pi = \zeta$ and $\varepsilon_{\pi+i} = \rho_i$ (for $i = 1, \ldots, f$). Then $\{\varepsilon_1, \ldots, \varepsilon_{\pi+f}\}$ are $p$-independent. By Kummer's theorem, there exist prime ideals $Q_1, \ldots, Q_{\pi+f}$, distinct from $A\lambda$, such that $Q_i$ does not divide any of the ideals $J_1, \ldots, J_f$ and for every

$i = 1, \ldots, \pi + f$, $\{\varepsilon_i/Q_i\} \neq 1$, $\{\varepsilon_i/Q_j\} = 1$ (for $j \neq i$). So the ideals $Q_i$ are all distinct.

I write

$$[Q_j] = [J_1]^{u_1^{(j)}} \cdots [J_f]^{u_f^{(j)}}$$

with $1 \leq u_i^{(j)} \leq p^{b_i}$. Let $w_i^{(j)} = p^{b_i} - u_i^{(j)}$, so $Q_j J_1^{w_1^{(j)}} \cdots J_f^{w_f^{(j)}} = A\kappa_j$ where $\kappa_j \in A \backslash A\lambda$, $0 \leq w_i^{(j)} \leq p^{b_i} - 1$.

Let $S$ be the set of all algebraic integers $\alpha$ of the form

$$\alpha = \varepsilon_1^{u_1} \cdots \varepsilon_{\pi+f}^{u_{\pi+f}} \kappa_1^{v_1} \cdots \kappa_{\pi+f}^{v_{\pi+f}},$$

$0 \leq u_i \leq p - 1$, $0 \leq v_i \leq p - 1$ and such that the following $f$ congruences modulo $p$ are satisfied:

$$v_1 w_f^{(1)} + \cdots + v_{\pi+f} w_f^{(\pi+f)} \equiv 0 \pmod{p}.$$
$$\vdots$$
$$v_1 w_f^{(1)} + \cdots + v_{\pi+f} w_f^{(\pi+f)} \equiv 0 \pmod{p}.$$

Then $S \subset A \backslash A\lambda$, and $\#(S) \geq p^{\pi+f+\pi} = p^{p-1+f}$. Also, if $\alpha \in S$, then $A\alpha = J_1^{n_1} J_2^{n_2} \cdots J_f^{n_f} Q_1^{v_1} \cdots Q_{\pi+f}^{v_{\pi+f}}$, where $p | n_1, \ldots, p | n_f$.

In 1902, Hilbert defined $\alpha, \beta \in A \backslash A\lambda$ to be of the *same kind* if there exists $\gamma \in A$ such that $\alpha \equiv \beta\gamma^p \pmod{A\lambda^p}$. This implies that $\gamma \notin A\lambda$. The above relation is an equivalence.

Furtwängler then defined $\alpha \in A \backslash A\lambda$ to be a *primary integer* if there exists $\beta \in A$ such that $\alpha \equiv \beta^p \pmod{A\lambda^p}$. It follows that there exists $b \in \mathbb{Z}$ such that $\alpha \equiv b^p \pmod{A\lambda^p}$.

It is not difficult to show:

**Lemma 6.2.**

1. *All primary integers are of the same kind.*
2. *If $\beta \in A \backslash A\lambda$ is of the same kind as the primary integer $\alpha$, then $\beta$ is also primary.*
3. *There are $p^{p-1}$ distinct kinds of integers in $A \backslash A\lambda$.*

Since $S \subset A \backslash A\lambda$ has at least $p^{p-1+f}$ elements, there exist $\omega_1, \omega_2 \in S$, distinct but of the same kind. Let $\omega = \omega_1 \omega_2^{p-1}$. Then $\omega$ is a primary integer and it may be written in the form:

$$\omega = \varepsilon_1^{u_1} \cdots \varepsilon_{\pi+f}^{u_{\pi+f}} \kappa_1^{v_1} \cdots \kappa_{\pi+f}^{v_{\pi+f}} \alpha^p$$

with $\alpha \in A$, $0 \leq u_i, v_j \leq p - 1$ and at least one of the integers $u_i, v_j$ is different from 0. In particular, $\omega$ is not the $p$th power of an element of $K$.

The determination of $\omega$ depends on the choice of the prime ideals $Q_1, \ldots, Q_{\pi+f}$. With another choice $Q'_1, \ldots, Q'_{\pi+f}$, another element $\omega'$ is obtained. It may be shown that $\omega, \omega'$ are $p$-independent.

Since $\omega$ is not a $p$th power in $K$, $K(\sqrt[p]{\omega})|K$ is a cyclic extension of degree $p$, that is, a Kummer extension with Galois group generated by $\sigma: \sqrt[p]{\omega} \mapsto \zeta \sqrt[p]{\omega}$.

With every unramified cyclic extension of degree $p$ of $K$, $K_\beta = K(\sqrt[p]{\beta})$ (where $\beta \in A$, $\beta$ not a $p$th power in $K$) Furtwängler associated a subgroup

$\mathscr{G}_\beta$ of the $p$-class group $\mathscr{C}_p$ as follows: $\mathscr{G}_\beta$ is generated by all ideal cla
where $P$ is a prime ideal, $P \neq A\lambda$, $P \nmid A\beta$ and $\{\beta/P\} = 1$; that is, $P$ splits in
$K_\beta|K$. Thus $\#(\mathscr{G}_\beta) \leq p^{a+a_1-1}$.

In fact, it will be shown ultimately that $\mathscr{G}_\beta$ has order $p^{a+a_1-1}$. Pending this
result, let $\mathscr{H}_\beta$ be a subgroup of $\mathscr{C}_p$ such that $\mathscr{G}_\beta \subset \mathscr{H}_\beta$ and $\mathscr{H}_\beta$ has order
$p^{a+a_1-1}$.

To study the correspondence $K_\beta \to \mathscr{H}_\beta$, Furtwängler used the following
general lemma:

**Lemma 6.3.** *Let $L$ be any algebraic number field with class group $\mathscr{C}\ell(L)$
and class number $h_L = p^m r$, $m \geq 0$, $p \nmid r$. Let $\mathscr{H}$ be a subgroup of order $p^{m-1}$
of $\mathscr{C}\ell(L)$. Then*

$$\sum_{[P] \in \mathscr{H}} \frac{1}{N(P)^s} \leq \frac{1}{p} \log \frac{1}{s-1} + f(s),$$

*where $f(s)$ remains bounded in a neighborhood of $s = 1$.*

Applying this to $K$, as well as to $K^+$, Furtwängler showed that if $K_\beta$, $K_{\beta'}$
are distinct unramified cyclic extensions of degree $p$ of $K$, then $\mathscr{H}_\beta \neq \mathscr{H}_{\beta'}$.

Before stating the main theorem, I recall that $K_1|K, \ldots, K_m|K$ are
called *independent extensions* if none is contained in the composite of the
others. Furtwängler's main theorem is the following:

**(6B)**

1. *$K$ has $f$ independent unramified cyclic extensions of degree $p$.*
2. *Any $f + 1$ unramified cyclic extensions of degree $p$ must be dependent.*

It follows from this theorem that $\mathscr{H}_\beta = \mathscr{G}_\beta$. More specifically:

**(6C)** *$K$ has $e$ independent unramified cyclic extensions of degree $p$, say
$K_i = K(\sqrt[p]{\omega_i})$, with $\omega_i \bar{\omega}_i = \beta_i^p$, where $\beta_i \in A^+$.*

Based on this theory, Hecke could prove (1910)—and this was certainly
easier than what Furtwängler did—the following:

**(6D)**

1. *There exist ideal classes $[H_1], \ldots, [H_e] \in \mathscr{C}_p$ such that*
   a. *$N_{K|K^+}([H_i]) = [A^+]$ for $i = 1, \ldots, e$.*
   b. *$H_i^p = A\omega_i$, where $\omega_i$ is a primary integer.*
   c. *If $[H_1]^{m_1} \cdots [H_e]^{m_2} = [A]$, then $p|m_1, \ldots, p|m_e$.*
2. *$e \leq e_1$.*

I observe that this generalizes Kummer's theorem that if $p$ divides $h^+$,
then $p$ divides $h^*$.

It is interesting to see the implications of this theorem to Fermat's problem. Assume that there exists $\alpha$, $\beta$, $\gamma \in A \backslash A\lambda$ such that $\alpha^p + \beta^p + \gamma^p = 0$. Hecke showed that the above implies that $e_1 - e \geq 1$. Previously, in 1857, Kummer had shown that $p^2$ divides $h = h^* h^+$. It follows that $p^2 | h^*$. Indeed, if $p | h^+$, then $\text{rank}(\mathscr{C}_p^+) = e \geq 1$ and $e_1 \geq e + 1 \geq 2$, thus $p^2 | h^*$; while, if $p \nmid h^+$, then $p^2 | h^*$.

Furtwängler proved that under the above hypothesis (failure of the first case in the ring $A$ for the exponent $p$) then $e_1 - e \geq 4$ and therefore $p^4 | h^*$.

These results have been much improved by a totally different method invented by Vandiver. I will discuss this later in this lecture.

To conclude this section, I shouldn't pass over the beautiful reflection theorem of Leopoldt (1958). It is in fact a generalization of Scholz's theorem of 1932. In Lecture XIII, I'll specifically need this theorem. I think it worthwhile to describe, even though succinctly, Leopoldt's theorem. In this, I follow the "exposé" by F. Bertrandias (1969).

Let $\mathscr{G}$ be the Galois group of $K | \mathbb{Q}$, let $\hat{\mathscr{G}}$ be the group of characters with values in the group of (nonzero) $p$-adic numbers, namely, all the homomorphisms $\chi : \mathscr{G} \to \mathbb{Q}_p$. The group ring $\mathbb{Q}_p[\mathscr{G}]$ contains a system of orthogonal idempotents:

$$ l_\chi = \frac{1}{p-1} \sum_{\sigma \in \mathscr{G}} \chi(\sigma^{-1}) \sigma $$

(for all $\chi \in \hat{\mathscr{G}}$) and moreover

$$ 1 = \sum_{\chi \in \hat{\mathscr{G}}} l_\chi. $$

This induces a decomposition

$$ \mathbb{Q}_p[\mathscr{G}] = \bigoplus_{\chi \in \hat{\mathscr{G}}} \mathbb{Q}_p l_\chi, $$

and similarly

$$ \mathbb{Z}_p[\mathscr{G}] = \bigoplus_{\chi \in \hat{\mathscr{G}}} \mathbb{Z}_p l_\chi. $$

Suppose now that $V$ is any abelian $p$-group, say of order $p^n$, on which $\mathscr{G}$ operates; I write $v^\sigma$ for the image of $v$ by $\sigma$. This may be extended to an operation of $\mathbb{Z}_p[\mathscr{G}]$ in the following way: if $\alpha = \sum_{\sigma \in \mathscr{G}} \alpha_\sigma \sigma$, with $\alpha_\sigma \in \mathbb{Z}_p$, let $a_\sigma \in \mathbb{Z}$, $0 \leq a_\sigma \leq p^n - 1$ be such that $\alpha_\sigma \equiv a_\sigma \pmod{p^n}$. Then, the action of $\alpha$ on $v$ is defined by

$$ v^\alpha = \prod_{\sigma \in \mathscr{G}} (v^\sigma)^{a_\sigma}. $$

Letting $V_\chi = \{v^{l_\chi} | v \in V\}$, then $V_\chi$ is a subgroup of $V$; $v \in V_\chi$ if and only if $v^{l_\chi} = v$, and in this case, $v^\sigma = v^{\chi(\sigma)}$. It follows also that $V = \prod_{\chi \in \hat{\mathscr{G}}} V_\chi$.

Let $\tau \in \mathscr{G}$ complex conjugation. A character $\chi$ such that $\chi(\tau) = \mathscr{G}$ is even, while if $\chi(\tau) = -1$, it is odd. I note that if $\chi$ is even, $v^\tau = v$, while if $\chi$ is odd, then $v^\tau = v^{-1}$ for all $v \in V_\chi$. An element $v \in V$ is invariant by $\tau$ exactly if $v \in V^+ = \prod_{\chi \text{ even}} V_\chi$, while $v^\tau = v^{-1}$ if and only if $v \in V^* = \prod_{\chi \text{ odd}} V_\chi$.

The group $\mathscr{G}$ operates on various interesting abelian $p$-groups. First, it operates on the group $W$ of all $p$th roots of 1. This is a cyclic group of order $p$, so there is a character $\chi^*$ of $\mathscr{G}$ such that $W = W_{\chi^*}$. Namely, $\chi^*(\sigma) = \lim_{n \to \infty} g^{p^n}$ (limit in the $p$-adic sense) where $\zeta^\sigma = \zeta^g$, $1 \leq g \leq p - 1$, $\zeta$ being a primitive $p$th root of 1. I note, therefore, that $\zeta^\sigma = \zeta^{\chi^*(\sigma)}$.

Given any character $\chi$ of $\mathscr{G}$, the character $\bar{\chi} = \chi^{-1}\chi^*$ is called the *reflected character* of $\chi$. The reflection $\chi \mapsto \bar{\chi}$ in the group $\hat{\mathscr{G}}$ is an involution, that is, $\bar{\bar{\chi}} = \chi$. Since $\chi^*(\tau) = -1$, $\chi^*$ is an odd character and $\chi, \bar{\chi}$ have different parity.

Another abelian $p$-group on which $\mathscr{G}$ operates is $\mathscr{C}_p$. So, for every character $\chi$ of $\mathscr{G}$, I may consider the subgroup $(\mathscr{C}_p)_\chi$, whose rank is denoted by $r_\chi$.

Leopoldt's theorem relates the ranks $r_\chi$ and $r_{\bar{\chi}}$, for reflected characters $\chi$ and $\bar{\chi}$. To bound the difference $r_{\bar{\chi}} - r_\chi$, Leopoldt considered the action of $\mathscr{G}$ on the $\mathbb{F}_p$-vector space $\mathscr{U} = U/U^p$, where $U$ is the group of units of $A$ (the ring of integers of $K$). The rank of $\mathscr{U}$ is $(p - 1)/2$. Consider now all units $\varepsilon$, such that the extension $K(\sqrt[p]{\varepsilon})|K$ is unramified. Their cosets $\varepsilon\mathscr{U}^p$ form a subgroup $\mathscr{U}_u$ of $\mathscr{U}$. For each character $\chi$ of $\mathscr{G}$ let $s_{u,\chi} = \mathrm{rank}(\mathscr{U}_{u,\chi})$.

Leopoldt's theorem may now be stated:

**(6E)**

1. *For every character $\chi$ of $\mathscr{G}$:* $-s_{u,\bar{\chi}} \leq r_{\bar{\chi}} - r_\chi \leq s_{u,\chi}$.
2. *If $\chi$ is an even character, then* $r_\chi \leq r_{\bar{\chi}} \leq r_\chi + 1$ *and*
3. $e \leq e_1 \leq e + (p - 3)/2$.

Of course the inequality $e \leq e_1$ is Hecke's theorem.

# 7. Criteria for $p$-Divisibility of the Class Number

The general idea is quite simple: If the first case is false for the exponent $p$, to deduce that a very high power of $p$ divides the first (or second) factor of the class number. Then by some means, like explicit computation, rule out this possibility, thereby assuring that the first case actually holds for the exponent in question.

There are various papers of this kind in the literature. To begin with, I consider statements involving the first factor of the class number. For example, as I mentioned earlier, if the first case fails for $p$, then $p^2 | h^*$ (Hecke) and, better, $p^4 | h^*$ (Furtwängler).

The next results in the same line depend on a congruence for the first factor which was discovered in 1919 by Vandiver:

$$h^* \equiv \frac{(-1)^{(p-1)/2} p}{2^{(p-3)/2}} \prod_{\substack{1 \leq i \leq p-2 \\ i \text{ odd}}} B_{ip^n+1} \pmod{p^n}$$

for $n \geq 1$. This congruence was proved again in 1966 by Hasse, using $p$-adic methods. A still simpler proof was given by Slavutskii in 1968.

In 1925, using the above congruence in appropriate ways, Vandiver showed that if the first case fails for the exponent $p$, then $p^8$ divides $h^*$. Morishima (1932) sharpened this to $p^{12}\mid h^*$ (with two undecided primes) and Lehmer (1932) ruled out the exceptions.

Whatever the virtue of these results, it would be much better to have some power $p^{n(p)}$ dividing $h^* = h_p^*$, where $n(p)$ increases with $p$.

In 1965, Eichler proved the following result:

**(7A)** *If the first case fails for $p$, then $p^{[\sqrt{p}]-1}$ divides the first factor $h^*$.*

The proof is one that relies only on basic principles and may be explained without much effort. In spite of the danger of making this lecture too long, I will now give Eichler's clever proof.

PROOF. If $x$, $y$, $z$ are relatively prime integers, not multiples of $p$, and such that $x^p + y^p + z^p = 0$ then, as said many times, there are ideals $J_i$ of $A$ such that $A(x + \zeta^i y) = J_i^p$ (for $i = 0,1,\ldots,p-1$). Let $r = [\sqrt{p}] - 1$ and $h^* = p^s t$, $s \ge 0$, $p \nmid t$. Assume that $s < r$. We will show this leads to a contradiction. By the pigeon-hole principle and the fact that $p^s$ is the order of the group $\mathcal{N} \cong \mathscr{C}_p/\mathscr{C}_p^+$, there exist $a_1,\ldots,a_r$, $0 \le a_i \le p-1$, but not all $a_i$ equal to zero, such that $J_1^{a_1} \cdots J_r^{a_r} \sim AJ^+$, where $J^+$ is some ideal of $A^+$ (the ring of integers of $K^+ = \mathbb{Q}(\zeta + \zeta^{-1})$). Then

$$\prod_{i=1}^{r} A(x + \zeta^i y)^{a_i} = (J_1^{a_1} \cdots J_r^{a_r})^p = (A\rho \cdot AJ^+)^p,$$

where $\rho \in A$; it follows that $A(J^+)^p$ is principal, say $A(J^+)^p = A\alpha$ with $\alpha \in A^+$. Then

$$\prod_{i=1}^{r} (x + \zeta^i y)^{a_i} = \rho^p \alpha \zeta^j \varepsilon,$$

where $\varepsilon$ is a real unit (from the nature of units of the cyclotomic field). Since $A\rho = (A\lambda)^{p-1}$ and $A/A\lambda \cong \mathbb{F}_p$, there exists $b \in \mathbb{Z}$ such that $\rho \equiv b \pmod{\lambda}$. Then $\rho^p \equiv b^p \pmod{p}$. Therefore $\prod_{i=1}^{r} (x + \zeta^i y)^{a_i} \equiv b^p \alpha \zeta^j \varepsilon \pmod{p}$. Taking the complex conjugates and dividing,

$$\prod_{i=1}^{r} \left(\frac{x + \zeta^i y}{x + \zeta^{-i} y}\right)^{a_i} \equiv \zeta^{2j} \pmod{p}$$

therefore

$$\prod_{i=1}^{r} \left(\frac{x + \zeta^i y}{y + \zeta^i x}\right)^{a_i} \equiv \zeta^n \pmod{p}$$

with $n = 2j - \sum_{i=1}^{r} i a_i$. Let

$$x_i = \begin{cases} x & \text{when } a_i \ge 0 \\ y & \text{when } a_i < 0 \end{cases}$$

$$y_i = \begin{cases} y & \text{when } a_i \ge 0 \\ x & \text{when } a_i < 0. \end{cases}$$

If $Z$ is an indeterminate, let

$$F(Z) = \prod_{i=1}^{r} (x_i + Z^i y_i)^{|a_i|}$$

and

$$G(Z) = \prod_{i=1}^{r} (y_i + Z^i x_i)^{|a_i|}.$$

Then $F(\zeta) \equiv \zeta^n G(\zeta) \pmod{p}$ so there exists a polynomial $P(Z)$ with coefficients in $\mathbb{Z}$, degree at most $p - 2$, such that

$$F(\zeta) = \zeta^n G(\zeta) + pP(\zeta).$$

This means that

$$F(Z) = Z^n G(Z) + pP(Z) + \Phi_p(Z)Q(Z),$$

where $\Phi_p(Z)$ is the $p$th cyclotomic polynomial and $Q(Z)$ has coefficients in $\mathbb{Z}$, because of Gauss's lemma on factorization of polynomials.

Multiplying by $1 - Z$, taking the derivatives, and setting $Z$ equal to $\zeta$ leads to the congruence

$$(1 - \zeta)F'(\zeta) - F(\zeta) \equiv -\zeta^n G(\zeta) + n(1 - \zeta)\zeta^{n-1}G(\zeta) + (1 - \zeta)\zeta^n G'(\zeta) \pmod{\lambda}.$$

Dividing by the preceding congruence and computing explicitly the logarithmic derivatives yield:

$$(1 - \zeta) \sum_{i=1}^{r} |a_i| i \frac{y_i \zeta^{i-1}}{x_i + \zeta^i y_i} \equiv \frac{n(1 - \zeta)}{\zeta} + (1 - \zeta) \sum_{i=1}^{r} |a_i| i \frac{x_i \zeta^{i-1}}{y_i + \zeta^i x_i} \pmod{\lambda}.$$

From here, it follows that

$$(1 - \zeta) \sum_{i=1}^{r} \zeta^i a_i i \left( \frac{y}{x + \zeta^i y} - \frac{x}{y + \zeta^i x} \right) \equiv n(1 - \zeta) \pmod{p}.$$

Let $k \geq 1$ be the smallest index such that $a_k \neq 0$. Upon multiplication with the product

$$\prod_{i=1}^{r} (x + \zeta^i y)(y + \zeta^i x)$$

a congruence is obtained between polynomial expressions in $\zeta$, where the highest term is of the form

$$a_k k (xy)^m (x^2 - y^2) \zeta^{r(r+1)+1-k}$$

($m$ being a positive integer). Since $r(r + 1) \leq p - 2$, by the assumption $r = [\sqrt{p}] - 1$ and since $\{1, \zeta, \ldots, \zeta^{p-2}\}$ is an integral basis, the congruence in question gives $x^2 \equiv y^2 \pmod{p}$, i.e., $x \equiv \pm y \pmod{p}$. By symmetry, $y \equiv \pm z \pmod{p}$.

Hence from $x + y + z \equiv x^p + y^p + z^p \equiv 0 \pmod{p}$ it follows that $p \mid x$, against the hypothesis.                                                              $\square$

With a slight modification in the proof, the following is also true (but has not been explicitly published in this form):

**(7B)** *If the first case fails for p, then the p-rank of the class group* $\mathscr{C} = \mathscr{C}\ell(K)$ *is greater than* $\sqrt{p} - 2$.

Despite the interest of the above results, it is, in practice, not easy to determine the p-part of the first factor $h^*$, or of the p-rank of the class group. Brückner obtained in 1979 the following more amenable result:

**(7C)** *If the first case of Fermat's theorem fails for p, then the irregularity index of p,* $\mathrm{ii}(p) = \#\{2j = 2, 4, \ldots, p - 3 \,|\, p \text{ divides } B_{2j}\}$, *is greater than* $\sqrt{p} - 2$.

Unaware of the variant of Eichler's theorem, Skula proved in 1972 that if the p-rank of the class group of $K$ is 1, then the first case holds for $p$. A substantial simplification of Skula's proof was published by Brückner (1972). But, in any case, this result is no more than a special case of the variant of the beautiful theorem of Eichler.

Now, I would like to turn my attention to similar statements for the second factor of the class number.

Carlitz established in 1968 a congruence between the first and second factors of the class number. Let $m = (p - 1)/2$, let $g$ be a primitive root modulo $p$ and $G_0 = \det(g^{2nj})_{j,n}$ for $0 \leq j \leq m - 2$, $1 \leq n \leq m - 1$. Let $\varepsilon_1(Z), \ldots, \varepsilon_{m-1}(Z)$ be polynomials of degree at most $p - 2$, coefficients in $\mathbb{Z}$, such that $\{\varepsilon_1(\zeta), \ldots, \varepsilon_{m-1}(\zeta)\}$ is a fundamental system of units of $K$ and $\varepsilon_k'(x)$ denotes the derivative of the polynomial $\varepsilon_k(x)$. Then

$$\zeta \frac{\varepsilon_k'(\zeta)}{\varepsilon_k(\zeta)} = \sum_{s=0}^{p-2} c_{ks} \zeta^{g^s},$$

where $c_{ks} \in \mathbb{Z}$. Put

$$C_{kn} = \sum_{s=0}^{p-2} c_{ks} g^{(2n-1)s} \quad \text{for } k, n = 1, \ldots, m - 1$$

and let $C = \det(C_{kn})_{k,n}$. Finally, define $G = 2^{m+2}(C/G_0)$.

Carlitz showed that $G$ is independent of the choice of the fundamental system of units and moreover:

**(7D)** $h^+ G \equiv \pm h^* \pmod{p}$.

Note that this says that if $p \,|\, h^+$, then $p \,|\, h^*$ (as Kummer showed), and conversely, if $p \,|\, h^*$, but $p \nmid h^+$, then $p \,|\, G$.

In 1973, using p-adic methods, Metsänkylä proved a congruence for $h^+$, which is actually equivalent to the one by Carlitz.

In 1934, Vandiver stated and gave an abbreviated proof of the following theorem:

*If p does not divide $h^+$, then the first case holds for the exponent p.*

Referring to this theorem, Vandiver wrote: "I now give a sketch of a proof of a theorem which appears to be the principal result I have so far found concerning the first case of [Fermat's] theorem."

Indeed, if this theorem is true, it is quite remarkable. However, L. C. Washington communicated to me in a letter that "Iwasawa has told me that there is an error in Vandiver's proof which he has not been able to fix. R. Greenberg has also looked at it and arrived at the same conclusion.... Anyway, the theorem should be regarded as questionable".

Therefore some of the results of Morishima's papers XI and XII of 1934 and 1935 are also unreliable (I have also discussed Gunderson's criticisms in another lecture).

If it can be established one day that, after all, Vandiver's theorem is true, then the following simple proof of Skula's result, as communicated to me by Hasse, will be justified:

Suppose that the $p$-rank of the class group is 1, that is, $\mathscr{C}_p$ is cyclic. Since $\mathscr{C}_p \cong \mathscr{C}_p^+ \times \mathscr{N}$, then either $\mathscr{C}_p^+$ or $\mathscr{N}$ is trivial. Recall that if $p$ divides $h^+$, then $p$ divides $h^*$, as Kummer proved. Therefore, $p$ cannot divide $h^+$, otherwise both groups $\mathscr{C}_p^+$ and $\mathscr{N}$ are nontrivial. By Vandiver's theorem (if it is true) the first case holds for $p$.

I note that in the same year as Vandiver, 1934, Grün proved the following result, which though not as strong as what Vandiver claimed, at least is correct.

(7E) If $p > 3$ is a prime not dividing the second factor $h^+$, and if $B_{2kp} \not\equiv 0$ (mod $p^3$) for $k = 1, 2, \ldots, (p - 3)/2$, then the first case holds for the exponent $p$.

# 8. Properly and Improperly Irregular Cyclotomic Fields

Vandiver, following in the footsteps of Kummer, recognized the importance, for Fermat's theorem, of the condition that $p$ divides the second factor of the class number.

In 1929, he introduced the following definition. A prime $p$ is called *properly irregular* if $p \mid h_p^*$, but $p \nmid h_p^+$. It is *improperly irregular* if $p \mid h_p^+$, hence also $p \mid h_p^*$ by Kummer's theorem.

Already in 1870, Kummer established that 2 divides $h_{163}^+$ and $h_{937}^+$, and 3 divides $h_{229}^+$ and $h_{257}^+$. However, Kummer found no example where $p \mid h_p^+$. In fact, no such example has yet been found, however there is no evidence to support a conjecture that $p$ never divides $h_p^+$ (see Herbrand, 1932).

The conjecture that $p$ does not divide $h_p^+$ is commonly called Vandiver's conjecture. Yet, it appears already in Kummer's early work. Indeed, in a letter to Kronecker, dated December 28, 1849, Kummer stated that he thinks to base his proof of Fermat's theorem for the exponent $p$ on two properties.

The first one, "still to be established" amounts to the fact that $p$ does not divide the second factor of the class number. In another letter to Kronecker (April 24, 1853), Kummer recognized the failure of his attempts to prove that $p$ does not divide $h_p^+$.

In 1929 and 1930, Vandiver gave a necessary and sufficient condition for $h_p^+$ to be a multiple of $p$. This is expressed, as one would expect, in terms of the units of $K$. I recall my previous notations.

Let $g$ be a primitive root modulo $p$, let $\sigma : \zeta \mapsto \zeta^g$ be the corresponding generator of the Galois group of $K|\mathbb{Q}$. For every $j \in \mathbb{Z}$, let $g_j$ be the unique integer such that $1 \le g_j \le p - 1$ and $g_j \equiv g^j \pmod{p}$ (with the obvious convention when $j < 0$). Let

$$R_i(\sigma) = \sum_{j=0}^{(p-3)/2} g_{-2ij}\sigma^j$$

(for $i = 1, 2, \ldots, (p-3)/2$). Let

$$\delta_g = \sqrt{\frac{1 - \zeta^g}{1 - \zeta} \times \frac{1 - \zeta^{-g}}{1 - \zeta^{-1}}}$$

and finally, let $\varepsilon_i = \delta_g^{R_i(\sigma)}$, for $i = 1, 2, \ldots, (p-3)/2$. Then:

**(8A)** $p$ divides $h_p^+$ if and only if one of the units $\varepsilon_i$ $(1 \le i \le (p-3)/2)$ is the pth power of a unit in $K$.

Vandiver also proved that the units $\varepsilon_1, \ldots, \varepsilon_{(p-3)/2}$ form an independent system of units in $K$.

A simpler proof of the above result was given by Inkeri in 1955.

The following theorem of Vandiver, was first announced in a special case in 1930. The proof was published later in 1939. It is quite interesting, since it relates the irregularity index of $p$ with the invariants of the $p$-class group of $K$.

Morishima had essentially claimed this result already in 1933, but as Vandiver points out, Morishima's proof is incomplete.

**(8B)** *Let $p$ be a properly irregular prime number, let the $p$-class group of $K$ be $\mathscr{C}_p = \mathscr{X}_1 \times \cdots \times \mathscr{X}_f$, where each $\mathscr{X}_i$ is a cyclic group, of order $p^{b_i}$ ($i = 1, \ldots, f$), $b_i \ge 1$. Let $m_1, m_2, \ldots, m_{f'}$ be the indices such that $1 \le m_i \le (p-3)/2$ and $p$ divides $B_{2m_i}$. Then:*

1. *$f' = f$ and $p^{b_i} \| B_{2m_i}$ (for $i = 1, \ldots, f$).*
2. *$\mathscr{C}_p$ has a basis $\{[J_1], \ldots, [J_f]\}$ such that $[J_i]$ has order $p^{b_i}$ and $J_i^{g - g^{p - 2m_i}} \sim A$ (for $i = 1, \ldots, f$).*

These results are deep, the proofs quite involved.

In 1976, using the theory of modular forms, Ribet proved a result which is intimately related with (1) of (8B) above. However, he did not assume that $p$ divides $h^+$.

Let $\Gamma = \mathscr{C}_p/\mathscr{C}_p^p$, where $\mathscr{C}_p$ is the $p$-primary component of the class group of $K$. Let $\chi:\mathrm{Gal}(K|\mathbb{Q}) \to \mathbb{F}_p$ be the character, with modulus $p - 1$, defined as follows. If $g$ is a primitive root modulo $p$, $\sigma:\zeta \mapsto \zeta^g$ the corresponding generator of $\mathrm{Gal}(K|\mathbb{Q})$, then $\chi(\sigma^i) = g^i$ for $i = 0, 1, \ldots, p - 2$. Let $\Gamma(\chi^j) = \{\gamma \in \Gamma \,|\, \sigma^i(\gamma) = \chi^j(\sigma^i)\gamma$ for $i = 0,1, \ldots, p - 2\}$. Then $\Gamma(\chi^j)$ is a subspace of the $\mathbb{F}_p$-vector space $\Gamma$, and if $j \equiv j' \pmod{p - 1}$, then $\Gamma(\chi^j) = \Gamma(\chi^{j'})$. Moreover $\Gamma = \bigoplus_{j=0}^{p-1} \Gamma(\chi^j)$.

Ribet proved:

**(8C)** *If $1 \le k \le (p - 3)/2$, then $p$ divides $B_{2k}$ if and only if $\Gamma(\chi^{1 - 2k}) \neq 0$.*

A better understanding of the structure of the class group and units of the cyclotomic fields is now being reached by means of the theory of $p$-adic $L$-functions, as illustrated in the recent work of Washington.

I will now describe the method used by Washington to give a more systematic and transparent proof of various results by Pollaczek and Dénes.

Borrowing from Iwasawa's beautiful book (1972), I present the $p$-adic $L$-functions. Let $\chi$ be a primitive modular character, with conductor $f$ (see Lecture VI, §1). So $\chi:\mathbb{Z} \to \mathbb{C}$, $\chi(a) = 0$ if and only if $\gcd(a,f) = 1$, $\chi(ab) = \chi(a)\chi(b)$, and $a \equiv b \pmod{f}$ implies $\chi(a) = \chi(b)$.

The generalized Bernoulli numbers $B_{n,\chi}$ are defined as follows. If

$$F_\chi(T) = \sum_{i=1}^{f} \frac{\chi(i)Te^{iT}}{e^{fT} - 1},$$

the formal power series development of $F_\chi(T)$ is written as:

$$F_\chi(T) = \sum_{n=0}^{\infty} B_{n,\chi} \frac{T^n}{n!}.$$

In particular, if $\chi = \chi_0$, the principal character (with conductor $f = 1$), then $B_{n,\chi_0} = B_n$ (for $n \neq 1$), $B_{1,\chi_0} = -B_1 = \frac{1}{2}$.

If $L(s,\chi) = \sum_{n=1}^{\infty} (\chi(n)/n^s)$ (for $s > 1$) and if $\chi \neq \chi_0$, the above $L$-series may be extended to a meromorphic function on the whole plane. For $n = 1, 2, 3, \ldots$ it may be shown that $L(1 - n, \chi) = -B_{n,\chi}/n$. This is similar to a formula connecting the Bernoulli numbers $B_{2s}$ and the values of the Riemann zeta-function $\zeta(2s)$, once the functional equation for the Riemann zeta-function is taken into account.

The aim in defining the $p$-adic $L$-functions is to have at our disposal, in the $p$-adic number field, a function which behaves like the ordinary $L$-function. Kubota and Leopoldt (1964) were able to construct such $p$-adic $L$-functions.

I assume $p > 2$, but everything holds with slight modification when $p = 2$.

The group $U$ of units of the ring $\mathbb{Z}_p$ of $p$-adic integers has the following structure: $U \cong W \times U_1$ (group isomorphism and homeomorphism), where $U_1$ is the subgroup of all units $\alpha$ such that $\alpha \equiv 1 \pmod{p}$ and $W$ is the group of order $p - 1$ of all $(p - 1)$th roots of $1$ in $\mathbb{Q}_p$.

Let $\omega(\alpha) = \lim_{n \to \infty} \alpha^{p^n}$ for every $\alpha \in U$. Then $\omega(\alpha) \in W$ and $\alpha^{(1)} = \alpha/\omega(\alpha) \in U_1$.

In particular, if $a \in \mathbb{Z} \subset \mathbb{Z}_p$, $p \nmid a$, then $a \in U$, so $\omega(a)$ is well defined; for the integers $a$, such that $p \mid a$, just put $\omega(a) = 0$. Since the values of $\omega(a)$ are algebraic numbers, hence complex numbers, the mapping $\omega : \mathbb{Z} \to \mathbb{C}^\cdot$ is a primitive modular character with conductor $p$.

If $\chi$ is a primitive modular character with conductor $f$, for every $n \geq 1$ let $\chi_n = \chi\omega^{-n}$. This is the product modular character, which is the unique primitive modular character with conductor dividing $fp$ such that $\chi_n(a) = \chi(a)\omega(a)^{-n}$ for every integer $a$ with $\gcd(a, fp) = 1$.

The theorem of Kubota and Leopoldt asserts the existence of a unique $p$-adic meromorphic function $L_p(s,\chi)$ such that the following two conditions are satisfied:

i.

$$L_p(s,\chi) = \frac{c_{-1}}{s-1} + \sum_{n=0}^{\infty} c_n(s-1)^n,$$

where each $c_i \in \mathbb{Q}_p(\chi)$ (the field generated over $\mathbb{Q}_p$ by the values $\chi(a)$, for $a = 1, 2, \ldots, p-1$),

$$c_{-1} = \begin{cases} 1 - \dfrac{1}{p} & \text{if } \chi = \chi_0, \\ 0 & \text{if } \chi \neq \chi_0, \end{cases}$$

and the series converges in the domain

$$\left\{ s \in \bar{\mathbb{Q}}_p = \text{algebraic closure of } \mathbb{Q}_p \,\middle|\, |s - 1|_p < p^{1-\frac{1}{p-1}} \right\}$$

($| \ |_p$ denotes the $p$-adic absolute value, normalized by $|p|_p = 1/p$).

ii. For $n = 1, 2, 3, \ldots$

$$L_p(1 - n, \chi) = (1 - \chi_n(p)p^{n-1})L(1 - n, \chi_n) = -(1 - \chi_n(p)p^{n-1})\frac{B_{n,\chi_n}}{n}.$$

The functions $L_p(s,\chi)$ defined above are the $p$-adic $L$-functions. Thus, if $\chi \neq \chi_0$, $L_p(s,\chi)$ is regular at $s = 1$, while $L_p(s,\chi_0)$ has a pole of order 1, with residue $1 - (1/p)$ at $s = 1$.

Now I shall require the $p$-adic regulator of an algebraic number field, which is defined in a way similar to the ordinary regulator. For this purpose, I first consider the $p$-adic logarithm. Let $\alpha \in \bar{\mathbb{Q}}_p$ (algebraic closure of $\mathbb{Q}_p$) be such that $|\alpha - 1|_p < 1$. Then the series

$$\sum_{n=1}^{\infty} \frac{(-1)^{n-1}}{n}(\alpha - 1)^n$$

is convergent and its sum is an element of $\mathbb{Q}_p$ which is defined to be $\log_p(\alpha)$, the $p$-adic logarithm of $\alpha$ (I have defined in Lecture VII, §5, the $\lambda$-adic logarithm in similar way). The mapping $\log_p$ may be extended in a unique way to the whole multiplicative group $\mathbb{Q}_p$, so as to be continuous and satisfies $\log_p(\tau(\alpha)) = \tau(\log_p(\alpha))$ for every $\tau \in \text{Gal}(\bar{\mathbb{Q}}_p | \mathbb{Q}_p)$ and $\log_p p = 0$. Note that $\bar{\mathbb{Q}}$ is algebraic over $\mathbb{Q}_p$, hence it may be imbedded into $\bar{\mathbb{Q}}_p$; I fix one such imbedding.

Let $L$ be a totally real number field of degree $n$, let $\varphi_1, \ldots, \varphi_n$ be the imbeddings of $L$ into $\bar{\mathbb{Q}} \subseteq \bar{\mathbb{Q}}_p$. Let $\{\varepsilon_1, \ldots, \varepsilon_r\}$ be a fundamental system of units of $L$, where $r = n - 1$.

The determinant of the matrix $(\log_p \varphi_i(\varepsilon_j))_{i,j=1,\ldots,r}$ belongs to $\mathbb{Q}_p$ and, up to a factor $\pm 1$, is independent of the choice of the fundamental system of units. It is, by definition, the $p$-*adic regulator* of $L$, denoted by $R_p$.

Leopoldt conjectured that $R_p$ is always nonzero. This has been proved by Brumer (1967) when $L | \mathbb{Q}$ is a totally real abelian extension, and it is quite a deep theorem. In particular, if $L = K^+$, its $p$-adic regulator is not equal to 0.

Leopoldt's $p$-adic class number formula puts together the various values of the $p$-adic $L$-functions at 1, the $p$-adic regulator and the class number $h$ of $K = \mathbb{Q}(\zeta)$:

$$h = \frac{p^{(p-3)/4}}{2^{(p-3)/2}R_p} \prod_{i=1}^{(p-3)/2} L_p(1,\omega^{2i}).$$

In particular, $L_p(1,\omega^{2i}) \neq 0$ for each $i = 1, \ldots, (p-3)/2$.

Dénes defined in 1954 the $p$-characters of the Bernoulli numbers as follows. If $1 \leq i \leq (p-3)/2$, the $p$-character of $B_{2i}$ is the smallest integer $u_i \geq 0$ such that $B_{2ip^{u_i}} \not\equiv 0 \pmod{p^{2u_i+1}}$. In particular, $p$ is regular if and only if the $p$-character of every Bernoulli number $B_{2i} (1 \leq i \leq (p-3)/2)$ is equal to 0.

Dénes raised the question whether such an integer $u_i$, as above, necessarily exists. He promised a proof, but this promise remained unfulfilled. It was only in 1977 that Washington succeeded in showing that the $p$-character is always defined (as a finite integer). Precisely

$$u_i = v_p(L_p(1,\omega^{2i})) < \infty,$$

where $v_p$ is the $p$-adic valuation. The inequality reflects the fact that $L_p(1,\omega^{2i}) \neq 0$.

The $p$-adic method of Washington was sufficient to obtain new proofs of results of Pollaczek (1924) and Dénes (1954) concerning fundamental systems of units for the real cyclotomic field. Because of its interest, I wish to mention the following result of Dénes (1954, 1956).

**(8D)** *There exists a fundamental system of units* $\{\eta_2, \eta_4, \ldots, \eta_{p-3}\}$ *for* $K^+$ *of the form*:

$$\eta_{2i} \equiv a_{2i} + b_{2i}\lambda^{c_{2i}} \pmod{\lambda^{c_{2i}+1}}$$

*with* $a_{2i}, b_{2i} \in \mathbb{Z}$ *not divisible by* $p$, *and*

$$c_{2i} = 2i + (p-1)u'_{2i},$$

*where*

$$0 \leq u'_{2i} \leq u_{2i} = v_p(L_p(1,\omega^{2i}))$$

*for* $2i = 2, 4, \ldots, p - 3$.

(8E) *If* $p^a$ *is the exact power of* $p$ *dividing* $h^+$, *then*

$$a = \sum_{i=1}^{(p-3)/2} (u_{2i} - u'_{2i}).$$

In particular

(8F) *If* $p$ *is properly irregular, then* $u'_{2i} = u_{2i}$ *for* $i = 1, \ldots, (p - 3)/2$.

# Bibliography

1844   Eisenstein, F. G.
Beweis des Reciprocitätssatzes für die cubischen Reste in der Theorie der aus dritten Wurzeln der Einheit zusammengesetzten complexen Zahlen. *J. reine u. angew. Math.*, **27**, 1844, 289–310. Reprinted in *Mathematische Werke*, vol. I, Chelsea Publ. Co., New York, 1975, 59–80.

1850   Eisenstein, F. G.
Beweis der allgemeinen Reziprozitätsgesetz zwischen reellen und komplexen Zahlen. *Monatsber. Akad. d. Wiss.*, Berlin, 1850. Reprinted in *Mathematische Werke*, vol. II, Chelsea Publ. Co., New York, 1975, 712–721.

1850   Kummer, E. E.*
Allgemeine Reciprocitätsgesetze für beliebig hohe Potenzreste. *Monatsber. Akad. d. Wiss.*, Berlin, 1850. 154–165.

1852   Kummer, E. E.
Über die Ergänzungssätze zu den allgemeinen Reciprocitätsgesetzen. *J. reine u. angew. Math.*, **44**, 1852, 93–146.

1857   Kummer, E. E.
Einige Sätze über die aus den Wurzeln der Gleichung $\alpha^\lambda = 1$ gebildeten complexen Zahlen, für den Fall dass die Klassenzahl durch $\lambda$ theilbar ist, nebst Anwendungen derselben auf einen weiteren Beweis des letztes Fermat'schen Lehrsatzes. *Math. Abhandl. Akad. d. Wiss.*, Berlin, 1857, 41–74.

1858   Kummer, E. E.
Über die allgemeinen Reciprocitätsgesetz der Potenzreste. *Monatsber. Akad. d. Wiss.*, Berlin, 1858, 158–171.

1859   Kummer, E. E.
Über die Ergänzungssätze zu den allgemeinen Reciprocitätsgesetzen. *J. reine u. angew. Math.*, **56**, 1859, 270–279.

1859   Kummer, E. E.
Über die allgemeinen Reciprocitätsgesetze unter den Resten und Nichtresten der Potenzen, deren Grad eine Primzahl ist. *Math. Abhandl. Akad. d. Wiss.*, Berlin, 1859, 19–159.

* The papers of Kummer are now easily accessible in Kummer's *Collected Papers*, vol. I, edited by A. Weil, Springer-Verlag, Berlin, 1975.

1870   Kummer, E. E.
Über eine Eigenschaft der Einheiten der aus den Wurzeln der Gleichung $\alpha^\lambda = 1$
gebildeten complexen Zahlen und über den Zweiten Faktor der Klassenzahl.
Monatsber. Akad. d. Wiss., Berlin, 1870, 855–880.

1887   Kummer, E. E.
Zwei neue Beweise der allgemeinen Reciprocitätsgesetze unter den Resten und
Nichtresten der Potenzen, deren Grad eine Primzahl ist. *J. reine u. angew. Math.*,
**100**, 1887, 10–50.

1897   Hilbert, D.
Die Theorie der algebraischen Zahlkörper. *Jahresber. d. Deutschen Math. Verein.*,
**4**, 1897, 175–546. Also in *Gesammelte Abhandlungen*, vol. I, Springer-Verlag,
Berlin, 1932. Reprinted by Chelsea Publ. Co., New York, 1965.

1898   Hilbert, D.
Über die Theorie der relativ-Abelschen Zahlkörper. Nachr. Ges. d. Wiss. zu
Göttingen, 1898, 370–379. Also in *Acta Math.*, **26**, 1902, 99–132. Reprinted in
*Gesammelte Abhandlungen* vol. I, Chelsea Publ. Co., New York, 1965, 483–509.

1904   Furtwängler, P.
Über die Reziprozitätsgesetze zwischen *l*ten Potenzresten in algebraischen
Zahlkörpern, wenn *l* eine ungerade Primzahl bedeutet. *Math. Annalen*, **58**, 1904,
1–50.

1907   Furtwängler, P.
Allgemeiner Existenzbeweis für den Klassenkörper eines beliebigen algebraischen
Zahlkörpers. *Math. Annalen*, **63**, 1907, 1–37.

1909   Furtwängler, P.
Die Reziprozitätsgesetze für Potenzreste mit Primzahlexponenten in algebraischen
Zahlkörpern, I. *Math. Annalen*, **67**, 1909, 1–31.

1910   Furtwängler, P.
Untersuchungen über die Kreisteilungskörper und den letzten Fermat'schen Satz.
Nachr. Akad. d. Wiss. Göttingen, 1910, 554–562.

1910   Hecke, E.
Über nicht-reguläre Primzahlen und den Fermatschen Satz. Nachr. Akad. d.
Wiss. Göttingen, 1910, 420–424.

1912   Furtwängler, P.
Die Reziprozitätsgesetze fur Potenzreste mit Primzahlexponenten in algebraischen
Zahlkörpern, II. *Math. Annalen*, **72**, 1912, 346–386.

1912   Furtwängler, P.
Letzter Fermatschen Satz und Eisensteins'ches Reziprozitätsgesetz. *Sitzungsber.
Akad. d. Wiss. Wien.*, Abt. IIa, **121**, 1912, 589–592.

1914   Vandiver, H. S.
A note on Fermat's last theorem. *Trans. Amer. Math. Soc.*, **15**, 1914, 202–204.

1919   Vandiver, H. S.
A property of cyclotomic integers and its relation to Fermat's last theorem. *Annals
of Math.*, **21**, 1919, 73–80.

1922   Fueter, R.
Kummer's Kriterien zum letzten Theorem von Fermat. *Math. Annalen*, **85**, 1922,
11–20.

1922   Takagi, T.
Über das Reciprocitätsgesetz in einem beliebigen algebraischen Zahlkörper.
*J. Coll. of Science, Imp. Univ. Tokyo*, **44**, art. 5, 1922.

1922 Takagi, T.
On the law of reciprocity in cyclotomic corpus. *Proc. Phys. Math. Soc. Japan*, **4**, 1922, 173–182.

†1923 Čebotarev, N. G.
Ob odnoi teoreme Hilberta (On a theorem of Hilbert). *Visti Ukrain. Akad. Nauk*, 1923, 3–7.

1924 Pollaczek, F.
Über die irregulären Kreiskörper der *l*-ten und *l*²-ten Einheitswurzeln. *Math. Zeits.*, **21**, 1924, 1–37.

1925 Artin, E. and Hasse, H.
Über den zweiten Ergänzungssatz sum Reziprozitätsgesetz der *l*-ten Potenzreste in Körper $k_\zeta$ der *l*-ten Einheitswurzeln und in Oberkörpern von $k_\zeta$. *J. reine u. angew. Math.*, **154**, 1925, 143–148. Reprinted in Hasse's *Mathematische Abhandlungen*, vol. I, W. de Gruyter, Berlin, 1975, 247–252.

1925 Hasse, H.
Über das allgemeine Reziprozitätsgesetz der *l*-ten Potenzreste in Körper $k_\zeta$ der *l*-ten Einheitswurzeln und in Oberkörpern von $k_\zeta$. *J. reine u. angew. Math.*, **154**, 1925, 96–109. Reprinted in *Mathematische Abhandlungen*, vol. I, W. de Gruyter, Berlin, 1975, 233–246.

1925 Vandiver, H. S.
A property of cyclotomic integers and its relation to Fermat's last theorem (2nd paper). *Annals of Math.*, **26**, 1925, 217–232.

1926 Hasse, H.
Bericht über neuere Untersuchungen und Probleme aus der Theorie der algebraischen Zahlköper (2 volumes) Part I: Jahresber. d. Deutschen Math. Verein. **35**, 1926, 1–55. Part Ia: Jahresber. d. Deutschen Math. Verein. **36**, 1927, 233–311. Part II: Jahresber. d. Deutschen Math. Verein. Supplement, 1930, 204 pages. Reprinted by Physica-Verlag, Würzburg, 1965.

1927 Artin, E.
Beweis des allgemeinen Reziprozitätsgesetzes. Hamburg Abhandl., **5**, 1927, 353–363. Reprinted in *Collected Papers*, Addison-Wesley, Reading, Mass., 1965, 131–141.

1927 Hasse, H.
Über das Reziprozitätsgesetz der *m*-ten Potenzreste. *J. reine u. angew. Math.*, **158**, 1927, 228–259. Reprinted in *Mathematische Abhandlungen*, vol. I, W. de Gruyter, Berlin, 1975, 294–325.

1927 Landau, E.
*Vorlesungen über Zahlentheorie*, vol. III, S. Hirzel, Leipzig, 1927. Reprinted by Chelsea Publ. Co., New York, 1969.

1927 Takagi, T.
Zur Theorie des Kreiskörpers. *J. reine u. angew. Math.*, **157**, 1927, 230–238.

1928 Morishima, T.
Über die Fermatsche Vermutung. *Proc. Imp. Acad. Japan*, **4**, 1928, 590–592.

1929 Vandiver, H. S.
Some theorems concerning properly irregular cyclotomic fields. *Proc. Nat. Acad. Sci. U.S.A.*, **15**, 1929, 202–207.

1929 Vandiver, H. S.
A theorem of Kummer's concerning the second factor of the class number of a cyclotomic field. *Bull. Amer. Math. Soc.*, **35**, 1929, 333–335.

1930   McDonnell, J.
New criteria associated with Fermat's last theorem. *Bull. Amer. Math. Soc.*, **36**, 1930, 553–558.

1930   Stafford, E. and Vandiver, H. S.
Determination of some properly irregular cyclotomic fields. *Proc. Nat. Acad. Sci. U.S.A.*, **16**, 1930, 139–150.

1930   Vandiver, H. S.
On the second factor of the class number of a cyclotomic field. *Proc. Nat. Acad. Sci. U.S.A.*, **16**, 1930, 743–749.

1932   Herbrand, J.
Sur les classes des corps circulaires. *J. Math. Pures et Appl.*, **11**, 1932, 417–441.

1932   Lehmer, D. H.
A note on Fermat's last theorem. *Bull. Amer. Math. Soc.*, **38**, 1932, 723–724.

1932   Morishima, T.
Über die Fermatsche Vermutung, VII. *Proc. Imp. Acad. Japan*, **8**, 1932, 63–66.

1932   Morishima, T.
Über die Fermatsche Vermutung, VIII. *Proc. Imp. Acad. of Japan*, **8**, 1932, 67–69.

1932   Scholz, A.
Über die Beziehung der Klassenzahlen quadratischer Zahlkörper zueinander. *J. reine u. angew. Math.*, **166**, 1932, 201–203.

1933   Morishima, T.
Über die Einheiten und Idealklassen des Galoisschen Zahlkörpers und die Theorie des Kreiskörper de $l^r$-ten Einheitswurzeln. *Jpn. J. Math.*, **10**, 1933, 83–126.

1933   Moriya, M.
Über die Fermatsche Vermutung. *J. reine u. angew. Math.*, **169**, 1933, 92–97.

1934   Grün, O.
Zur Fermatsche Vermutung. *J. reine u. angew. Math.*, **170**, 1934, 231–234.

1934   Morishima, T.
Über die Fermatsche Vermutung, XI. *Jpn. J. Math.*, **11**, 1934, 241–252.

1934   Vandiver, H. S.
Fermat's last theorem and the second factor in the cyclotomic class number. *Bull. Amer. Math. Soc.*, **40**, 1934, 118–126.

1935   Holzer, L.
Takagische Klassenkörpertheorie, Hassesche Reziprozitätsformel und Fermatsche Vermutung, *J. reine u. angew. Math.*, **173**, 1935, 114–124.

1935   Morishima, T.
Über die Fermatsche Vermutung, XII. *Proc. Imp. Acad. Japan*, **11**, 1935, 307–309.

1938   Vandiver, H. S.
On criteria concerning singular integers in cyclotomic fields. *Proc. Nat. Acad. Sci. U.S.A.*, **24**, 1938, 330–333.

1939   Vandiver, H. S.
On the composition of the group of ideal classes in a properly irregular cyclotomic field. *Monatshefte f. Math. u. Phy.* **48**, 1939, 369–380.

1939   Vandiver, H. S.
On basis systems for groups of ideal classes in a properly irregular cyclotomic field. *Proc. Nat. Acad. Sci. U.S.A.*, **25**, 1939, 586–591.

1941   Vandiver, H. S.
On improperly irregular cyclotomic fields. *Proc. Nat. Acad. Sci. U.S.A.*, **27**, 1941, 77–82.

1946   Inkeri, K.
Untersuchungen über die Fermatsche Vermutung. *Annales Acad. Sci. Fennicae*, Ser. A, I, 1946, No. **33**, 60 pages.

1948   Inkeri, K.
Some extensions of criteria concerning singular integers in cyclotomic fields. *Annales Acad. Sci. Fennicae*, Ser. A, I, 1948, No. **49**, 15 pages.

1949   Inkeri, K. A.
On the second case of Fermat's last theorem. *Ann. Acad. Sci. Fennicae*, Ser. A, I, 1949, No. **60**, 32 pages.

1954   Dénes, P.
Über irreguläre Kreiskörper. *Publ. Math. Debrecen*, **3**, 1954, 17–23.

1954   Dénes, P.
Über Grundeinheitssysteme der irregulären Kreiskörper von besonderen Kongruenzeigenschaften. *Publ. Math. Debrecen*, **3**, 1954, 195–204.

1955   Inkeri, K.
Über die Klassenzahl des Kreiskörpers der $l$ ten Einheitswurzeln. *Annales Acad. Sci. Fennicae*, Ser. A, I, 1955, No. **199**, 12 pages.

1956   Dénes, P.
Über den zweiten Faktor der Klassenzahl und den Irregularitätsgrad der irregulären Kreiskörper. *Publ. Math. Debrecen*, **4**, 1956, 163–170.

1958   Leopoldt, H. W.
Zur Struktur der $l$-Klassengruppe galoisscher Zahlkörper. *J. reine u. angew. Math.*, **199**, 1958, 165–174.

1964   Kubota, T. and Leopoldt, H. W.
Eine $p$-adische Theorie der Zetawerte, I. *J. reine u. angew. Math.*, **214/5**, 1964, 328–339.

1965   Eichler, M.
Eine Bemerkung zur Fermatschen Vermutung. *Acta. Arith.*, **11**, 1965, 129–131. (Errata) p. 261.

1966   Hasse, H.
Vandiver's congruence for the relative class number of the $p$th. cyclotomic field. *J. Math. Anal. and Applic.*, 15, 1966, 87–90.

1967   Brumer, A.
On the units of algebraic number fields. *Mathematika*, **14**, 1967, 121–124.

1968/9   Bertrandias, F.
Sur la Structure du $p$-Groupe des Classes du Corps Cyclotomique $\mathbb{Q}^p$ (d'après Leopoldt). Bordeaux Sém. Th. Nombers (J. Lesca), 1968/9, No. 7, 10 pages.

1968   Carlitz, L.
A congruence for the second factor of the class number of a cyclotomic field. *Acta Arith.*, **14**, 1968, 27–34. Corrigendum: *Acta Arith.* **16**, 1969, 437.

1968   Slavutskii, I. Š.
The simplest proof of Vandiver's theorem. *Acta Arithm.*, 15, 1968, 117–118.

1969   Kaplan, P.
Démonstration des lois de réciprocité quadratique et biquadratique. *J. Fac. Sci. Tokyo Univ.*, 1969, 115–145.

1971   Gandhi, J. M.
On Fermat's last theorem. *J. reine u. angew. Math.*, **250**, 1971, 49–55.

1972  Brückner, H.
Zum Beweis des ersten Falles der Fermatschen Vermutung fur pseudo-reguläre
Primzahlen *l* (Bemerkungen zur vorstehenden Arbeit von L. Skula). *J. reine u.
angew. Math.*, **253**, 1972, 15–18.

1972  Iwasawa, K.
*Lectures on p-adic L-Functions*, Annals of Math. Studies, Princeton Univ. Press,
Princeton, 1972.

1972  Skula, L.
Eine Bemerkung zu dem ersten Fall der Fermatschen Vermutung. *J. reine u.
angew. Math.*, **253**, 1972, 1–14.

1973  Metsänkylä, T.
A class number congruence for cyclotomic fields and their subfields. *Acta Arith.*,
**23**, 1973, 107–116.

1974  Hayashi, H.
On a simple proof of Eisenstein's reciprocity law. *Mem. Fac. Sci. Kyushu Univ.*,
Ser. A, **28**, 1974, 93–99.

1975  Brückner, H.
Zum Ersten Fall der Fermatschen Vermutung. *J. reine u. angew Math.*, **274/5**,
1975, 21–26.

1976  Ribet, K.
A modular construction of unramified $p$-extensions of $\mathbb{Q}(\mu_p)$. *Invent. Math.*, **34**,
1976, 151–162.

1977  Schinzel, A.
Abelian binomials, power residues and exponential congruences. *Acta Arith.*, **32**,
1977, 245–274.

1978  Washington, L. C.
Units of irregular cyclotomic fields. To appear.

1978  Washington, L. C.
A note on $p$-adic $L$-functions. *J. Number Theory* (to appear).

1979  Brückner, H. *Explizit Reziprozitätsgesetz und Anwendungen.* Univ. Essen. 1979,
83 pages.

# LECTURE X

# Fresh Efforts

In this lecture, I'll present results obtained by various new methods. My choice is rather encompassing. There are some attempts, which belong among those described in my Lecture IV, on the naïve approach. Others involve penetrating studies of the class group. And entirely new avenues are opening with ideas from the theory of algebraic functions.

Whether or not these methods will solve the problem, they at any rate stimulate interesting investigations.

## 1. Fermat's Last Theorem Is True for Every Prime Exponent Less than 125000

I announced in my first lecture that Fermat's last theorem is true for every prime exponent $p < 125000$. Though, as I said earlier in Lecture VIII, the first case holds for $p < 3 \times 10^9$.

Of course, the above verification has been made using modern computers. This is the result of the patient work of Wagstaff, but—as he assured me—not using the most modern computers available today. Instead, the majority of the programs were run on a ten-year-old IBM 360/65. He also used an IBM 370 computer, which is between three and five years old, in doing this work. These computers do not have either the speed or the features of the most advanced machines available today. But such advanced machines were not made available for this kind of work. Whatever computers used, the methods applied are quite reliable, so the outcome is not in question.

Before I start explaining the method, it is perhaps a good idea to review what was already done.

When Kummer proved his main theorem in 1847, and subsequently recognized that 37 is the smallest irregular prime, FLT was proved for $p < 37$. In his 1857 paper, Kummer proved a theorem which assured that FLT holds for a certain class of irregular primes (see Lecture VII). Applying his criterion to the irregular primes $p = 37, 59, 67$ less than 100, he concluded that FLT holds for all $p < 100$. However, in 1920 Vandiver noted that Kummer's proofs of 1857 contained various gaps. Vandiver succeeded in 1922, 1926 in repairing these imperfections guaranteeing already by 1926 that FLT holds for $p < 157$.

The prime 157, the smallest prime $p$ with $p^2$ dividing the first factor of the class number $h_p^*$, was not covered by Kummer's criterion.

In 1929, Vandiver published a long paper where he gave various new criteria. Let $\zeta$ be a primitive $p$th root of 1, $K = \mathbb{Q}(\zeta)$, and let $A$ be the ring of integers of $K$.

**(1A)** *Assume*:

I. *The second factor $h^+$ of the class number of $K$ is not a multiple of $p$.*
II. *None of the Bernoulli numbers $B_{2np}$ ($n = 1, 2, \ldots, (p-3)/2$) is a multiple of $p^3$.*

*Then the second case of FLT is true for the exponent $p$.*

The above criterion is not easily applicable because the computation of the second factor $h^+$ is so tricky.

The next criterion is more practical:

**(1B)** *Assume*:

III. *There exists one and only one index $2s$, $2 \le 2s \le p - 3$, such that $p$ divides $B_{2s}$.*
IV. *With the above index $2s$, $p^3$ does not divide $B_{2sp}$.*

*Then FLT holds for the exponent $p$.*

All irregular primes $p < 211$, except 157, are covered by the above criterion. The third of Vandiver's theorems was:

**(1C)** *If $p \equiv 1 \pmod 4$ and if $p \nmid B_{2s}$ for every odd index $s$, $2 \le 2s \le p - 3$, then FLT holds for the exponent $p$.*

This criterion is not satisfied by 157, because 157 divides $B_{62}$ and $B_{110}$. It is only his fourth theorem which applies to the prime 157. I introduce the following notation. Let $a_1, a_2, \ldots, a_t$ be the indices such $1 \le a_i \le (p-3)/2$ and $p \mid B_{2a_i}$. As in Lecture VII, let $g$ be a primitive root modulo $p$ and let $\sigma$ be the automorphism of $K = \mathbb{Q}(\zeta)$ such that $\sigma(\zeta) = \zeta^g$. Let

$$\delta = \sqrt{\frac{1 - \zeta^g}{1 - \zeta} \cdot \frac{1 - \zeta^{-g}}{1 - \zeta^{-1}}}$$

and

$$\theta_a = \prod_{i=0}^{(p-3)/2} \sigma^i(\delta^{g-2ia})$$

for $a = 1, 2, \ldots, (p-3)/2$.

**(1D)** *Assume*:

V. *There exists a prime* $l$, $l \equiv 1 \pmod{p}$, $l < p^2 - p$, *such that if* $1 \le a_i \le (p-3)/2$, $p|B_{2a_i}$, *then the unit* $\theta_{a_i}$ *is not congruent to the* $p$th *power of an integer of* $K = \mathbb{Q}(\zeta)$, *modulo* $L$, *where* $L$ *is a prime ideal dividing* $Al$.

*Then FLT holds for the exponent* $p$.

Using this theorem, Vandiver proved, in the same article, the more practical criterion:

**(1E)** *Assume*:

VI. *There exists a prime* $l$, $l \equiv 1 \pmod{p}$, $l \not\equiv 1 \pmod{p^2}$ *such that the congruence*

$$X^p + Y^p + Z^p \equiv 0 \pmod{l}$$

*has only the trivial solution.*

VII. *With the above prime* $l$, *if* $1 \le a_i \le (p-3)/2$, $p|B_{2a_i}$, *then*

$$\left\{\frac{\theta_{a_i}}{L}\right\} \ne 1,$$

*where* $L$ *is a prime ideal of* $K = \mathbb{Q}(\zeta)$ *dividing* $Al$ *and* $\{-\}$ *denotes the* $p$th *power residue symbol.*

*Then FLT holds for the exponent* $p$.

Finally this is the first of the criteria which may be applied to $p = 157$. Vandiver chose $l = 1571$, the primitive root $g = 139$ modulo 157, $a_1 = 31$, $a_2 = 55$ and $\operatorname{ind}(\theta_{31}) \equiv 150 \pmod{157}$, $\operatorname{ind}(\theta_{55}) \equiv 39 \pmod{157}$.

Vandiver also concluded that $h_p^+$ is not a multiple of $p$ for every $p < 211$.

In 1930, with the same methods, Vandiver proved FLT for every prime $p$, $211 \le p < 269$. This was extended in 1931 up to 307 and again in 1937, up to 617 by Vandiver and his assistants. At this point the computations became much too laborious for desk calculators.

In an important paper of 1954, Lehmer, Lehmer, and Vandiver introduced a new criterion more appropriate for computations.

I have already mentioned, in Lecture VI, how Stafford and Vandiver gave practical computational methods to decide whether a given prime is irregular; this remains after Kummer the only case which requires further work.

The following lemma is basic:

**Lemma 1.1.** *Let $l$ be a prime, $l = kp + 1 < p^2 - p$. Let $t$ be a natural number such that $t^k \not\equiv 1 \pmod{l}$. For $a \geq 1$, let*

$$d = \sum_{j-1}^{(p-1)/2} j^{p-2a} \quad and \quad Q_a = \frac{1}{t^{kd/2}} \prod_{i=1}^{(p-1)/2} (t^{ki} - 1)^{i^{p-1-2a}}.$$

*With the notation already introduced, the unit $\theta_a$ is congruent to the $p$th power of an integer in $K = \mathbb{Q}(\zeta)$, modulo a prime ideal $L$ dividing $Al$, if and only if $Q_a^k \equiv 1 \pmod{l}$.*

Using this lemma, Lehmer, Lehmer, and Vandiver proved the following criterion:

**(1F)** *Assume that $p$ is an irregular prime, let $a_i$, $1 \leq a_i \leq (p-3)/2$, be the indices such that $p \mid B_{2a_i}$. With above notations, if $2^k \not\equiv 1 \pmod{l}$ and $Q_{a_i}^k \not\equiv 1 \pmod{l}$ for all above indices $a_i$, then FLT holds for the exponent $p$.*

Using the SWAC calculating machine, the above authors proved that FLT holds for every prime exponent $p < 2003$. Soon after, Vandiver (1954) extended this to 2521.

This only began a long series of computations by various mathematicians:

Selfridge, Nicol, and Vandiver, in 1955—up to 4001,
Selfridge and Pollack, in 1967—up to 25000,
Kobelev, in 1970—up to 5500,
Johnson, in 1975—up to 30000,
Wagstaff, in 1975—up to 58150,
Wagstaff, in 1976—up to 100000,
Wagstaff, in 1977—up to 125000.

Much valuable experimental information about irregular primes has been obtained along with these computations. This I have described in Lecture VI.

In 1956, Inkeri extended the above criterion so as to guarantee that no solutions exist in integers of the real cyclotomic field. However, the proof is based on a result of Morishima (1935), which was criticized in Gunderson's thesis (1948).

## 2.  Euler Numbers and Fermat's Theorem

The Bernoulli numbers seem like cousins to other numbers, considered first by Euler in the series development of the secant function:

$$\sec x = 1 - \frac{E_2}{2!} x^2 + \frac{E_4}{4!} x^4 - \frac{E_6}{6!} x^6 + \cdots \quad \left( \text{convergent for } |x| < \frac{\pi}{2} \right). \quad (2.1)$$

The secant coefficients were later called *Euler numbers* by Scherk (1825). They have been extensively studied, just like the Bernoulli numbers.

For example, they satisfy the basic recurrence relation:

$$E_{2n} + \binom{2n}{2n-2}E_2 + \binom{2n}{2n-4}E_4 + \cdots + \binom{2n}{2}E_{2n-2} + 1 = 0. \quad (2.2)$$

Thus,

$$E_2 = -1, \quad E_4 = 5, \quad E_6 = -61, \quad E_8 = 1385, \quad E_{10} = -50521,$$
$$E_{12} = 2702765, \quad E_{14} = -199360981, \quad \text{etc.}$$

It is also customary to write

$$E_0 = 1, \quad E_{2n+1} = 0 \quad \text{for } n \geq 0.$$

It is easily seen that the Euler numbers are integers. Moreover it may be proved that they are, in fact, odd integers, with alternating signs, and their last digit is alternately 1, 5 (a fact which is irrelevant here).

More useful is to establish their relationship with the Bernoulli numbers. This was indicated by Scherk (1825):

$$E_{2k} = 1 - \sum_{i=1}^{k} \binom{2k}{2i-1} \frac{2^{2i}(2^{2i}-1)B_{2i}}{2i}$$

$$= \frac{1}{2k+1}\left[1 - \sum_{i=1}^{k}\binom{2k+1}{2i}2^{2i+1}(2^{2i-1}-1)B_{2i}\right]. \quad (2.3)$$

On the other hand,

$$B_{2k} = \frac{2k}{2^{2k}(2^{2k}-1)}\left[\sum_{i=1}^{k-1}\binom{2k-1}{2i}E_{2i} + 1\right]. \quad (2.4)$$

If I have to say only a few more words about Euler numbers, then I choose to give the formula which is analogous to the relation between the Bernoulli numbers and the values $\zeta(2k)$ of the Riemann zeta-function. This formula, also discovered by Euler, is the following:

$$\sum_{n=0}^{\infty}(-1)^n \frac{1}{(2n+1)^{2k+1}} = \frac{\pi^{2k+1}(-1)^k}{2^{2k+2}(2k)!}E_{2k}. \quad (2.5)$$

I will pass in silence over all the numerous identities, congruences and other properties of Euler numbers and I will just say a word about Euler polynomials: they are defined and studied in a way analogous to Bernoulli polynomials. The classical books by Saalschütz (1893), Nielsen (1923) and Nörlund (1923) contain much interesting information.

It is not all surprizing that the connection, via Kummer's theorem, between the primes dividing certain Bernoulli numbers and the truth of Fermat's theorem, would suggest a similar theorem using the Euler numbers.

A prime $p$ is called *E-regular* if $p$ does not divide the Euler numbers $E_2, E_4, \ldots, E_{p-3}$.

In 1940, Vandiver proved:

**(2A)** *If $p$ does not divide $E_{p-3}$, then the first case of Fermat's theorem holds for the exponent $p$.*

Thus, if $p$ is $E$-regular, the above conclusion is still true.

The basic fact used in the proof was a congruence obtained by E. Lehmer in 1938:

$$\sum_{j=1}^{[p/4]} \frac{1}{j^2} \equiv \sum_{j=1}^{[p/4]} j^{p-3} \equiv (-1)^{(p-1)/2} 4 E_{p-3} \pmod{p}. \tag{2.6}$$

In his paper of 1954 about irregular primes, Carlitz also showed:

**(2B)** *There exist infinitely many $E$-irregular primes $p$, i.e., $p \mid E_2 E_4 \cdots E_{p-3}$.*

As in the case of regular primes, it is not known whether there are infinitely many $E$-regular primes.

The distribution of $E$-irregular primes was the object of two recent papers, by Ernvall (1975) and Ernvall and Metsänkylä (1978). I quote the following results. The first one sharpens Carlitz's theorem.

**(2C)** *There exist infinitely many $E$-irregular primes $p$ such that $p \not\equiv \pm 1$ (mod 8).*

Let $\mathrm{ii}_E(p)$ denote the *E-irregularity index* of the prime $p$; that is, the number of integers $2n$, $2 \le 2n \le p - 3$, such that $p$ divides $E_{2n}$. Ernvall and Metsänkylä computed all $E$-irregular primes $p < 10^4$ and found examples of primes $p$ with $\mathrm{ii}_E(p) = 1, 2, 3, 4, 5$ in the range in question.

Let $x$ be a positive real number, let

$\pi(x) = $ number of primes $p$, $p \le x$

$\pi_E(x) = $ number of $E$-irregular primes $p$, $p \le x$

$\pi_{BE}(x) = $ number of primes $p$, $p \le x$, which are both irregular and $E$-irregular.

Under the heuristic assumption that the Euler numbers are randomly distributed modulo $p$, it follows that

$$\lim_{x \to \infty} \frac{\pi_E(x)}{\pi(x)} = 1 - \frac{1}{\sqrt{e}},$$

$$\lim_{x \to \infty} \frac{\pi_{BE}(x)}{\pi(x)} = 1 - \frac{2}{\sqrt{e}} + \frac{1}{e}.$$

The observed and predicted values of these ratios agree pretty well.

Gut derived, in 1950, Euler number criteria, analogous to those of Kummer for the first case, when the exponent is equal to $2p$, $p$ an odd prime.

He showed:

**(2D)** *If there exist* $x$, $y$, $z$ *such that* $p \nmid xyz$ *and* $x^{2p} + y^{2p} = z^{2p}$, *then* $p$ *divides* $E_{p-3}, E_{p-5}, E_{p-7}, E_{p-9}, E_{p-11}$.

These results, involving Euler numbers, should only be viewed as an alternate way of expressing equivalent facts involving Bernoulli numbers. These results are now superseded by Terjanian's theorem (page 68).

# 3. The First Case Is True for Infinitely Many Pairwise Relatively Prime Exponents

Up to now, it has not been possible to show that the first case of Fermat's theorem is true for infinitely many distinct prime exponents. The situation, however, is different for pairwise relatively prime exponents. Indeed, this was established by Maillet in 1897:

**(3A)** *Let* $p > 3$, *let* $h = p^u k$, *with* $u \geq 0$, $p \nmid k$, *be the class number of the cyclotomic field* $K = \mathbb{Q}(\zeta)$ ($\zeta^p = 1, \zeta \neq 1$). *Then if* $n \geq u + 1$ *and* $\alpha$, $\beta$, $\gamma$ *are nonzero integers in* $K$ *satisfying* $\alpha^{p^n} + \beta^{p^n} + \gamma^{p^n} = 0$, *it follows that* $\lambda = 1 - \zeta$ *divides* $\alpha$, $\beta$ *or* $\gamma$.

In particular, for $n \geq u + 1$ the first case of Fermat's theorem holds for the exponent $p^n$. In 1972, Hellegouarch gave a shorter proof for this theorem. In fact, his result applies for $n \geq t + 1$, where $p^t$ is the exponent of the $p$-primary component of the class group of $K$ (note that $u \geq t$, so this is an improvement over Maillet's result).

As a consequence, for every prime $p$ the first case holds for some exponent of the form $p^n$ and therefore, the first case is true for infinitely many pairwise relatively prime exponents.

It is worthwhile saying more about the smallest exponent $M(p)$ such that the first case is true for every $p^n$, $n \geq M(p)$. Hellegouarch treated this question by generalizing the theorems of Furtwängler (see Lecture IX) for prime-power exponents. As I have previously said, this had been achieved earlier by Moriya. With this method Hellegouarch proved, among other things, the following generalizations of the theorems of Wieferich and Mirimanoff:

**(3B)** *If* $p > 3$, $n \geq 1$ *and the order of* 2 *modulo* $p$ *is smaller than the order of* 2 *modulo* $p^{2n}$, *or the order of* 3 *modulo* $p$ *is smaller than the order of* 3 *modulo* $p^{2n}$, *then the first case holds for the exponent* $p^n$.

In other words, if the first case fails for the exponent $p^n$ then $2^{p-1} \equiv 1 \pmod{p^{2n}}$ and also $3^{p-1} \equiv 1 \pmod{p^{2n}}$.

These congruences are stronger than the congruence $2^{p-1} \equiv 1 \pmod{p^{n+1}}$ obtained by Inkeri as indicated in Lecture IX.

Thus the constant $M(p)$ is at most equal to the smallest integer $n$ such that $o(2 \bmod p) < o(2 \bmod p^{2n})$ or $o(3 \bmod p) < o(3 \bmod p^{2n})$.

Another result on the value of $M(p)$ has been published recently by Washington (1977). Following the method of Eichler and writing the first factor of the class number as $h^* = p^{u^*} k^*$, where $p \nmid k$ and $u^* \geq 0$, then:

(3C) *Let $p \geq 3$; if $n \geq \max\{1, u^* - \sqrt{p} + 3\}$ and $x$, $y$, $z$ are nonzero integers such that $x^{p^n} + y^{p^n} + z^{p^n} = 0$, then $p$ divides $xyz$.*

In some instances this result of Washington is better than Maillet's and Hellegouarch's. If $p = 3511$, then by Maillet's theorem $M(p) \leq 3$, Hellegouarch's theorem gives $M(p) \leq 2$, while Washington's theorem gives $M(p) = 1$.

For the exponents of the form $2^n$, Bohníček proved the following theorem in 1912:

(3D) *If $n \geq 3$, then there do not exist integers $\alpha$, $\beta$, $\gamma$, different from 0, in the cyclotomic field $\mathbb{Q}(\zeta_{2^n})$ (where $\zeta_{2^n}^{2^n} = 1$, $\zeta_{2^n}^{2^{n-1}} \neq 1$), such that*

$$\alpha^{2^{n-1}} + \beta^{2^{n-1}} + \gamma^{2^{n-1}} = 0.$$

For example, if $n = 3$, this theorem says more than Theorem 169 of Hilbert's *Zahlbericht*. Indeed it asserts that $X^4 + Y^4 + Z^4 = 0$ has only the trivial solution in $\mathbb{Q}(\zeta_8)$, and not merely in $\mathbb{Q}(i)$.

More recently, in 1964, Kapferer, quite unaware of Maillet's theorem, proved in a rather cumbersome way the existence of infinitely many pairwise relatively prime exponents for which the first case holds. His basic result is the following:

(3E) *Let $a \geq 2$, $n \geq 1$, $e \geq 2$ be integers such that $2 \nmid e$, $3 \nmid e$ and $a^n - 1 = ef$, where $\gcd(e, f) = 1$. If $x$, $y$, $z$ are nonzero pairwise relatively prime integers such that $x^e + y^e + z^e = 0$ and $a \mid x$, then $e \mid x$.*

The above result may be applied with $a = 2$. According to a theorem of Bang–Zsigmondy–Birkhoff and Vandiver (1886, 1892, 1904), each integer $2^n - 1$ (when $n > 6$) has a prime factor $p_n$ such that $p_n$ does not divide $2^m - 1$ for every $m < n$. Let $2^n - 1 = p_n^{r_n} f_n$, where $p_n \nmid f_n$ and $e_n = p_n^{r_n}$. Then, by Kapferer's theorem, the first case holds for all exponents $e_n$.

If it can be shown that there are infinitely many indices $n$ such that $r_n = 1$, then the first case would hold for infinitely many prime exponents $p_n$.

Odlyzko called my attention to a recent paper by Powell, where there is a very elementary proof of the following result:

(3F) *There exist infinitely many relatively prime exponents $n$ such that the equation $X^n + Y^n = Z^n$ has no solution in nonzero integers $x$, $y$, $z$ such that $\gcd(n, xyz) = 1$, that is, the first case holds for $n$.*

As a matter of fact, if $p$ is any prime and $n = p(p-1)/2$, then the first case holds for $n$.

The result (3F) is weaker than what was proved earlier, by analytic means by Ankeny (1952) and Ankeny and Erdös (1954). Namely:

**(3G)** *The density of the set of exponents for which the first case is false is necessarily zero.*

# 4. Connections between Elliptic Curves and Fermat's Theorem

Let $K$ be an algebraic number field (of finite degree over $\mathbb{Q}$). Let $\mathscr{C}$ be a (nonsingular) elliptic curve, with a fixed point 0, both defined over $K$. Let $\mathscr{G}$ be the abelian group of $K$-rational points of $\mathscr{C}$. The celebrated theorem of Mordell (1922) later generalized by Weil (1928), states that the group $\mathscr{G}$ is finitely generated. So the subgroup $\mathscr{F}$, consisting of the points of finite order, is finite.

More precisely, if $K = \mathbb{Q}$ and $P = (x,y)$ with $x$, $y \in \mathbb{Q}$ is a point of finite order of the curve $Y^2 = X^3 + aX + b$, now in Weierstrass form ($a$, $b \in \mathbb{Z}$, $4a^3 + 27b^2 \neq 0$), then $x$, $y \in \mathbb{Z}$ and $y = 0$ or $y^2$ divides $4a^3 + 27b^2$.

Much more is known and may be found in Cassels's excellent survey article (1966). I single out the following fact: $\mathscr{F}$ is either cyclic or the product of a cyclic group with the group of two elements.

Cassels also mentions the "folklore" conjecture, that there is a uniform bound for the orders of all points in $\mathscr{F}$, which depends only on the field of definition $K$ of $\mathscr{C}$, and not on $\mathscr{C}$.

In 1969, Manin proved the local analogue of the folklore conjecture:

**(4A)** *For every algebraic number field $K$ and every prime $p$, there exists a constant $M = M(p,K) > 0$ such that for every (nonsingular) elliptic curve $\mathscr{C}$, defined over $K$, the orders of points in $\mathscr{C}$, of order a power of $p$, are bounded by $M$.*

However, no upper bound for $M$ is effectively determined.

In 1971, Demjanenko published a "proof" of the following result:

**(4B)** *For every algebraic number field $K$ there exists a positive constant $M = M(K)$ such that for any elliptic curve $\mathscr{C}$, defined over $K$, the order of every point of finite order is at most $M$.*

However, Cassels' review in 1972 should be taken into account:

> The author purports to prove the long-standing conjecture that the Mordell–Weil rank of an elliptic curve (=abelian variety of dimension one)

over an algebraic number field is bounded by a constant depending only on the field. Unfortunately, the exposition is so obscure that the reviewer has yet to meet someone who would vouch for the validity of the proof; on the other hand, he has yet to be shown a mistake that unambiguously and irretrievably vitiates the argument . . . .

So, whether or not the above statement of Demjanenko is true, since $M(K)$ is not effective, the question still remains open: Which are the possible orders of the points of finite order of an elliptic curve over $K$?

At this point, I would like to mention the work of Ogg (1971) and his conjecture. Ogg associated with every $n \geq 1$ a curve $\mathscr{X}_n$, defined over $\mathbb{Q}$, with the following property: For each couple $(\mathscr{C}, P)$, where $\mathscr{C}$ is an elliptic curve defined over $K$, and $P$ is a $K$-rational point of $\mathscr{C}$ of order $n$, there corresponds a $K$-rational point of $\mathscr{X}_n$. Moreover, "almost" all points of $\mathscr{X}_n$ are so obtained.

The genus $g_n$ of $\mathscr{X}_n$ is computed to be; when $n \geq 5$:

$$g_n = 1 + \frac{\varphi(n)}{4}\left[\frac{n}{6}\prod_{p|n}\left(1 + \frac{1}{p}\right) - \sum_{d|n}\frac{\varphi(t(d))}{t(d)}\right],$$

where $\varphi$ denotes Euler's totient function and $t(d) = \gcd(d, n/d)$. Thus $g_n = 0$ if and only if $n \leq 10$, $n = 12$. Also $g_n = 1$ if and only if $n = 11, 14, 15$.

Ogg's conjecture was the following: If $g_n \geq 1$, no elliptic curve $\mathscr{C}$ defined over $K$ has a rational point of order $n$.

A recent paper by Mazur (1978) settles the problem, confirming Demjanenko's theorem, over the field $\mathbb{Q}$ (see Serre, 1977, 1978):

**(4C)** *The possible orders of points of finite order of an elliptic curve over* $\mathbb{Q}$ *are* $n = 1, \ldots, 10$ *and* $n = 12$.

This confirms Ogg's conjecture. The question remains open over general algebraic number fields.

Since these lectures focus on Fermat's theorem, I wish to report on Hellegouarch's work, and its surprising connection with Fermat's equation. Among his theorems, I mention:

**(4D)** *There exists an integer* $N_0 \geq 17$ *and an integer* $N \geq 4$ *(effectively computable) such that if* $p$ *is a prime,* $p \geq N_0$, *if* $h \geq 2$ *and if* $\mathscr{C}$ *is any (nonsingular) elliptic curve over* $\mathbb{Q}$ *with a rational point* $P$ *of order* $p^t$, *then for every* $u = 1, 2, \ldots, t$ *the equation*

$$\sum_{i=0}^{N} T_i^{p^{t-u}} = 0$$

*has at least* $(p^u - 1)/2$ *distinct solutions in* $\mathbb{Q}$ *(not counting the trivial solution* $(0, 0, \ldots, 0)$*).*

For elliptic curves having a rational point of order 2, the above results may be sharpened as follows:

**(4E)** *If $p > 3$ (respectively $p = 3$), and if the elliptic curve $\mathscr{C}$ over $\mathbb{Q}$ has a rational point of order $2p^t$ with $t \geq 1$ (respectively $t \geq 2$), then for $u = 1, 2, \ldots, t$ (respectively $u = 2, \ldots, t$) the equations*

$$T_1^{p^{t-u}} + T_2^{p^{t-u}} + T_3^{p^{t-u}} = 0$$

*and*

$$T_1^{p^{t-u}} + T_2^{p^{t-u}} + 2T_3^{p^{t-u}} = 0$$

*have at least $(p^u - 1)/2$ distinct solutions in $\mathbb{Z}$ [not counting the trivial solution $(0,0,0)$].*

The above results, which were obtained before Mazur's, may be viewed as giving information about the solutions of Fermat's equation and analogous equations.

On the other hand, without the newer results, they imply already the nonexistence of a rational point of order $2 \times 3^3 = 54$, as well as one of order $2^2 \times 3^2 = 36$. But they wouldn't be strong enough to rule out the existence of a point of order 18.

Hellegouarch also showed, with similar methods, that if $p > 3$, there are no points of $\mathscr{C}$ (defined over $\mathbb{Q}$) of order $2p^2$. But again, this has been superseded by the recent more powerful result of Mazur.

# 5. Iwasawa's Theory

At this point, I'm tempted to present some material which is very attractive and goes deep into the study of cyclotomy. I don't know whether this theory will have any direct connection with Fermat's theorem. I consulted Iwasawa on this matter, but he wasn't sure either.

The idea is simply to take, for a given prime $p$, the tower of cyclotomic extensions $K_n = \mathbb{Q}(\zeta_{p^{n+1}})$, where $\zeta_{p^{n+1}}$ is a primitive $p^{n+1}$th root of 1 and to examine these fields, together with some associated Galois groups and their ideal class groups. Since $p$ is prominent, the $p$-adic valuation plays an important role. Moreover, it is only natural to consider the $p$-primary components of the ideal class groups. Passing to the limit, the description becomes smoother and more natural. There will be certain numerical invariants attached to the whole system, which ultimately depend only on the given prime and reflect its special properties.

I may trace two classical theorems as roots of this theory. First, let $p = 2$. In 1886, Weber proved (see also his *Algebra*, volume II, 1899):

**(5A)** *For $p = 2$, and every $n \geq 1$, the class number of $\mathbb{Q}(\zeta_{2^n})$ divides the* *class number of* $\mathbb{Q}(\zeta_{2^{n+1}})$ *and the quotient is odd. In particular, for every* $n \geq 1$, *the class number of* $\mathbb{Q}(\zeta_{2^n})$ *is odd.*

In 1911, Furtwängler extended Weber's theorem for the case of odd primes:

**(5B)** *Let* $p$ *be an odd prime, let* $n \geq 1$. *The class number of* $K_n$ *is a multiple of* $p$ *if and only if the class number of* $K_0$ *is a multiple of* $p$.

In other words, if $p$ is a regular prime, then $p$ does not divide $h(K_n)$ for every $n \geq 1$. However, if $p$ is irregular, let $e_n \geq 1$ satisfy $p^{e_n} \| h(K_n)$. How does $e_n$ behave with respect to $n$? This question is answered by Iwasawa's theory, which, also applies to other more general situations.

Serre's lecture at Bourbaki's Seminar (December 1958) is a very good guide to this theory. As usual, he is crystal clear, and I can't do better than to follow his lead. However, due to lack of space, I'll not enter into too many details.

Let $\mathbb{Q} \subset K_0 \subset K_1 \subset \cdots \subset K_n \subset \cdots$ be the tower of cyclotomic fields, with $K_n = \mathbb{Q}(\zeta_{p^{n+1}})$, and $K_\infty = \bigcup_{n=0}^\infty K_n$.

Let $G_n = \mathrm{Gal}(K_n|\mathbb{Q})$, so $G_n \cong (\mathbb{Z}/p^{n+1})^\times$, the multiplicative group of invertible residue classes modulo $p^{n+1}$. Since $p \neq 2$, $G_n$ is a cyclic group; it has order $p^n(p-1)$. Let $\Gamma_n = \mathrm{Gal}(K_n|K_0)$, so $\Gamma_n$ is a cyclic group of order $p^n$. The Galois group $\Gamma_\infty = \mathrm{Gal}(K_\infty|K_0)$ is the inverse limit of the groups $\Gamma_n$ (relative to the natural restriction of automorphisms). $\Gamma_\infty$ is a compact totally disconnected group, which is in fact isomorphic (both algebraically and topologically) to the additive group of $p$-adic integers: $\Gamma_\infty \cong \hat{\mathbb{Z}}_p$. As such, $\Gamma_\infty$ is a (topologically) cyclic group.

Another aspect of the situation concerns the class group. Let $\mathscr{C}(K_n)$ denote the class group of $K_n$; this is an abelian group of order $h(K_n)$. Let $\mathscr{C}_p(K_n)$ be its $p$-primary component. If $h(K_n) = p^{e_n} k_n$, with $p \nmid k_n$, then $\mathscr{C}_p(K_n)$ has order $p^{e_n}$. The group $\Gamma_n$ acts on the class group $\mathscr{C}(K_n)$ and also on $\mathscr{C}_p(K_n)$.

For every $n \geq 1$, taking norms of ideals, gives a homomorphism from $\mathscr{C}_p(K_{n+1})$ into $\mathscr{C}_p(K_n)$, which is obviously compatible with the operations of the Galois groups $\Gamma_{n+1}$, $\Gamma_n$. Let $\mathscr{C}_p$ be the inverse limit of the abelian $p$-groups $\mathscr{C}_p(K_n)$, with respect to the norm homomorphisms.

Since $\Gamma_n$ operates on $\mathscr{C}_p(K_n)$, the group algebra over the ring of $p$-adic integers, $A_n = \hat{\mathbb{Z}}_p[\Gamma_n]$ also operates on $\mathscr{C}_p(K_n)$, that is, $\mathscr{C}_p(K_n)$ is a module over $A_n$. If $A = \varprojlim A_n$ then $\mathscr{C}_p = \varprojlim \mathscr{C}_p(K_n)$ is also an $A$-module. The first structure theorem asserts:

**(5C)** *For every odd prime* $p$, *there exist integers* $\lambda$, $\mu$, $\nu$ *and* $n_0 \geq 0$ *such that if* $n \geq n_0$, *then*

$$e_n = \lambda n + \mu p^n + \nu.$$

The integers $\lambda$, $\mu$, $\nu$ depend only on the prime $p$.

The proofs of the above theorems of Iwasawa (1959) essentially use class field theory and I refer to Serre's "exposé".

If $\mu \neq 0$, the growth of $e_n$ is very rapid, so it is interesting to know whether this may happen. Iwasawa has conjectured that for every $p \geq 3$, the corresponding $\mu_p = 0$.

In successive papers, Iwasawa and Sims (1966) showed that $\mu_p = 0$ for all primes $p \leq 4001$. In 1973, Johnson established the same for $p < 8000$, in 1976, he went to 30000 and in 1977, Wagstaff pushed it further to $p < 125000$. Some theoretical work by Metsänkylä (1974) gave the estimate $\mu_p < (p-1)/2$, improving a previous bound of Iwasawa. Finally, in a 1978 paper, Ferrero and Washington proved that $\mu_p = 0$. Actually, the same result holds also when the ground field is any abelian extension of finite degree of the rational field.

There is a similar theory concerning the growth of the $p$-primary part of the first factor of the class number of $K_n = \mathbb{Q}(\zeta_{p^{n+1}})$, which was also developed by Iwasawa, using $p$-adic $L$-functions (1972).

The theory may be generalized for certain ground fields different from $\mathbb{Q}$ and some other types of extensions, called $\Gamma$-extensions.

There is now a flurry of research in this area. However, as yet, no explicit connection has been made with Fermat's last theorem.

# 6. The Fermat Function Field

Fermat's equation $X^n + Y^n + Z^n = 0$ (for $n \geq 3$) defines a curve $\mathscr{F}_n$, in the projective plane over the field $\bar{\mathbb{Q}}$ of all algebraic numbers. Hasse was the first to investigate the rational points of $\mathscr{F}_n$, in the light of the theory of algebraic functions.

If $P$ is a point of $\mathscr{F}_n$, let its coordinates be $P = (x(P), y(P), z(P))$ with $x(P)$, $y(P)$, $z(P) \in \bar{\mathbb{Q}}$ (they are defined up to a proportionality factor). Let $\mathscr{F}_n^0$ be the affine portion of the curve $\mathscr{F}_n$, consisting of all points $P$ such that $z(P) \neq 0$. Let $\xi, \eta$ be the functions from $\mathscr{F}_n^0$ to $\bar{\mathbb{Q}}$ defined by $\xi(P) = x(P)/z(P)$, $\eta(P) = y(P)/z(P)$. Let $\zeta$ be a primitive $2n$th root of 1, $K_0 = \mathbb{Q}(\zeta)$, and $K = K_0(\xi, \eta)$. This is the quotient field of the ring $K_0[T, U]$ (where $T$, $U$ are indeterminates) modulo the prime ideal generated by $T^n + U^n + 1$. $K$ is called the function field of the affine curve $\mathscr{F}_n^0$. $\xi$ is transcendental over $K_0$ and $\eta = \sqrt[n]{-(1 + \xi^n)}$, so it is algebraic over $K_0(\xi)$. Thus $K$ is an algebraic function field of one variable, having field of constants equal to $K_0$. This last statement reflects the fact that $T^n + U^n + 1$ is an absolutely irreducible polynomial over $K_0$.

The arithmetic theory of algebraic function fields of one variable, when the constant field is not necessarily the field of complex numbers, was developed by Hasse and Chevalley, among others.

The arithmetic Riemann surface $\mathfrak{S}$ of $K$ is the collection of all prime divisors $\not\!p$ of $K \mid K_0$. These may be viewed as symbols which correspond bijectively to the valuations of $K$, which are trivial on $K_0$. I write $w_{\not\!p}$ for the valuation corresponding to the divisor $\not{p}$. Let $\mathcal{O}_{\not{p}}$ be the valuation ring of $w_{\not{p}}$, that is, the ring of all elements $\alpha \in K$ such that $w_{\not{p}}(\alpha) \geq 0$. The ring $\mathcal{O}_{\not{p}}$ has a unique maximal ideal, $\mathcal{M}_{\not{p}} = \{\alpha \in K | w_{\not{p}}(\alpha) > 0\}$. The quotient field $K_{\not{p}} = \mathcal{O}_{\not{p}}/\mathcal{M}_{\not{p}}$ is an algebraic extension of finite degree of $K_0$. This degree $f_{\not{p}} = [K_{\not{p}} : K_0]$ is called the degree of $w_{\not{p}}$, or of the divisor $\not{p}$. The residue class of $\alpha \in \mathcal{O}_{\not{p}}$ modulo $\mathcal{M}_{\not{p}}$ is written $\alpha(\not{p})$, and it is called the value of $\alpha$ at the divisor $\not{p}$.

It may be shown that the prime divisors of $K | K_0$ correspond to the points $P$ of the affine Fermat curve $\mathscr{F}_n^0$. Those of degree 1 correspond to the points $P$ with coordinates $x(P)/z(P)$, $y(P)/z(P)$ in $K_0$. More generally, each prime divisor $\not{p}$ corresponds to a point $P$ with coordinates in $K_{\not{p}}$.

The free abelian group $\mathscr{D}$ generated by all the prime divisors is called the group of divisors. So the divisors are the formal products $\mathfrak{a} = \prod_{\not{p}} \not{p}^{a(\not{p})}$ with $a(\not{p}) \in \mathbb{Z}$ and $a(\not{p}) = 0$ except for finitely many prime divisors $\not{p} \in \mathfrak{S}$. The degree of $\mathfrak{a}$ is, by definition, $\deg(\mathfrak{a}) = \sum_{\not{p}} a(\not{p})\deg(\not{p})$.

The unit divisor is $\mathfrak{e} = \prod_{\not{p}} \not{p}^{e(\not{p})}$ with $e(\not{p}) = 0$ for every $\not{p}$. The divisor $\mathfrak{a}$ is called integral if $a(\not{p}) \geq 0$ for every $\not{p}$. The divisor $\mathfrak{b}$ divides the divisor $\mathfrak{a}$ if there exists an integral divisor $\mathfrak{c}$ such that $\mathfrak{bc} = \mathfrak{a}$. With obvious notation, this amounts to saying $b(\not{p}) \leq a(\not{p})$ for every $\not{p}$.

Any divisors $\mathfrak{a}_1, \ldots, \mathfrak{a}_n$ have a unique greatest common divisor $\mathfrak{d} = \gcd(\mathfrak{a}_1, \ldots, \mathfrak{a}_n)$, with $d(\not{p}) = \min\{a_1(\not{p}), \ldots, a_n(\not{p})\}$ for every $\not{p}$. They also have a unique least common multiple. If $\gcd(\mathfrak{a}_1, \ldots, \mathfrak{a}_n) = \mathfrak{e}$ (unit divisor) then the (integral) divisors $\mathfrak{a}_1, \ldots, \mathfrak{a}_n$ are said to be relatively prime.

For each element $\alpha \in K$, $\alpha \neq 0$, it is true that $w_{\not{p}}(\alpha) = 0$ except for finitely many prime divisors $\not{p}$. Then $\operatorname{div}(\alpha) = \prod_{\not{p}} \not{p}^{w_{\not{p}}(\alpha)}$ is a divisor, called the divisor of $\alpha$. If $\not{p}_1, \ldots, \not{p}_m$ $(m \geq 0)$ are the prime divisors such that $w_{\not{p}_i}(\alpha) > 0$ and if $\not{p}_{m+1}, \ldots, \not{p}_n$ $(n \geq m)$ are those such that $w_{\not{p}_i}(\alpha) < 0$, then $\operatorname{div}(\alpha) = \mathfrak{a}/\mathfrak{b}$, where $\mathfrak{a} = \not{p}_1^{w_{\not{p}_1}(\alpha)} \cdots \not{p}_m^{w_{\not{p}_m}(\alpha)}$, $\mathfrak{b} = \not{p}_{m+1}^{-w_{\not{p}_{m+1}}(\alpha)} \cdots \not{p}_n^{-w_{\not{p}_n}(\alpha)}$. Thus $\mathfrak{a}$, $\mathfrak{b}$ are integral divisors, $\gcd(\mathfrak{a},\mathfrak{b}) = \mathfrak{e}$, $\mathfrak{a}$ is called the divisor of zeroes, $\mathfrak{b}$ the divisor of poles of $\alpha$. It is true that $\mathfrak{a}$, $\mathfrak{b}$ have the same degree, i.e., $\deg(\operatorname{div}(\alpha)) = 0$.

In particular, considering $\xi$, $\eta \in K$, I may write $\operatorname{div}(\xi) = \mathfrak{a}/\mathfrak{c}$, $\operatorname{div}(\eta) = \mathfrak{b}/\mathfrak{c}$ with $\mathfrak{a}$, $\mathfrak{b}$, $\mathfrak{c}$ integral divisors and $\gcd(\mathfrak{a},\mathfrak{b},\mathfrak{c}) = \mathfrak{e}$. Moreover, $\deg(\mathfrak{c}) = n$ because the field extension $K | K_0(\xi)$ has degree $n$. So $\deg(\mathfrak{b}) = n$ and I may write $\mathfrak{b} = \not{p}_1 \cdots \not{p}_n$, where the $\not{p}_i$ are prime divisors (a priori not necessarily distinct). From the relation $\xi^n + \eta^n + 1 = 0$ it follows that $\operatorname{div}(\xi^n + 1) = \mathfrak{b}^n/\mathfrak{c}^n = (\not{p}_1^n/\mathfrak{c}) \times \cdots \times (\not{p}_n^n/\mathfrak{c})$.

On the other hand, from

$$\xi^n + 1 = (\xi - \zeta_1)(\xi - \zeta_2) \cdots (\xi - \zeta_n),$$

where $\zeta_i^n = -1$ for $i = 1, 2, \ldots, n$, writing $\operatorname{div}(\xi - \zeta_i) = \mathfrak{a}_i/\mathfrak{c}$ gives $\deg(\mathfrak{a}_i) = \deg(\mathfrak{c}) = n$, and $\mathfrak{a}_1 \cdots \mathfrak{a}_n = \not{p}_1^n \cdots \not{p}_n^n$. After renumbering, $\mathfrak{a}_i = \not{p}_i^n$, so $\deg(\not{p}_i) = 1$ and hence $\not{p}_1, \ldots, \not{p}_n$ are distinct.

The prime divisors of degree 1 correspond to the points $P_i$ with coordinates $(\zeta_i, 0, 1)$ belonging to $K_0 = \mathbb{Q}(\zeta)$. In the same way, other trivial solutions $P_i'$, with coordinates $(0, \zeta_i, 1)$ correspond to the prime divisors dividing $\mathfrak{a}$.

Searching for nontrivial solutions of Fermat's equation in $K_0$ corresponds to discovering other prime divisors of degree 1. The idea behind this approach to the problem is to make use of the vast arithmetic theory of algebraic functions and see whether some of its powerful theorems would imply or exclude nontrivial solutions. This approach has as yet produced no conclusive results. I want however to describe what has been achieved, without attempting to explain all the terms and facts used.

The differential $d(1/\xi) = -d\xi/\xi^2$ has the divisor

$$\operatorname{div}\left(d\left(\frac{1}{\xi}\right)\right) = \frac{\operatorname{diff}(K\,|\,K_0(\xi))}{\mathfrak{a}^2}.$$

The numerator is the different of the extension $K\,|\,K_0(\xi)$. Because $\eta = \sqrt[n]{-(1 + \xi^n)}$, the prime divisors $\not{p}_i$, with $\mathfrak{b} = \not{p}_1 \cdots \not{p}_n$, are the only ramified primes; so $\operatorname{diff}(K\,|\,K_0(\xi)) = \mathfrak{b}^{n-1}$.

The genus $g$ is computed by the formula: $2g - 2 = $ degree of $\operatorname{div}(d(1/\xi))$. So $2g - 2 = (n - 1)\operatorname{deg}(\mathfrak{b}) - 2\operatorname{deg}(\mathfrak{a}) = n(n - 3)$ and therefore

$$g = \frac{(n - 1)(n - 2)}{2}.$$

Thus if $n = 2$, then $g = 0$, so the Fermat curve is a rational curve. If $n = 3$, then $g = 1$ and the associated field $K$ is an elliptic function field. For $n \geq 4$, $g \geq 2$. It is known that, for any divisor $\mathfrak{a}$, the set $\{\alpha \in K\,|\,\operatorname{div}(\alpha)\mathfrak{a}$ is an integral divisor$\}$ is a vector space of finite dimension over $K_0$. This dimension is, by definition, called the dimension of the divisor $\mathfrak{a}$.

Now, I'll discuss the Weierstrass points of the Fermat function field. Let $\mathfrak{d} = \operatorname{div}(d(1/\xi))$ and let $\not{p}$ be any prime divisor. I consider the sequence of integers

$$1 = r_1 < r_2 < \cdots < r_g \leq 2g - 1$$

which is defined as follows. From the Riemann–Roch theorem, $\dim(\mathfrak{d}) = g$. Let

$$\dim\left(\frac{\mathfrak{d}}{\not{p}^{r_1}}\right) = \cdots = \dim\left(\frac{\mathfrak{d}}{\not{p}^{r_2 - 1}}\right) = g - 1,$$

$$\dim\left(\frac{\mathfrak{d}}{\not{p}^{r_2}}\right) = \cdots = \dim\left(\frac{\mathfrak{d}}{\not{p}^{r_3 - 1}}\right) = g - 2,$$

$$\vdots$$

$$\dim\left(\frac{\mathfrak{d}}{\not{p}^{r_g}}\right) = \cdots = \dim\left(\frac{\mathfrak{d}}{\not{p}^{2g - 1}}\right) = 0.$$

The weight of $\not{p}$ is by definition

$$W(\not{p}) = \sum_{i=1}^{g} (r_i - i) = \sum_{i=1}^{g} r_i - \tfrac{1}{2}g(g + 1).$$

Hence $W(\not{p}) = 0$ if and only if $r_i = i$ for $i = 1, \ldots, g$.

The prime divisor $\not\!\!\!\!/$ is called a *Weierstrass point* if $W(\not\!\!\!\!/) > 0$. The theory tells that $W(\not\!\!\!\!/) < (g - 1)g$ when $g > 1$, and that there are only finitely many, say $N$, Weierstrass points. The sum of their weights is $W = \sum_{\not\!\!\!\!/} W(\not\!\!\!\!/) = (g - 1)g(g + 1)$.

If $g > 1$, then $g + 1 < N \le (g - 1)g(g + 1)$. In case the exponent in Fermat's equation is $n > 3$, so $g > 1$, then the 3n trivial solutions of $X^n + Y^n + Z^n = 0$ correspond to Weierstrass points, all with the same weight

$$\frac{g(n - 3)(n + 4)}{12} = \frac{(g - 1)g(n + 4)}{6n}$$

so the sum of these weights is $W_0 = \frac{1}{2}(g - 1)g(n + 4)$.

If $n = 4$, $g = 3$, then $W_0 = W$ and the above are therefore all the Weierstrass points.

However, if $n > 4$ there are many more Weierstrass points, the sum of their weights being equal to $W_0 = \frac{1}{2}(g - 1)g(2g - n - 2)$. The corresponding points of Fermat's curve have to be distinguished points (with coordinates algebraic, but not necessarily in $\mathbb{Q}(\zeta)$). Therefore, it should be of interest to determine the Weierstrass points.

# 7. Mordell's Conjecture

Let $K$ be an algebraic number field and let $\mathscr{C}$ be a nonsingular projective curve over $K$, with genus $g \ge 2$. In 1922, Mordell gave his now famous conjecture: The set of points of $\mathscr{C}$ which are $K$-rational (that is, have coordinates in $K$) is necessarily finite.

For example, if this turns out to be true, taking $K = \mathbb{Q}$ and the Fermat curve $\mathscr{F}_n: X^n + Y^n + Z^n = 0$, with $n \ge 4$, then there would be only finitely many solutions of Fermat's equation in rational numbers, or equivalently, in integers. This is not as strong as what Fermat's last theorem asserts. Still less so, if no bound on the number or size of the solutions is obtained. At any rate, for the moment a proof of Mordell's conjecture seems, with good reason, very remote.

I will report here on some work done recently to connect Mordell's conjecture and Fermat's equation.

In 1972, Hellegouarch noted that if $f(X,Y)$ is a polynomial with coefficients in $K$ and the equation $f(X,Y) = 0$ has only finitely many solutions $(x_i,y_i)$, with $x_i, y_i \in K$ then, for any integer $m \ge 2$, there exists an integer $h_0 \ge 1$ such that for every $h \ge h_0$ $f(X^{m^h}, Y^{m^h}) = 0$ has only the trivial solution in $K$. This remark led to the formulation of the following weaker Mordell conjecture:

Let $K$ be an algebraic number field, let $p \ge 2$ be a prime, $f(X,Y) \in K[X,Y]$. If there exists $h \ge 1$ such that the curve defined by $f(X^{p^h}, Y^{p^h}) = 0$ has genus

$g \geq 2$, then there exists $k \geq h$ such that $f(X^{p^k}, Y^{p^k}) = 0$ has only the trivial solution in $K$.

For example, if this conjecture is true, then it would give at once a weaker form of Maillet's theorem of 1897. Namely, let $K = \mathbb{Q}$, let $f(X,Y) = X + Y - 1$ and $p \geq 5$, $h = 1$. Then the curve of equation $X^p + Y^p - 1 = 0$ has genus $g = (p-1)(p-2)/2 \geq 6$. So there would exist $k \geq 1$ such that the equation $X^{p^k} + Y^{p^k} - 1 = 0$ has only the trivial solution, as Maillet proved. Hellegouarch gave some more support to this weaker conjecture.

In 1965, Mumford examined the countable set of $K$-rational points of a nonsingular projective curve $\mathscr{C}$ of genus $g \geq 2$. Short of proving that there are only finitely many solutions, Mumford wrote the solutions in order, by increasing height: $P_1, P_2, P_3 \cdots$ and he was able to prove:

**(7A)** *There are real numbers $a$, $b$, with $a > 0$, such that the height of $P_i$ satisfies*: $ht(P_i) \geq e^{ai+b}$.

The concept of height of a point is explained in Mumford's paper. In the special case of Fermat's curve and $K = \mathbb{Q}$, this result becomes:

**(7B)** *Let $n \geq 4$ and let $(x_i, y_i, z_i)$ be an infinite set of distinct positive integral solutions of $X^n + Y^n = Z^n$, such that $\gcd(x_i, y_i, z_i) = 1$ and $z_1 < z_2 < z_3 < \cdots$. Then there exist real numbers $a$, $b$, $a > 0$, such that $z_i \geq \exp\{\exp(ai + b)\}$.*

To conclude this brief excursion into Mordell's conjecture, I would like to mention the work initiated by Lang and enhanced by Kubert and Rohrlich (1975, 1977). Kubert and Lang have already published three papers, in what is announced to be a long series. The papers are very sophisticated and I dare not explain their contents. Perhaps, a paragraph which is indicative of the tenor of the first paper should be quoted here:

> ... given a projective nonsingular curve $V$ and an affine open subset $V_0$, does there exist an unramified covering $W$ of $V$, and units $u$, $v$ in the corresponding affine coordinate ring of $W$, such that $u + v = 1$? Whenever this is the case, we get a correspondence

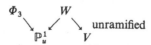

> which reduces the study of integral points on $V$ to integral points on [the Fermat curve] $\Phi_3 \cdots$. We see that the Fermat curve $\Phi_n$ has such a correspondence with $\Phi_3$.

Rohrlich considers the group $\mathscr{D}^\infty$ of divisors of the function field of Fermat's equation which are on the points at infinity, that is, the trivial solutions of $X^n + Y^n + Z^n = 0$ ($n \geq 3$).

Let $\mathscr{F}^{\infty}$ be the subgroup of principal divisors. In his thesis, Röhrlich computes the quotient group, arriving at the following result:

$$\mathscr{D}^{\infty}/\mathscr{F}^{\infty} \cong \begin{cases} (\mathbb{Z}/n)^{3n-7} & \text{if } n \text{ is odd,} \\ (\mathbb{Z}/n)^{3n-7} \times \mathbb{Z}/2 & \text{if } n \text{ is even.} \end{cases}$$

It is not explicitly stated where these studies will lead, and for this we all have to wait and see.

## 8. The Logicians

What do the logicians say about Fermat's last theorem? They are concerned about the truth, the possibility of proving the theorem starting from a given set of axioms, and about the question of undecidability. I wish to explain briefly the meaning of these expressions, avoiding any technicalities.

Fermat's last theorem concerns a sentence of the *universal* type, because all its variables are preceded by the quantifier "for every": "For every $x$, for every $y$, for every $z$, for every $n$: either $xyz = 0$ or $x^{n+3} + y^{n+3} \neq z^{n+3}$." It is understood that this sentence is to be interpreted with $x$, $y$, $z$, $n$ as natural numbers (including zero). If the sentence is true with this interpretation, then Fermat's last theorem is true.

The negation of this sentence is an *existential* sentence: "There exists $x$, there exists $y$, there exists $z$, there exists $n$, such that: $xyz \neq 0$ and $x^{n+3} + y^{n+3} = z^{n+3}$."

The axioms of Peano's arithmetic involve the symbol 0, the unary operation $\sigma$ (successor), and the binary operations addition and multiplication. They are very well-known, so I need not repeat them here. Let me just recall that one of the axioms, the principle of finite induction, is in fact an infinite set of axioms, namely one axiom for each formula in the formal language.

The set $\mathbb{N}$ of natural numbers, with the symbol 0 interpreted as the number "zero," the operations $+$, $\cdot$, as addition and multiplication of natural numbers, and the successor operation interpreted as the successor of a natural number, becomes a model for Peano arithmetic. This means that every axiom of Peano arithmetic is true in $\mathbb{N}$, with the above interpretation.

More generally, any set $A$, with a constant, two binary operations, one unary operation, such that Peano's axioms are true in $A$, with an appropriate interpretation, is also called a model of Peano arithmetic.

The theory of Peano arithmetic consists of all sentences which may be proved (with the usual rules of inference) from the axioms of Peano arithmetic. Clearly, every theorem of Peano arithmetic is true in any model of Peano arithmetic. In other words, if a sentence may be proved "syntactically" (i.e., with the rules of inference) from the axioms, all its interpretations in any model are true, so it is "semantically" true.

In 1934, Skolem provided specific models of Peano's arithmetic which are not isomorphic to $\mathbb{N}$, but have exactly the same theorems as $\mathbb{N}$. As one says, they are elementary equivalent to $\mathbb{N}$. Such models are called non-standard models of $\mathbb{N}$.

A theory is *consistent* if, for any sentence $S$ in the language of the theory, it is not possible that both $S$ and its negation $\neg S$ be theorems. If a theory has a model then it is necessarily consistent.

On the other hand, a theory is *complete* if, for any sentence $S$ in the language of the theory, either $S$ is a theorem or its negation $\neg S$ is a theorem. Thus, if a theory is complete, any sentence $S$ in the language of the theory is either true in every model, or false in every model of the theory.

One of the most fundamental theorems in logic is Gödel's completeness theorem: *If a theory is consistent, then it has a model*. Another way of expressing this theorem is the following: *Every consistent theory may be extended to a consistent and complete theory*. Still another way to put it, closer to our concern, is the following: *A sentence is a theorem in a given theory, that is, it may be proved from the axioms if (and only if) it is true in every model of the theory*.

If the theory is not complete, there are sentences $S$ which are true in some model and false (that is, $\neg S$ is true) in some other model. Any such sentence is called an *undecidable* sentence. Both $S$ and $\neg S$ cannot be proved from the axioms.

Considering specifically the theory of Peano arithmetic, Gödel proved that this theory is incomplete. And much more. In fact, roughly speaking, every formal extension of Peano's arithmetic is still incomplete. So, there are sentences in Peano arithmetic which are undecidable.

Is the statement $F$ (Fermat's last theorem) undecidable in Peano's arithmetic? I show that if $F$ is undecidable, then $F$ is true in the model $\mathbb{N}$. Indeed, since $F$ is undecidable, $\neg F$ is not a theorem in Peano arithmetic. Thus, $\neg F$ is not true in some model $M$ of Peano arithmetic. So $F$ is true in the model $M$. But the model $\mathbb{N}$ has the special feature of being a *prime model*, which means that if a universal sentence $S$ is true in some model $M$, then it is true in $\mathbb{N}$. In particular, $F$ (Fermat's last theorem) is true in $\mathbb{N}$.

There are other interesting considerations about the logical aspect of Fermat's last theorem. The fact that the sentence $F$ is a theorem depends essentially on how rich is the given collection of axioms. In a sense which I'll not make precise, Shepherdson (1965) has shown that, with methods current at the time of Fermat, and without "auxiliary operations" (like, say, Legendre's symbol, and this is explained in Shepherdson's paper) for no value of $n \geq 3$ it is possible to prove the sentence

$F_n$: "For every $x$, for every $y$, for every $z$ either $xyz = 0$ or $x^n + y^n \neq z^n$."

The recent solution of Hilbert's tenth problem has also some bearing on Fermat's last theorem. This is explained well in the paper of Davis, Matijasevič, and Julia Robinson (1976). The new idea is the possibility of transforming the proof of Fermat's last theorem into the verification that a polynomial equation has no solution in nonnegative integers.

The method is based on the following fact. There is a polynomial $P(X, Y, Z, W_1, \ldots, W_k)$, with coefficients in $\mathbb{Z}$, such that the equation $Y^Z = X$ has solution in natural numbers $X = a$, $Y = b$, $Z = c$ if and only if the Diophantine equation $P(a,b,c,W_1, \ldots, W_k) = 0$ has a solution in natural numbers $W_1 = w_1, \ldots, W_k = w_k$.

Thus, Fermat's theorem is true if and only if the system of equations

$$P(A, X + 1, N + 3, U_1, \ldots, U_k) = 0$$
$$P(B, Y + 1, N + 3, V_1, \ldots, V_k) = 0$$
$$P(A + B, Z, N + 3, W_1, \ldots, W_k) = 0$$

with indeterminates $A$, $B$, $X$, $Y$, $Z$, $N$, $U_1, \ldots, U_k$, $V_1, \ldots, V_k$, $W_1, \ldots, W_k$, has no solution in natural numbers $a$, $b$, $x$, $y$, $z$, $n$, $u_1, \ldots, w_k$. Considering the square of the polynomial $P$, this is equivalent to the condition that the polynomial equation

$$P^2(A, X + 1, N + 3, U_1, \ldots, U_k) + P^2(B, Y + 1, N + 3, V_1, \ldots, V_k)$$
$$+ P^2(A + B, Z, N + 3, W_1, \ldots, W_k) = 0$$

has no solution in natural numbers.

At present, it is possible to construct explicitly a polynomial like the polynomial $P$, with less than 12 indeterminates. Can we hope for more with this method?

Recently, I came across a still unpublished manuscript by Zinoviev, which was circulated in the Symposium on Modern Logic, organized by the Istituto dell'Enciclopedia Italiana, in 1977, in Roma. The last sentence of that text is: "It means that GFT (the Great Fermat Theorem) is unprovable."

Having been unable to understand this paper, I consulted Professor Kreisel, who was present at that particular meeting and who replied to me (1978): "In my considered opinion, Zinoviev's paper on Fermat's conjecture is worthless."

## Bibliography

1825  Scherk, H. F.
Von den numerische Coefficienten der Secantereihe ihrem Zussammenhange, und ihrer Analogie mit den Bernoullischen Zahlen. *Gesammelte Mathematische Abhandlungen*, Reimer, Berlin, 1825, 1–30.

1886  Bang, A. S.
Taltheoretiske Undersøgelser. *Tidskrift for Math.*, series 5, **4**, 1886, 70–80, 130–137.

1886  Weber, H.
Theorie der Abelschen Zahlkörper. *Acta Math.* **8**, 1886, 193–263.

1892  Zsigmondy, K.
Zur Theorie der Potenzreste. *Monatsh f. Math.*, **3**, 1892, 265–284.

1893  Saalschütz, L.
*Vorlesungen über die Bernoullischen Zahlen, ihren Zusammenhang mit den Secanten-Coefficienten und ihre wichtigeren Anwendungen*, Springer-Verlag, Berlin, 1893.

1897  Hilbert, D.
Die Theorie der algebraischen Zahlkörper. *Jahresbericht d. Deutschen Math. Vereinigung* **4**, 1897, 175–546, Reprinted in *Gesammelte Abhandlungen*, vol. I, Chelsea Publ. Co., New York, 1965, 63–363.

1897  Maillet, E.
Sur l'équation indéterminée $ax^{\lambda^t} + by^{\lambda^t} = cz^{\lambda^t}$. *Assoc. Française Avancement Science, St. Etienne*, **26**, 1897, II, 156–168.

1899  Weber, H.
*Lehrbuch der Algebra*, vol. II, 2nd edition, Braunschweig, 1899. Reprinted by Chelsea Publ. Co., New York, 1961.

1901  Maillet, E.
Sur les équations indéterminées de la forme $x^\lambda + y^\lambda = cz^\lambda$. *Acta Math.* **24**, 1901, 247–256.

1904  Birkhoff, G. D. and Vandiver, H. S.
On the integral divisors of $a^n - b^n$. *Annals of Math.*, (2), **5**, 1904, 173–180.

1911  Furtwängler, P.
Über die Klassenzahlen der Kreisteilungskörper. *J. reine und angew. Math.*, **140**, 1911, 29–32.

1912  Bohniček, S.
Über die Unmöglichkeit der Diophantischen Gleichung $\alpha^{2^{n-1}} + \beta^{2^{n-1}} + \gamma^{2^{n-1}} = 0$ im Kreiskörper der $2^n$-ten Einheitswurzeln, wenn $n$ grösser als 2 ist. *Sitzungsber. Akad. d. Wiss. Wien.*, Abt. IIa, **121**, 1912, 727–742.

1923  Nielsen, N.
*Traité Elémentaire des Nombres de Bernoulli*, Gauthier-Villars, Paris, 1923.

1923  Nörlund, N. E.
Vorlesungen über Differenzenrechnung, Springer-Verlag, Berlin, 1923. Reprinted by Chelsea Publ. Co., New York, 1954.

1928  Weil, A.
L'arithmétique sur les courbes algébriques. *Acta Math.*, **52**, 1928, 281–315.

1929  Vandiver, H. S.
Summary of results and proofs concerning Fermat's last theorem (3rd paper). *Proc. Nat. Acad. Sci. U.S.A.*, **15**, 1929, 43–48.

1929  Vandiver, H. S.
Summary of results and proofs concerning Fermat's last theorem (4th paper). *Proc. Nat. Acad. Sci. U.S.A.*, **15**, 1929, 108–109.

1929  Vandiver, H. S.
On Fermat's last theorem. *Trans. Amer. Math. Soc.*, **31**, 1929, 613–642 and Corrections, *Trans. Amer. Math. Soc.*, **33**, 1931, 998.

1930  Vandiver, H. S.
Summary of results and proofs on Fermat's last theorem (5th paper). *Proc. Nat. Acad. Sci. U.S.A.*, **16**, 1930, 298–304.

1931  Vandiver, H. S.
Summary of results and proofs on Fermat's last theorem (6th paper). *Proc. Nat. Acad. Sci. U.S.A.*, **17**, 1931, 661–673.

1934   Skolem, T.
Über die nicht-Charakterisierbarkeit der Zahlentheorie mittels endlich oder
abzählbar unendlich vieler Aussagen, mit aussschliesslich Zahlenvariabeln. *Fund.
Math.*, **23**, 1934, 150–161.

1935   Morishima, T.
Über die Fermatsche Vermutung, XI. *Jpn. J. Math.*, **11**, 1934, 241–252.

1937   Vandiver, H. S.
On Bernoulli numbers and Fermat's last theorem. *Duke Math. J.*, **3**, 1937, 569–584.

1938   Lehmer, E.
On congruences involving Bernoulli numbers and the quotients of Fermat and
Wilson. *Annals of Math.*, **39**, 1938, 350–359.

1940   Vandiver, H. S.
Note on Euler number criteria for the first case of Fermat's last theorem. *Amer. J.
Math.*, **62**, 1940, 79–82.

1948   Gunderson, N. G.
Derivation of Criteria for the First Case of Fermat's Last Theorem and the Com-
bination of the these Criteria to produce a new lower Bound for the Exponent.
Thesis, Cornell University, 1948, 111 pages.

1950   Hasse, H.
Über den algebraischen Funktionenkörper der Fermatschen Gleichung. *Acta
Szeged*, **13**, 1950, 195–207. Reprinted in *Mathematische Abhandlungen*, vol. 2,
422–434. De Gruyter, Berlin, 1975.

1950   Gut, M.
Eulersche Zahlen und grösser Fermat'scher Satz. *Comm. Math. Helv.* **24**, 1950,
73–99.

1952   Ankeny, N. C.
The insolubility of sets of diophantine equations in the rational numbers. *Proc.
Nat. Acad. Sci. U.S.A.*, **38**, 1952, 880–884.

1954   Ankeny, N. C. and Erdös, P.
The insolubility of classes of diophantine equations. *Amer. J. Math.*, **76**, 1954,
488–496.

1954   Carlitz, L.
Note on irregular primes. *Proc. Amer. Math. Soc.* **5**, 1954, 329–331.

1954   Hasse, H.
Zetafunktion und *L*-Funktionen zu einem arithmetischen Funktionenkörper vom
Fermatschen Typus. *Abhandl. d. Deutschen Akad. d. Wiss. Berlin, Math.-Nat. kl.*
**4**, 1954, No. 4, 5–70. Reprinted in *Mathematische Abhandlungen*, vol. 2, 450–515.
De Gruyter, Berlin, 1975.

1954   Kreisel, G.
Applications of mathematical logic to various branches of mathematics. Colloque
de Logique Mathématique, Paris, 1954, 37–49.

1954   Lehmer, D. H., Lehmer, E., and Vandiver, H. S.
An application of high speed computing to Fermat's last theorem. *Proc. Nat. Acad.
Sci. U.S.A.*, **40**, 1954, 25–33.

1954   Vandiver, H. S.
Examination of methods of attack of the second case of Fermat's last theorem.
*Proc. Nat. Acad. Sci. U.S.A.*, **40**, 1954, 732–735.

1955   Selfridge, J. L., Nicol, C. A., and Vandiver, H. S.
Proof of Fermat's last theorem for all exponents less than 4002. *Proc. Nat. Acad.
U.S.A.*, **41**, 1955, 970–973.

1956 Inkeri, K.
Über eine Verallgemeinerung des letzten Fermatschen Satzes. *Annales Univ. Turku,* Ser. A, I, 1956, No. **23**, 13 pages.

1957 Hasse, H.
Über die Charakterführer zu einem Arithmetischen Funktionen Körper vom Fermatschen Typus. *Wissenschaftliche Veröffentlichungen der Nationalen Technischen Universität Athens,* **12**, 1957, 3–50.

1958 Kemeny, I. G.
Undecidable problems of elementary number theory. *Math. Ann.,* **135**, 1958, 160–169.

1958 Serre, J.-P.
Classes des corps cyclotomiques. Séminaire Bourbaki, 1958, No. 174, 11 pages.

1959 Iwasawa, K.
On Γ-extensions of algebraic number fields. *Bull. Amer. Math. Soc.,* **65**, 1959, 183–226.

1962 Lang, S.
*Diophantine Geometry,* Wiley-Interscience, New York, 1962.

1964 Kapferer, H.
Verifizierung des symmetrisches Teils der Fermatschen Vermutung fur unendlichen viele paarweise teilerfremde Exponenten E. *J. reine u. angew. Math.,* **214/5**, 1964, 360–372.

1965 Hellegouarch, Y.
Une propriété arithmétique des points exceptionnels rationnels d'ordre pair d'une cubique de genre 1. *C. R. Acad. Sci. Paris,* **260**, 1965, 5989–5992.

1965 Hellegouarch, Y.
Application d'une propriété arithmétique des points exceptionnels d'ordre pair d'une cubique de genre 1. *C. R. Acad. Sci. Paris,* **260**, 1965. 6256–6258.

1965 Mumford, D.
A remark on Mordell's conjecture. *Amer. J. Math.,* **87**, 1965, 1007–1016.

1965 Shepherdson, J. C.
Non-standard models for fragments of number theory. *Symposium on the Theory of Models* (edited by J. W. Addison, L. Henkin and A. Tarski) North-Holland, Amsterdam, 1965, 342–358.

1966 Cassels, J. W. S.
Diophantine Equations with special reference to elliptic curves. *J. London Math. Soc.,* **41**, 1966, 193–291.

1966 Iwasawa, K. and Sims, C.
Computation of invariants in the theory of cyclotomic fields. *J. Math. Soc. Japan,* **18**, 1966, 86–96.

1966 Samuel, P.
*Lectures on Old and New Results on Algebraic Curves,* Tata Institute, Bombay, 1966.

1967 Selfridge, J. L. and Pollack, B. W.
Fermat's last theorem is true for any exponent up to 25000. *Notices Amer. Math. Soc.,* **11**, 1967, 97 Abstract 608–638.

1969 Manin, Yu. I.
The p-torsion of elliptic curves is uniformly bounded. *Izvestija Akad. Nauk. SSSR,* ser. Mat. 33, 1969, 459–465. Translated in *Math. USSR Izv.,* **3**, 1969, 433–438.

1969 Mordell, L. J.
*Diophantine Equations,* Academic Press, New York, 1969.

1970   Demjanenko, B. A.
O Totchkax kenietchnovo poriadka Elliptitcheskix Krivix. *Mat. Zametki*, 7, 1970, 563–567.

1970   Demjanenko, B. A.
O Totchkax Krutchenia Elliptitscheskix Krivix (On torsion points of elliptic curves). *Izvestija Akad. Nauk. SSSR*, ser. Mat., **34**, 1970, 757–774. Translated in *Math. USSR Izv.*, **4**, 1970, 765–783.

1970   Kobelev, V. V.
Proof of Fermat's last theorem for all prime exponents less than 5500. *Soviet Math. Dokl.* **11**, 1970, 188–190.

1971   Demjanenko, B. A.
O Krutcheni Elliptitcheskix Krivix (On torsion of elliptic curves). *Izvestija Akad. Nauk SSSR*, ser. Mat. **35**, 1971, 280–307. Translated in *Math. USSR Izv.*, **5**, 1971, 286–318.

1971   Hellegouarch, Y.
Sur un théorème de Maillet. *C. R. Acad. Sci. Paris*, **273**, 1971, 477–478.

1971   Hellegouarch, Y.
Points d'ordre fini sur les courbes elliptiques. *C. R. Acad. Sci. Paris*, **273**, 1971, 540–543.

1971   Ogg, A.
Rational points of finite order on elliptic curves. *Invent. Math.*, **12**, 1971, 105–111.

1972   Cassels, J. W. S.
Review of Demjanenko's paper "The torsion of elliptic curves." *Math. Reviews*, **44**, 1972, review #2755, p. 519.

1972   Hellegouarch, Y.
Courbes Elliptiques et Equation de Fermat. Thèse, 1972, Besançon.

1972   Iwasawa, K.
*Lectures on p-adic L-Functions*, Princeton Univ. Press, Princeton, 1972.

1972   Iwasawa, K.
On the $\mu$-invariants of cyclotomic fields. *Acta Arithm.*, **21** 1972, 99–101.

1973   Chang, C. C. and Keisler, H. J.
*Model Theory*, North-Holland, Amsterdam, 1973.

1973   Johnson, W.
On the vanishing of the Iwasawa invariant $\mu_p$, for $p < 8000$. *Math. Comp.* **27**, 1973, 387–396.

1973   Mazur, B. and Tate, J.
Points of order 13 on elliptic curves. *Invent. Math.*, **22**, 1973, 41–49.

1974   Metsänkylä, T.
Class numbers and $\mu$-invariants of cyclotomic fields. *Proc. Amer. Math. Soc.*, **43**, 1974, 299–300.

1975   Ernvall, R.
On the distribution mod 8 of the E-irregular primes. *Annales Acad. Sci. Fennicae*, ser. A, I, **1**, 1975, 195–198.

1975   Johnson, W.
Irregular primes and cyclotomic invariants. *Math. Comp.*, **29**, 1975, 113–120.

1975   Kubert, D. and Lang, S.
Units in the modular function field, I. *Math. Ann.*, **218**, 1975, 67–96.

1975   Kubert, D. and Lang, S.
Units in the modular function field, II. A full set of units. *Math. Ann.*, **218**, 1975, 175–189.

1975   Kubert, D. and Lang, S.
Units in the modular function field, III. Distribution relations. *Math. Ann.*, **218**, 1975, 273–285.

1975   Wagstaff, S. S.
Fermat's last theorem is true for all exponents less than 58150. *Notices Amer. Math. Soc.*, **22**, 1975, A-507.

1976   Davis, M., Matijasevič, Yu., and Robinson, J.
Hilbert's tenth problem. Diophantine equations: positive aspects of a negative solution. *Proc. Symposia Pure Math.*, **28**, 1976, 323–378. American Math. Soc. Providence, R. I.

1976   Wagstaff, S. S.
Fermat's last theorem is true for any exponent less than 100,000. *Notices Amer. Math. Soc.*, **23**, 1976, A-53.

1977   Manin, Yu. I.
*A Course in Mathematical Logic*, Springer-Verlag, New York, 1977.

1977   Mazur, B.
*Rational Points on Modular Curves.* Lect. Notes in Math., No. 601, 107–148. Springer-Verlag, Berlin, 1977.

1977   Rohrlich, D. E.
Points at infinity on the Fermat curves. *Invent. Math.*, **39**, 1977, 95–127.

1977/8   Serre, J.-P.
Points rationnels des courbes modulaires $X_0(N)$ (d'après B. Mazur). Séminaire Bourbaki, $30^e$ année, 1977/8, No. 511, 3 pages.

1978   Wagstaff, S. S.
The irregular primes to 125000. *Math. Comp.*, **32**, 1978, 583–592.

1977   Washington, L. C.
On Fermat's last theorem. *J. reine. u. angew. Math.*, **289**, 1977, 115–117.

1978   Ernvall, R. and Metsänkylä, T.
Cyclotomic invariants and $E$-irregular primes. *Math. Comp.*, **32**, 1978, 617–629.

1978   Ferrero, B. and Washington, L. C.
The Iwasawa invariant $\mu_p$ vanishes for abelian number fields. Preprint (to appear).

1978   Kreisel, G.
Letter to the author (6 June 1978). Atlantis Hotel, Zurich.

1978   Lang, S.
*Cyclotomic Fields*, Springer-Verlag, New York, 1978.

1978   Mazur, B.
Modular curves and the Eisenstein ideal. *Publ. Math. I.H.E.S.*, **47**, 1978, 35–193.

1978   Mazur, B.
Rational isogenies of prime degree. *Invent. Math.*, **44**, 1978, 129–162.

1978   Powell, B.
Proof of a special case of Fermat's last theorem. *Amer. Math. Monthly*, 1978, **85**, 750–751.

1978   Zinoviev, A. A.
Complete induction and great Fermat's theorem. Atti Convegno Logiche Moderne, Istituto delle Enciclopedia Italiana, Roma, 1977 (to appear). This paper will also appear in *J. Logique et Analyse* together with a report by G. Kreisel.

# LECTURE XI

# Estimates

In the preceding lectures, I have always mentioned the efforts of mathematicians to prove Fermat's last theorem. Their attitude was, in the main, the following:

Fermat's last theorem is true; let's try to find a proof. Any progress towards the complete proof will be worthwhile, whether it deals only with the first case, or if only covers certain exponents.

With this rationale, more and more complicated methods have been brought into the battle with, I must admit, only relatively minor success. This seemingly hopeless struggle prompted some mathematicians to begin to doubt the truth of the theorem. But their work might even turn out to be useful in proving the theorem.

I'll begin evaluating, with elementary methods, the size of the smallest possible solution, for any given exponent $p$. Some more sophisticated estimates for the first case are based on the criteria involving the Fermat quotients.

Recently, the methods of diophantine approximation and linear forms of logarithms have provided a new tool to attack the problem. It has still to be fully exploited.

## 1. Elementary (and Not So Elementary) Estimates

The approach is the following. Let $n$ be any natural number, $n > 2$, and let $x$, $y$, $z$ be *real* numbers, such that

$$0 < x < y < z \tag{1.1}$$

and

$$x^n + y^n = z^n. \tag{1.2}$$

In 1856, Grünert proved:

**(1A)** *If* $0 < x < y < z$ *are integers and* $x^n + y^n = z^n$, *then* $x > n$.

PROOF.

$$x^n = z^n - y^n = (z - y)(z^{n-1} + z^{n-2}y + \cdots + y^{n-1}) > (z - y)ny^{n-1}.$$

Hence

$$0 < z - y < \frac{x^n}{ny^{n-1}} < \frac{x}{n}$$

and

$$y + 1 \le z < y + \frac{x}{n}$$

so $n < x$.                                                                    □

This shows that any counterexample to Fermat's last theorem must involve large integers. For example, if $n = 101$ then $x > 102$ and the numbers involved would be at least like $102^{101}$. As a matter of fact, it is rather easy, without any powerful methods, to show that any would-be solution must be substantially larger.

From the above proof, it follows that

$$y < z < y + \frac{x}{n} < y\left(1 + \frac{1}{n}\right). \tag{1.3}$$

Hence $z$, $y$ are relatively close together and therefore the size of $x$ should be much smaller.

On the other hand, Perisastri showed in 1969 that $x$ cannot be much smaller than $z$:

$$z < x^2. \tag{1.4}$$

To obtain further estimates, it is convenient to introduce the positive real numbers $r$, $s$, $t$, $r_1$, $s_1$, $t_1$ defined as follows: If $n = p$ is an odd prime, if $0 < x < y < z$ are real numbers and $x^p + y^p = z^p$, let

$$x + y = t^p, \qquad \frac{x^p + y^p}{x + y} = t_1^p, \qquad z = tt_1,$$

$$z - y = r^p, \qquad \frac{z^p - y^p}{z - y} = r_1^p, \qquad x = rr_1, \tag{1.5}$$

$$z - x = s^p, \qquad \frac{z^p - x^p}{z - x} = s_1^p, \qquad y = ss_1.$$

These relations are reminiscent of the ones obtained by Abel, as explained in my fourth Lecture. Now, it is not required that these numbers be integers. However, if $x$, $y$, $z$ are nonzero integers and $p$ does not divide $xyz$, then all the above numbers are integers, as Abel proved.

From (1.5), it follows that

$$x = \frac{r^p - s^p + t^p}{2} \qquad y = \frac{-r^p + s^p + t^p}{2} \qquad z = \frac{r^p + s^p + t^p}{2} \qquad (1.6)$$

and

$$x + y - z = -\frac{r^p + s^p - t^p}{2}. \qquad (1.7)$$

It is also obvious that

$$0 < r < s < t. \qquad (1.8)$$

A better bound for the first case is the following:

**(1B)** *If $x$, $y$, $z$ are integers, not multiples of $p$ and $x^p + y^p = z^p$, then $x > 6p^3$. If $p$ divides $xyz$, then $x > 6p^2$.*

PROOF. I will illustrate in this proof how some of the naïve results of Lecture IV may be put to use. There it was proved that if $p \nmid xyz$, then $p^3 rst$ divides $x + y - z$ [see Lecture IV, (3B)]. Since $0 < r < s < t$, at worst $r = 1$, $s = 2$, $t = 3$, so $6p^3 \leq x + y - z < x$.
The other case is similar. $\qquad \square$

The search for improved lower bounds for eventual solutions of Fermat's equation was the object of numerous papers. I will not narrate step by step all these improvements. Conceptually, these results do not throw any more light onto the problem.

However, as in giving the latest news, I will mention Inkeri's estimate, which is at the present time the best one known. His method depends not only on inequalities which would be generally true, but also on precise arithmetical properties of the numbers $r$, $s$, $t$. Inkeri showed successively (1946):

$$0 < t_1 < s_1 < r_1. \qquad (1.9)$$

If $2 \leq y$ (for example, if $x$, $y$ are integers) then $t < t_1$. $\qquad (1.10)$

These two inequalities are quite obvious from the definitions. More work is required to show that

$$t_1 > \frac{t^{p-1}}{2} \quad \text{and} \quad t - s > \frac{r}{2p}. \qquad (1.11)$$

The next inequality is fundamental. It was first shown by James in 1934 and rediscovered by Inkeri, who based his proof on the following lemma:

**Lemma 1.1.** *Let $a$, $b$ be distinct positive real numbers, $a + b \leq 1$ and $n \geq 5$. Then*

$$(1 + a^n - b^n)^n + (1 - a^n + b^n)^n > (1 + a^n + b^n)^n.$$

From relations (1.2), it follows that

$$(r^p - s^p + t^p)^p + (-r^p + s^p + t^p)^p = (r^p + s^p + t^p)^p. \qquad (1.12)$$

The fundamental inequality is

$$r + s > t. \qquad (1.13)$$

To prove it, let $a = r/t$, $b = s/t$ hence $0 < a < b < 1$. If $r + s \leq t$, then $a + b \leq 1$ and by Lemma 1.1 this would contradict (1.12).

If $x$, $y$, $z$ are *integers*, not multiples of $p$, $0 < x < y < z$, $x^p + y^p = z^p$, then $r$, $s$, $t$ have the following arithmetical property: each one of these numbers has a prime factor $q$ such that $q \equiv 1 \pmod{p^2}$.

The proof of this statement is not elementary. More precisely, using methods described in my Lecture IV (The Naïve Approach) it is only possible to show that $q^p \equiv 1 \pmod{p^2}$.

To conclude the proof, Furtwängler's theorem is required: Each prime factor $q$ dividing $xyz$ satisfies the congruence $q^{p-1} \equiv 1 \pmod{p^2}$. Hence $q \equiv q^p \equiv 1 \pmod{p^2}$. Since class field theory is needed to establish Furtwängler's theorem, the above proof is no longer elementary.

Finally, Inkeri showed (1953):

**(1C)** *If $p$ is an odd prime, $0 < x < y < z$ are relatively prime integers, such that $x^p + y^p = z^p$, and $p \nmid xyz$, then*

$$x > \left( \frac{2p^3 + p}{\log(3p)} \right)^p.$$

SKETCH OF THE PROOF. By the preceding result $r = r'q_1$, $s = s'q_2$, $t = t'q_3$, where $r'$, $s'$, $t'$ are natural numbers, and $q_1$, $q_2$, $q_3$ are primes and $q_1 \equiv 1 \pmod{2p^2}$, $q_2 \equiv 1 \pmod{2p^2}$, $q_3 \equiv 1 \pmod{2p^2}$. By Lecture IV, (3B),

$$r' + s' - t' \equiv r'q_1 + s'q_2 - t'q_3 \equiv 0 \pmod{p^2}.$$

*First Case.* If $\max\{r', s', t'\} \geq p^2/2$, then $t > (p^2/2)(2p^2 + 1) > p^4$. Since $z - y \geq 1$ and $z > (x + y)/2 = t^p/2$,

$$x^p = z^p - y^p > (z - y)z^{p-1} > \left( \frac{t^{p-1}}{2} \right)^p.$$

so $x > t^{p-1}/2 > \frac{1}{2}p^{4(p-1)} > [(2p^3 + p)/\log(3p)]^p$.

*Second Case.* If $\max\{r', s', t'\} < p^2/2$, then

$$-\frac{p^2}{2} < r' + s' - t' < p^2.$$

But $p^2$ divides $r' + s' - t'$ so $r' + s' = t'$. Hence $t - s \equiv t' - s' \equiv r' \pmod{2p^2}$ so $t - s = 2mp^2 + r'$, where $m$ is an integer. Using (1.11) it follows that $m \geq 1$.

It may be shown, using (1.12) that

$$\frac{r}{t} < \xi = \sqrt[p]{\frac{\sqrt[p]{2} - 1}{2}} < \frac{s}{t}, \qquad (1.14)$$

hence

$$t > \frac{t - s}{1 - \xi}.$$  (1.15)

Since

$$x^p > (z - y)z^{p-1} = r^p z^{p-1} > \frac{r^p t^{(p-1)p}}{2^p},$$

it follows that

$$x > \frac{rt^{p-1}}{2} > \frac{r}{2}\left(\frac{t-s}{1-\xi}\right)^{p-1} > \frac{2p^2+1}{2}\left[\frac{2p^2+1}{\frac{1}{p}\log\left(\frac{2p}{\log 2}\right)}\right]^{p-1}$$

Assuming $p > 1190$ (otherwise there is no solution) then

$$2p\left[\log\left(\frac{2p}{\log 2}\right)\right]^{p-1} < [\log(3p)]^p$$

so

$$x > \left[\frac{2p^3 + p}{\log(3p)}\right]^p.$$   □

Similarly, Inkeri obtained bounds for the second case:

**(1D)** *If $p$ is an odd prime, $0 < x < y < z$ are relatively prime integers, such that $x^p + y^p = z^p$ and $p$ divides $xyz$, then*

$$x > p^{3p-4} \quad and \quad y > \tfrac{1}{2}p^{3p-1}.$$

Since the first case is true for every prime exponent $p < 57 \times 10^9$ then in this case $x$ would have at least $18 \times 10^{11}$ digits. Similarly, in the second case, $x$ would have at least $18 \times 10^5$ digits.

Another interesting estimate was obtained by Inkeri and van der Poorten (in 1977); it gives the following lower bound for the difference $z - x$ in terms of the exponent $p$:

$$z - x > 2^p p^{2p}.$$  (1.16)

## 2. Estimates Based on the Criteria Involving Fermat Quotients

I have said in Lecture VIII that if the first case fails for the exponent $p$, then the Fermat quotients of $p$ with bases $q = 2, 3, 5, 7, \ldots, 43$ (all primes at most 43) are congruent to zero modulo $p$. In other words

$$q^{p-1} \equiv 1 \pmod{p^2}$$  (2.1)

for $q$ prime, $q \le 43$.

In 1940 and 1941, Lehmer and Lehmer devised a very ingenious method to considerably extend the truth of the first case.

In his doctoral thesis (1948), written under the supervision of Rosser, Gunderson devised another method, along different lines from Lehmers', which actually provided stronger results. I cannot enter into great detail about these methods, but I'll expose their general lines.

Let $p_1 = 2, p_2 = 3, p_3 = 5, \ldots, p_{11} = 31, \ldots, p_{14} = 43, \ldots$ be the sequence of primes. For every $n \geq 1$ let $P_n$ be the set of natural numbers whose prime factors are at most $p_n$.

For every $x \geq 1$ let $P_n(x)$ be the number of elements $a$ of $P_n$, such that $1 \leq a \leq x$. Similarly, let $P_n^*(x)$ be the number of odd integers $a$, such that $1 \leq a \leq x$, $a \in P_n$.

For every prime $p$ let $W_p$ be the set of natural numbers $a \geq 1$ such that $a^{p-1} \equiv 1 \pmod{p^2}$, and for $x > 1$ let $W_p(x)$ be the number of elements $a \in W_p$, $1 \leq a \leq x$.

I rephrase the known criteria for the first case as follows: if the first case of Fermat's theorem fails for $p$ then $P_{14} \subset W_p$, because $W_p$ is closed under multiplication.

Rosser (1939, 1941) and Lehmer and Lehmer (1941) showed:

**Lemma 2.1.**

1. $W_p(p^2) = p - 1$, $W_p(p^2/2) = (p - 1)/2$.
2. *If $N > 1$ and $P_N \subset W_p$, then*

$$P_N\left(\frac{p^2}{3}\right) + P_N^*\left(\frac{p^2}{3}\right) \leq P_N\left(\frac{p^2}{2}\right) \leq \frac{p-1}{2}.$$

If $M$ is a good approximation to $P_{11}(p^2/2)$ from below, the lemma gives a lower bound $2M + 1 < p$. Hence if $p \leq 2M + 1$ then the first case holds for $p$.

This is what Gunderson worked out.

Let $n \geq 1$, $x \geq 1$, $y \geq 1$ and let $P_n(x,y)$ be the number of pairs $(a,b)$ with $\gcd(a,b) = 1$, $a, b \in P_n$, $1 \leq a \leq x$, $1 \leq b \leq y$.

**Lemma 2.2.** *If $P_N \subseteq W_p$ and if $1 \leq x \leq p^2/2$, then $P_N(x,p^2/2x) \leq (p - 1)/2$.*

The function $P_N(a,b)$ is not simple to evaluate. Gunderson proposed to replace it by a function with smaller values.

If $n = 2$, $x \geq 1$, $y \geq 1$, let

$$g_2(x,y) = \frac{2(\log x)(\log y)}{(\log 2)(\log 3)} \tag{2.2}$$

and for $n \geq 3$ let

$$g_n(x,y) = \frac{1}{\log p_n}\left[\int_1^y g_{n-1}(x,z)\frac{dz}{z} + \int_1^x g_{n-1}(y,z)\frac{dz}{z}\right]. \tag{2.3}$$

First Gunderson showed:

$$P_n(x,y) \geq g_n(x,y) \quad \text{(when max}\{x,y\} \geq 3).$$  (2.4)

Next, he gave the following expression of $g_n(x,y)$. Let $n \geq 2$ and

$$L_n = \frac{2}{n!} \frac{1}{(\log 2)(\log 3) \cdots (\log p_n)}.$$  (2.5)

Then

$$g_n(x,y) = L_n \sum_{i=1}^{n-1} (\log x)^i (\log y)^{n-i} \binom{n-2}{i-1}\binom{n}{i}.$$  (2.6)

With this preparation, Gunderson proved

**(2A)** *If $N \geq 3$, $p \geq 3$ and $P_N \subset W_p$, then*

$$\frac{p-1}{2} \geq \binom{2N-2}{N-1} \times \frac{2}{N!} \times \frac{\left(\log\dfrac{p}{\sqrt{2}}\right)^N}{(\log 2)(\log 3) \cdots (\log p_N)}.$$

From the Fermat quotient criteria up to 43, it follows that the first case of Fermat's theorem holds for every prime exponent $p < 57 \times 10^9$.

To illustrate the power of Gunderson's method, we compare with the recent computations of Brillhart, Tonascia, and Weinberger. As I have already said, they found no primes $p < 3 \times 10^9$ (except 1093 and 3511) satisfying the Wieferich congruence $2^{p-1} \equiv 1 \pmod{p^2}$; and the two exceptional primes did not satisfy the Mirimanoff congruence $3^{p-1} \equiv 1 \pmod{p^2}$. Unless the computing techniques are substantially improved, it will be very time-consuming to test Wieferich's congruence for primes $p$ up to $57 \times 10^9$.

I'll say only a few words about Lehmers' method. If $P_N \subset W_p$, then

$$p \geq 1 + 2P_N\left(\frac{p^2}{2}\right)$$

and also

$$p \geq 1 + 2\left\{P_N\left(\frac{p^2}{3}\right) + P_N^*\left(\frac{p^2}{3}\right)\right\}.$$

The idea of Lehmer and Lehmer was to construct recursively, polynomials $f_n(X)$, $f_n^*(X)$, with real coefficients, such that $f_n(\lambda) \leq P_n(10^\lambda)$, $f_n^*(\lambda) \leq P_n^*(10^\lambda)$ (for $n = 1,2,\ldots,N$ and $\lambda \geq 0$) and such that they provide a good approximation to $P_n(10^\lambda)$ and $P_n^*(10^\lambda)$.

If $1 \leq a \leq x$, $a \in P_n$, then $a = \prod_{i=1}^{n} p_i^{x_i}$ ($x_i \geq 0$) and to $a$ is associated the point $(x_1,\ldots,x_n) \in \mathbb{Z}^n$, with $0 \leq x_i$, $\sum x_i \log p_i \leq \log x$. This is a one-to-one mapping. Therefore the determination of $P_n(x)$ (and similarly of $P_n^*(x)$) is a special case of the following problem:

Given the positive real number $\lambda$ and any basis $\{\omega^1,\ldots,\omega^n\}$ of $\mathbb{R}^n$, where

$\omega^i = (\omega^i_1, \ldots, \omega^i_n)$, determine the number $N_n(\lambda | \omega^1, \ldots, \omega^n)$ of all lattice points $(x_i)_{1 \le i \le n} \in \mathbb{Z}^n$ such that $x_i \ge 0$ and $\sum_{i=1}^n x_i \omega^i_i \le \lambda$.

The following is the basic recursion formula:

$$N_n(\lambda | \omega^1, \ldots, \omega^n) = \sum_{k=0}^{s} N_{n-1}(\lambda - k\omega^n_n | \omega^1, \ldots, \omega^{n-1}), \qquad (2.7)$$

where $s = [\lambda/\omega^n_n]$.

Since $N_n(\lambda | \omega^1, \ldots, \omega^n)$ cannot be easily computed, the Lehmers determined polynomials $f_n(X), g_n(X)$, of degree $n$, such that

$$f_n(X) \le N_n(\lambda | \omega^1, \ldots, \omega^n) \le g_n(X)$$

and such that the difference could be easily evaluated. The matter is rather technical, so I avoid any further discussion.

As I already said, the computations were for $n = 14$ and gave the celebrated bound for the first case

$$p \le 253747889$$

# 3. Thue, Roth, Siegel and Baker

Short of proving that Fermat's equation has only the trivial solution, a good substitute would be to show that (for every exponent $p$) the equation has at most finitely many solutions.

Better still, would be to determine a number $C(p) > 0$ such that if $x^p + y^p + z^p = 0$, with nonzero relatively prime integers $x$, $y$, $z$, then

$$\max\{|x|, |y|, |z|\} < C(p).$$

Finally, the very best possible in this vein, would be to determine a number $C > 0$ such that if $p$ is a prime, and if $x^p + y^p + z^p = 0$, then

$$\max\{p, |x|, |y|, |z|\} < C.$$

Whether $C$ would be very large, or not so large, such a result would mean essentially a solution of Fermat's theorem. The remaining task would be the investigation of each prime $p$ less than $C$, and with the estimates already known (and others as yet undiscovered) it would appear feasible to solve this problem.

Alas, nothing like this is yet known.

An early incursion along these lines is due to Turán, and later to Dénes and Turán.

In 1951, with the standard methods of analytical number theory (in particular using the prime number theorem), Turán proved the following result.

Let $N > 1$ be an integer, let $v_p(N)$ denote the number of triples $(x,y,z)$ of integers such that $\gcd(x,y,z) = 1$, $x^p + y^p = z^p$ (where $p$ is an odd prime) and

$1 \leq x, y, z \leq N$. To prove Fermat's theorem for the exponent $p$ amounts to showing that $v_p(N) = 0$ for every $N \geq 1$. Turán's result, is far from this: There exists a constant $C > 0$ such that for every $N$

$$v_p(N) < cN \log^{(p/p-1)}N.$$

In their joint paper of 1955, Dénes and Turán first gave the following estimate, using elementary methods:

$$v_p(N) < p(1 + 3 \times 2^{1/p})N^{2/p}.$$

Using deeper analytical estimates, they obtained the better result:

$$v_p(N) < C(p)\frac{N^{2/p}}{(\log N)^{2-(2/p)}}.$$

They also conjectured that

$$v_p(N) < CN^{1/p}$$

and perhaps, for every $\varepsilon > 0$

$$v_p(N) < C(\varepsilon)N^{\varepsilon}.$$

In the above inequalities, $C$, $C(p)$, $C(\varepsilon)$ denote positive real numbers.

Turán's conjecture follows from a much stronger result of Mumford [quoted in Lecture X, (7B)]. Turán's conjecture, however, does not imply that Fermat's equation for the exponent $p$ has only finitely many solutions in integers.

The recent methods of diophantine approximation, based on the work of Thue, Roth, Siegel and more recently Baker, give some hope that new results may be in sight. Anyway, it is worth a try.

It is not possible, in a short space, to render justice to the beautiful and deep theorems which I will quote. Still worse will be my omissions.

Following an idea of Thue, Roth proved:

**(3A)** *Let $n \geq 3$, let*

$$F(X,Y) = a_0X^n + a_1X^{n-1}Y + \cdots + a_{n-1}XY^{n-1} + a_nY^n,$$

*where $a_i \in \mathbb{Z}$, $a_0 \neq 0$ and assume that the roots of $F(1,X)$ are distinct. Let $G(X,Y) \neq 0$ have coefficients in $\mathbb{Z}$, of total degree at most $n - 3$. Then the diophantine equation $F(X,Y) = G(X,Y)$ has at most finitely many solutions $(x,y) \in \mathbb{Z} \times \mathbb{Z}$.*

The proof follows an idea of Thue and requires the famous theorem of Roth (1955) of approximation of a nonrational algebraic number by rational numbers.

The special case $G(X,Y) = a \in \mathbb{Z}$, $a \neq 0$, was proved by Thue (in 1909). In particular

**(3B)** *If $n \geq 3$, if $a$, $b$, $c$ are integers, $a$, $c$ not zero, then the equation $aX^n + bY^n = c$ has at most finitely many solutions in integers.*

Taking $a = 1$, $b = 1$, and $c = z^n$, for each $z \neq 0$ there are at most finitely many integers $x$, $y$ such that $x^n + y^n = z^n$, when $n$ is odd. This could also be derived using the theorems of Zsigmondy (1892) or Birkhoff and Vandiver (1904).

The next result, due to Siegel (1929) has a geometric flavor.

Let $f(X,Y)$ be a polynomial with coefficients in $\mathbb{Z}$ which is absolutely irreducible (that is, cannot be written as a product of polynomials of smaller degree and coefficients from any algebraic number field). The set of points $(x,y)$, with complex coordinates, such that $f(x,y) = 0$ is a curve. Siegel's theorem states:

(3C) *If the curve defined by $f(X,Y) = 0$ is not a rational curve (in other words it has genus greater than 0), then there exist only finitely many pairs of integers $(x,y)$ such that $f(x,y) = 0$.*

In the case of genus 1, by means of an appropriate birational transformation, the above theorem reduces to a similar one for the hyperelliptic equation

$$a_0 X^n + a_1 X^{n-1} + \cdots + a_n = aY^2, \tag{3.1}$$

where $a_0, a_1, \ldots, a_n \in \mathbb{Z}$, $a_0 \neq 0$, $a \neq 0$, and $n \geq 3$. This special case had been obtained by Siegel in 1926 and in that paper Siegel mentioned the generalization to the curve $F(X) = Y^m$. An explicit and direct proof was given by Inkeri and Hyyrö in 1964, and this was extended by Schinzel and Tijdeman in 1976.

(3D) *If $m \geq 2$, $n \geq 2$ and $\max\{m,n\} \geq 3$, if $f(X) \in \mathbb{Z}[X]$ has degree $n$ and simple roots, and if $a \neq 0$ is an integer, then the equation $f(X) = aY^m$ has at most finitely many solutions in integers.*

The more recent method of Baker concerns effective positive lower bounds for linear forms of logarithms. Baker considered in 1966 the linear form

$$\Lambda = b_1 \log \alpha_1 + \cdots + b_n \log \alpha_n, \tag{3.2}$$

where $n \geq 1$, $\alpha_1, \ldots, \alpha_n$ are any algebraic integers and $b_1, \ldots, b_n$ are any integers, and log denotes the principal determination of the logarithmic function.

Let $d$ be the degree of the field $\mathbb{Q}(\alpha_1, \ldots, \alpha_n)$ over $\mathbb{Q}$. For every $\alpha_i$ let $H(\alpha_i)$ be its height defined as follows. Let $f(X) = c_0 X^m + c_1 X^{m-1} + \cdots + c_m$ be the only polynomial with coefficients in $\mathbb{Z}$ such that $\gcd(c_0, c_1, \ldots, c_m) = 1$ and $f(\alpha_i) = 0$, then $H(\alpha_i) = \max_{0 \leq i \leq m}\{|c_i|\}$.

Let $A_i = \max\{4, H(\alpha_i)\}$, for $i = 1, 2, \ldots, n$, numbered such that $A_1 \leq A_2 \leq \cdots \leq A_n$, and let $B = \max_{1 \leq i \leq n}\{|b_i|, 4\}$.

Assuming that $\Lambda \neq 0$. Baker showed (1977) that a convenient lower bound for $|\Lambda|$ is of the form

$$|\Lambda| > \exp\{-2^{\gamma_1 n + \gamma_0} n^{\gamma_2 n} d^{\gamma_3 n + \gamma_4}(\log B)(\log A_1) \cdots (\log A_n)(\log \log A_{n-1})\},$$
(3.3)

where $\gamma_0, \gamma_1, \gamma_2, \gamma_3, \gamma_4$ are positive real numbers, effectively computable, and not depending on $n$, $\alpha_i$, $b_i$.

For example, Baker showed:

$$\gamma_0 = 0, \qquad \gamma_1 = 800, \qquad \gamma_2 = 200, \qquad \gamma_3 = 200, \qquad \gamma_4 = 0, \qquad (3.4)$$

that is,

$$|\Lambda| > \exp\{-(16nd)^{200n}(\log B)(\log A_1) \cdots (\log A_n)(\log \log A_{n-1}).\} \quad (3.5)$$

van der Poorten and Loxton gave in 1977 the following better values for the constants:

$$\gamma_0 = 47, \qquad \gamma_1 = 61, \qquad \gamma_2 = 10, \qquad \gamma_3 = 10, \qquad \gamma_4 = 10. \qquad (3.6)$$

Some refinement in the form of the minorant and on the values of the constants may be expected. This will have an effect on the bounds obtained with the method.

The main application of Baker's method has been to provide effective bounds for solutions of certain types of diophantine equations. For Mordell's equation $Y^2 = X^3 + k$ $(k \neq 0)$, Baker proved in 1968: *If $x$, $y$ are integers and $y^2 = x^3 + k$, then*

$$\max\{|x|, |y|\} < \exp\{(10^{10}|k|)^{10^4}\}. \qquad (3.7)$$

He also considered the hyperelliptic and superelliptic equations in 1969. Let $f(X) = a_0 X^n + a_1 X^{n-1} + \cdots + a_n$ with $a_0 \neq 0$, $a_i \in \mathbb{Z}$, $n \geq 3$ and let

$$A = \max\{|a_0|, |a_1|, \ldots, |a_n|\}.$$

**(3E)** *If $f(X)$ has at least three simple roots and if $x, y \in \mathbb{Z}$ satisfy*

$$f(x) = y^2,$$

*then*

$$\max\{|x|, |y|\} < \exp \exp \exp(n^{10n^3} A^{n^2}). \qquad (3.8)$$

**(3F)** *If $f(X)$ has at least two distinct roots, if $m \geq 3$, and if $x, y \in \mathbb{Z}$ satisfy*

$$f(x) = y^m,$$

*then*

$$\max\{|x|, |y|\} < \exp \exp\{(5m)^{10} n^{10n^3} A^{n^2}\}. \qquad (3.9)$$

The most striking application of Baker's method was obtained by Tijdeman in 1976. It concerns Catalan's equation

$$X^m - Y^n = 1$$

for arbitrary integers $m, n \geq 2$. In 1844, Catalan conjectured that if $x$, $y$ are natural numbers and $x^m - y^n = 1$, then $x = 3$, $y = 2$, $m = 2$, $n = 3$.

This conjecture has not yet been proved in its full generality, despite the efforts of many distinguished mathematicians. In some sense, it is a problem reminiscent of Fermat's last theorem. It has been shown to be true in many special cases. Euler proved it, assuming that $m = 2, n = 3$. In 1850, Lebesgue disposed of the case $n = 2$. Nagell (in 1921) disposed of the cases $m = 3$, $n \geq 2$ and $m \geq 3$, $n = 3$. S. Selberg (1932) disposed of the case $m = 4$, while much later in 1964, Chao Ko settled the case $m = 2$.

In 1952, LeVeque proved that given the natural numbers $a, b > 1$, there exist at most one pair of natural numbers $m$, $n$ such that $a^m - b^n = 1$. This may also be seen using the theorems of Zsigmondy or Birkhoff and Vandiver, already mentioned. LeVeque proved also (in 1956) that given $m, n \geq 2$ there exist at most finitely many pairs of natural numbers $(x,y)$ such that $x^m - y^n = 1$.

This result was generalized in 1964 by Inkeri and Hyyrö for the equation $X^m - Y^n = c$ ($c \neq 0$, $m, n \geq 2$ given). But both results are a simple application of theorem (3D).

In 1953, Cassels conjectured that there are at most finitely many quadruples of natural numbers $(x,y,m,n)$ such that $x^m - y^n = 1$. The preceding results don't show quite as much. They illustrate the force of the theorems of Thue, Siegel, and Roth—which therefore ought to be useful for Fermat's equation. But, the main weakness is also apparent. Even though it might be proved that there are at most finitely many solutions, an upper bound for the solutions is not given.

And now Baker's method comes to the rescue. First, he proved, generalizing (3.7), that if $m, n \geq 2$ are given, if $x$, $y$ are natural numbers such that $x^m - y^n = 1$, then

$$\max\{x,y\} < \exp\exp\{(5n)^{10}m^{10m^3}\}. \tag{3.10}$$

Tijdeman somewhat improved Baker's lower bound for a linear form in logarithms and proved: *There exists a number $C \geq 3$, which is effectively computable, such that if $x$, $y$, $m$, $n$ are natural numbers, $m, n \geq 2$ and $x^m - y^n = 1$, then $\max\{x,y,m,n\} < C$.*

Tijdeman did not pause to explicitly compute the value, being assured that it was possible to do so. In 1975, Langevin proved that

$$C \leq \exp\exp\exp\exp 730.$$

I should add that this estimate may be—and it is being—substantially lowered, as van der Poorten communicated to me.

Tijdeman's theorem has essentially solved Catalan's conjecture—anyway he settled the one by Cassels. What remains, is the examination of the finitely many (albeit large) remaining cases; perhaps they may lead to other solutions. But these should be treated as being exceptional, or as some mathematicians like to say, sporadic.

Like a wanderer, I have deviated from my main path, to contemplate another landscape. By watching the fate of Catalan's problem, it is reasonable to think that a similar method will give new results for Fermat's equation.

## 4. Applications of the New Methods

Yes, there are several applications of these powerful methods. But, somehow they fall short of the expectations aroused, no doubt because Fermat's equation has four variables. The fact that Catalan's equation had two sets of two variables (some in the exponent) made it more vulnerable.

The first result I want to mention is due to Inkeri, in 1946. By considering solutions by integers not too far apart, the problem was reduced to one of two variables. Precisely:

**(4A)** *Let $n \geq 3$, let $M$ be a positive integer.*

a. *There exist only finitely many triples $(x,y,z)$ of integers such that $0 < x < y < z$, $x^n + y^n = z^n$ and $y - x < M$.*
b. *Idem, with $z - y < M$.*

*In both cases*

$$z < \exp \exp \{(5n)^{10}(n^{10n}A)^{n^2}\}, \qquad (4.1)$$

*where*

$$A = \max_{1 \leq i \leq n} \left\{ \binom{n}{i} M^i \right\}.$$

PROOF. Consider the diophantine equation

$$(T + a)^n + T^n = U^n,$$

where $a \geq 1$ is any given integer.

Let $f(T) = (T + a)^n + T^n$. It has distinct roots, so by (3D), the above equation has only finitely many solutions.

Taking $a = 1, 2, \ldots, M$, there are only finitely many integers $x$, $z$ such that $x^n + (x + a)^n = z^n$. This proves the first assertion. The other is proved by considering $(T + a)^n - T^n = U^n$.

The bound given follows from Baker's estimate indicated in (3F). $\square$

Let me note now that, for the first case, with purely elementary methods, Inkeri himself had obtained a far better bound. Namely, if $x^p + y^p = z^p$,

$0 < x < y < z$, $p$ does not divide $xyz$, and $y - x < M$, then

$$z < \frac{\sqrt[p]{2M^{p/(p-1)}}}{\sqrt[p]{2} - 1}.$$

If $z - y < M$, then

$$z < \frac{\sqrt[p]{2M^p}}{\sqrt[p]{2} - 1}.$$

This illustrates the fact that if it is possible to obtain a bound by elementary methods, it turns out to be incomparably better than the bounds given by Baker.

Inkeri also settled Abel's conjecture for the first case.

I recall, from my fourth Lecture, that if $n > 2$, $0 < x < y < z$ are relatively prime integers such that $x^n + y^n = z^n$, then $y$, $z$ are not prime-powers. If $x$ is a prime-power then $n = p$ is an old prime and $z = y + 1$.

Inkeri showed (1946):

**(4B)** *Under the above assumptions, with $n = p$ not dividing $xyz$, then $x$ is not a prime-power.*

PROOF. If it is, then $z - y = 1$. But $z - y = r^p$ (from Abel's relations) so $r = 1$. By (1.8) and (1.13)

$$s < t < r + s = 1 + s$$

and this is a contradiction.                                                    □

For the first case, nothing more than elementary methods were used. However, for the second case, with the powerful methods, Inkeri only reached partial results:

**(4C)** *Let $p \geq 3$. Then there exist at most finitely many triples of relatively prime integers $(x,y,z)$ such that $0 < x < y < z$, $x^p + y^p = z^p$ and $x$ is a prime-power. For each such triple $z = y + 1$, $p \mid y(y + 1)$ and*

$$y < \exp\exp\{2^p(p - 1)^{10(p-1)}\}^{(p-1)^2} < \exp\exp(2p^{10})^{p^3}.$$

PROOF. Since $z - y = 1$, taking $M = 2$, the first assertion follows from (4A), part (b). By (4B), $p$ divides $xyz$, but it is easily seen from $x^p = (y + 1)^p - y^p$ that $p \nmid x$. So $p \mid y(y + 1)$.

The bound results by applying Baker's estimate given in (3F).          □

In the above situation, in view of Wagstaff's computations, $p > 125000$. By Inkeri's lower bound for $y$, given in (1D),

$$\tfrac{1}{2}p^{3p-1} < y < \exp\exp(2p^{10})^{p^3}$$

so the above interval is much too large to derive any practical consequences.

The next result exhibits quite an important new feature, an effective bound for the exponent. It requires, however, a strong hypothesis, namely, it is assumed that $z - x$ is fairly large with respect to $y - x$. In some sense, this is not to be expected in view of (1.3).

The following theorem was proved by Stewart in 1977, in a slightly more general form. A variant was discovered independently by Inkeri and van der Poorten. I indicate a special case:

(4D) *Let $M > 0$ be a given real number, let $p$ be an odd prime and let $0 < x < y < z$ be relatively prime integers such that $x^p + y^p = z^p$ If the following condition*

$$y - x < M(z - x)^{1 - (1/\sqrt{p})} \qquad (4.2)$$

*is satisfied, then there is a positive real number $C$, depending on $M$, which may be explicitly computed, such that*

$$p < C.$$

PROOF. I will give a fairly detailed proof, which is an amalgam of those of Stewart and of Inkeri and van der Poorten, so as to illustrate the method.

First, it is shown that

$$z - x = b^p, \qquad z - y = p^{p-1}a^p \quad \text{when } p \,|\, x, \qquad (4.3\text{I})$$

$$z - x = p^{-1}b^p, \qquad z - y = a^p \qquad \text{when } p \,|\, y, \qquad (4.3\text{II})$$

$$z - x = b^p, \qquad z - y = a^p \qquad \text{when } p \nmid xy, \qquad (4.3\text{III})$$

where $a$, $b$ are rational numbers, here so arranged that $1 \leq a < b$. This follows at once from Abel's relations, in Lecture IV.

Note that since $z > y > x$, $z - x \geq 2$, hence $z - x \geq 2^p$, or in the second case, $z - x = p^{-1}b^p$, with $p \,|\, b$, hence also $z - x \geq p^{p-1} > 2^p$.

Let $r = 0, 1, 0$ and $q = p - 1, 1, 0$, respectively, in cases (4.3I), (4.3II), (4.3III). I claim that there exists a positive real number $c$ such that

$$(y - x)p^r b^{-p} > b^{-c(\log p)^3}.$$

Indeed, since $3 < 2^{p(\log p)^3} \leq b^{(\log p)^3}$, if $(y - x)p^r b^{-p} > \frac{1}{3}$, then

$$b^{-(\log p)^3} < \tfrac{1}{3} < (y - x)p^r b^{-p}.$$

So let $(y - x)p^r b^{-p} < \frac{1}{3}$. But

$$(y - x)p^r b^{-p} = \frac{y - x}{z - x} = 1 - \frac{z - y}{z - x} = 1 - p^q \left(\frac{a}{b}\right)^p \neq 0.$$

It is easily seen that

$$\left| \log\left( p^q \left(\frac{a}{b}\right)^p \right) \right| < \frac{3}{2} \left| 1 - p^q \left(\frac{a}{b}\right)^p \right|$$

so

$$(y - x)p^r b^{-p} = \left| 1 - p^q \left( \frac{a}{b} \right)^p \right| > \frac{2}{3} \left| q \log p + p \log \left( \frac{a}{b} \right) \right|.$$

Applying Baker's lower bound (3.5) to the above linear form in logarithms it follows that

$$(y - x)p^r b^{-p} > \tfrac{2}{3} \exp\{ -c(\log B)(\log A_1)(\log A_2)(\log \log A_1) \},$$

where $A_1 = p$, $A_2 = \max\{a,b\} = b$, and $B = p \max\{q,p\}$. Thus

$$(y - x)p^r b^{-p} > \exp\{ -c(\log p)^2 (\log b)(\log \log p) \} > b^{-c(\log p)^3}.$$

On the other hand, by hypothesis

$$y - x < M(z - x)^{1 - (1/\sqrt{p})} = M(p^{-r} b^p)^{1 - (1/\sqrt{p})},$$

so

$$b^{-c(\log p)^3} < (y - x)p^r b^{-p} < M(p^r b^{-p})^{1/\sqrt{p}}.$$

But $r = 0, 1$ so

$$b^{\sqrt{p}} - c(\log p)^3 < Mp^{r/\sqrt{p}} < M.$$

Either $\sqrt{p} \le c(\log p)^3$ or $c(\log p)^3 < \sqrt{p}$, hence $2^{\sqrt{p} - c(\log p)^3} < b^{\sqrt{p} - c(\log p)^3} < M$, so $\sqrt{p} - c(\log p)^3 < \log M / \log 2$. In both cases there exists a real positive number $C$ such that $p < C$. ☐

A simpler and more suggestive statement is the following earlier version of Stewart's result:

**(4E)** *Let $\varepsilon > 0$ be a real number, let $p$ be an odd prime, let $0 < x < y < z$ be relatively prime integers such that $x^p + y^p = z^p$ and $y - x < 2^{(1-\varepsilon)p}$. Then there exists an explicitly computable constant $C$ such that $p < C$.*

In fact, Stewart proves the results cited above for an exponent which is not necessarily prime; this constitutes a nontrivial generalization:

**(4F)** *Given $M > 0$, there exists a number $C > 0$ (which may be explicitly computed in terms of $M$) such that if $n \ge 3$ is an integer, if $x$, $y$, $z$ are relatively prime integers such that $x^n + y^n = z^n$ with $0 < x < y < z$, then: if $2 < z - y < M$ or if $y - x < M$, it follows that $x$, $y$, $z$, $n$ are less than $C$.*

Inkeri and van der Poorten proved the following rather technical result:

**(4G)** *Let $p$ be an odd prime, let $l_1, \ldots, l_m$ (with $m \ge 0$) be distinct primes, $l_i < p$; let $w_1, \ldots, w_m$ be natural numbers. If $x$, $y$, $z$ are relatively prime integers such that $0 < x < y < z$, $x^p + y^p = z^p$ and $\prod_{i=1}^{m} l_i^{w_i}$ divides $y - x$, then*

$$\frac{y - x}{\prod_{i=1}^{m} l_i^{w_i}} > (z - x)^{1 - \{L(\log p)^3/(p-1)\}},$$

where $L = C(1 + l_1 + \cdots + l_m)$, with

| | | |
|---|---|---|
| $z - x = b^p,$ | $z - y = p^{p-1}a^p$ | when $p \mid x,$ |
| $z - x = p^{-1}b^p,$ | $z - y = a^p$ | when $p \mid y,$ |
| $z - x = b^p,$ | $z - y = a^p$ | when $p \nmid xy.$ |

With this $l$-adic generalization, one cannot only show that $y - x < M(z - x)^{1 - (1/\sqrt{p})}$ implies that $p$ is bounded in terms of $M$, as above, but in fact that $p$ is bounded in terms of the greatest prime factor of $M$:

**(4H)** *Given the prime numbers $l_1, \ldots, l_m$ and an integer $M_0 > 0$, there is an effectively computable number $C > 0$ such that if $x, y, z$ are relatively prime integers, $0 < x < y < z$, $x^p + y^p = z^p$, and if*

$$y - x < M_0 l_1^{w_1} \cdots l_m^{w_m} (z - x)^{1 - (1/\sqrt{p})}$$

for any nonnegative integers $w_1, \ldots, w_m$, *then $p < C$.*

The common idea behind these attempts is to find an effectively computable number which bounds the exponent. If this is done without any assumption, it amounts, at least theoretically (as I already said), to the solution of Fermat's problem. Up to now the bounds on the exponent could only be obtained under various more or less technical hypothesis on the nature of the would-be solutions. It may be possible that this kind of information, coupled with results of some other type, will indeed be very important for the solution of the problem.

Who knows?

# Bibliography

1844   Catalan, E.
Note extraite d'une lettre adressée à l' éditeur. *J. reine u. angew. Math.*, **27**, 1844, 192.

1850   Lebesgue, V. A.
Sur l'impossibilité en nombres entiers de l'équation $x^m = y^2 + 1$. *Nouv. Ann. de Math.*, **9**, 1850, 178–181.

1856   Grünert, J. A.
Wenn $n > 1$, so gibt es unter den ganzen Zahlen von 1 bis $n$ nicht zwei Werte von $x$ und $y$, für welche, wenn $z$ einen ganzen Wert bezeichnet, $x^n + y^n = z^n$ ist. *Archiv Math. Phys.*, **27**, 1856, 119–120.

1892   Zsigmondy, K.
Zur Theorie der Potenzreste. *Monatsh. f. Math.*, **3**, 1892, 265–284.

1904   Birkhoff, G. D. and Vandiver, H. S.
       On the integral divisors of $a^n - b^n$. *Annals of Math.*, (2), **5**, 1904, 173–180.

1909   Thue, A.
       Über Annäherungswerte algebraische Zahlen. *J. reine u. angew. Math.*, **135**, 1909, 284–305.

1921   Nagell, T.
       Des équations indéterminées $x^2 + x + 1 = y^n$ et $x^2 + x + 1 = 3y^n$. *Norske Mat. Forenings Skrifter*, 1921, No. 2, 11–14.

1926   Siegel, C. L.
       The integer solutions of the equation $y^2 = ax^n + bx^{n-1} + \cdots + k$ (extract from a letter to Prof. L. J. Mordell). *J. London Math. Soc.*, **1**, 1926, 66–68. Also in *Gesammelte Abhandlungen*, vol. I, Springer-Verlag, Berlin, 1966, 207–208.

1929   Siegel, C. L.
       Über einige Anwendungen diophantischer Approximationen. Abhandl. Preuss. Akad. Wiss., No. 1, 1929. Reprinted in *Gesammelte Abhandlungen*, vol. I, Springer-Verlag, Berlin, 1966, 209–266.

1931   Morishima, T.
       Über den Fermatschen Quotienten. *Jpn. J. Math.*, **8**, 1931, 159–173.

1932   Selberg, S.
       Sur l'impossibilité de l'équation indéterminée $z^p + 1 = y^2$. *Norske Mat. Tidskrift*, **14**, 1932, 79–80.

1934   James, G.
       On Fermat's last theorem. *Amer. Math. Monthly*, **41**, 1934, 419–424.

1939   Rosser, J. B.
       On the first case of Fermat's last theorem. *Bull. Amer. Math. Soc.* **45**, 1939, 636–640.

1940   Lehmer, D. H.
       The lattice points of an $n$-dimensional tetrahedron. *Duke M. J.*, **7**, 1940, 341–353.

1940   Rosser, J. B.
       A new lower bound for the exponent in the first case of Fermat's last theorem. *Bull. Amer. Math. Soc.* **46**, 1940, 299–304.

1941   Lehmer, D. H. and Lehmer, E.
       On the first case of Fermat's last theorem. *Bull. Amer. Math. Soc.*, **47**, 1941, 139–142.

1941   Rosser, J. B.
       An additional criterion for the first case of Fermat's last theorem. *Bull. Amer. Math. Soc.*, **47**, 1941, 109–110.

1946   Inkeri, K.
       Untersuchungen über die Fermatsche Vermutung. *Annales Acad. Sci. Fennicae*, ser. A, I, 1946, **33**, 1–60.

1948   Gunderson, N. G.
       Derivation of Criteria for the First Case of Fermat's Last Theorem and the Combination of these Criteria to Produce a New Lower Bound for the Exponent. Thesis, Cornell University, 1948.

1951   Turán, P.
       A note on Fermat's conjecture. *J. Indian Math. Soc.*, **15**, 1951, 47–50.

1952   LeVeque, W. J.
       On the equation $a^x - b^y = 1$. *Amer. J. Math.*, **74**, 1952, 325–331.

1953   Cassels, J. W. S.
       On the equation $a^x - b^y = 1$. *Amer. J. Math.*, **75**, 1953, 159–162.

1953 Inkeri, K.
Abschätzungen für eventuelle Lösungen der Gleichung im Fermatschen Problem. *Ann. Univ. Turku*, Ser. A, 1953, No. 1, 3–9.

1955 Dénes, P. and Turán, P.
A second note on Fermat's conjecture. *Publ. Math. Debrecen.* **4**, 1955, 28–32.

1955 Roth, K. F.
Rational approximation to algebraic numbers. *Mathematika*, **2**, 1955, 1–20. Corrigendum, page 168.

1956 LeVeque, W. J.
*Topics in Number Theory*, vol. II. Addison-Wesley, Reading, Mass., 1956.

1961 Cassels, J. W. S.
On the equation $a^x - b^y = 1$, II. *Proc. Cambridge Phil. Soc.*, **56**, 1961, 97–103.

1964 Chao Ko
On the diophantine equation $x^2 = y^n + 1$, $xy \neq 0$. *Sciencia Sinica* (Notes), **14**, 1964, 457–460.

1964 Inkeri, K. and Hyyrö, S.
Über die Anzahl der Lösungen einiger Diophantischer Gleichungen. *Annales Univ. Turku*, Ser. A, 1964, No. 78, 3–10.

1966 Baker, A.
Linear forms in logarithms of algebraic numbers, I. *Mathematika*, **13**, 1966, 204–216.

1968 Baker, A.
The diophantine equation $y^2 = x^3 + k$. *Phil. Trans. R. Soc. London*, **263**, 1968, 193–208.

1969 Perisastri, M.
On Fermat's last theorem. *Amer. Math. Monthly*, **76**, 1969, 671–675.

1971 Brillhart, J., Tonascia, J., and Weinberger, P.
On the Fermat quotient, in *Computers in Number Theory*, Academic Press, New York, 1971, 213–222.

1975 Baker, A.
*Transcendental Number Theory*, Cambridge University Press, London, 1975.

1975 Langevin, M.
Sur la fonction plus grand facteur premier. Sém. Delange-Pisot-Poitou, 16$^e$ année. 1974/5, G 22, 29 pages. Inst. Henri Poincaré, Paris, 1975.

1976 Schinzel, A. and Tijdeman, R.
On the equation $y^m = P(x)$. *Acta Arithm.*, **31**, 1976, 199–204.

1976 Tijdeman, R.
On the equation of Catalan. *Acta Arithm.*, **29**, 1976, 197–209.

1977 Baker, A.
The theory of linear forms in logarithms, in *Transcendence Theory, Advances and Applications*, Academic Press, New York, 1977, 1–27.

1977 Inkeri, K. and van der Poorten, A. J.
Some remarks on Fermat's conjecture. (To appear).

1977 Stewart, C. L.
A note on the Fermat equation. *Mathematika*, **24**, 1977, 130–132.

1977 van der Poorten, A. J. and Loxton, J. H.
Multiplicative relations in number fields. *Bull. Austr. Math. Soc.*, **16**, 1977, 83–98.

# LECTURE XII

# Fermat's Congruence

In this lecture, I will turn my attention to an analogue of Fermat's theorem. Instead of the equation, it will be a question of a congruence. In addition to the intrinsic interest of this modified problem, I mentioned in my fourth lecture how Sophie Germain's criterion for the first case involves Fermat's congruence modulo some prime. Accordingly, I will begin by studying the Fermat equation over prime fields.

## 1. Fermat's Theorem over Prime Fields

Let $p$, $q$ be primes, and consider the congruence

$$X^p + Y^p + Z^p \equiv 0 \pmod{q} \tag{1.1}$$

or equivalently, Fermat's equation over the field $\mathbb{F}_q$ with $q$ elements.

If $p = q$, then since $a^p \equiv a \pmod{p}$ for every integer $a$, not a multiple of $p$, (1.1) obviously has solutions $(x, y, z)$ where $p \nmid xyz$. This case is uninteresting. The same thing happens if $p$ or $q$ is equal to 2.

So, I assume that $p$, $q$ are distinct odd primes. A nontrivial solution of (1.1) is a triple of integers $x$, $y$, $z$, such that they satisfy the congruence (1.1), and $1 \leq x, y, z \leq q - 1$.

Let $N(p,q)$ be the number of nontrivial solutions. It is equal to the number of pairs $(x, y)$ such that $1 \leq x, y \leq q - 1$ and $x^p + y^p + 1 \equiv 0 \pmod{q}$. I will state theorems which guarantee that $N(p,q) > 0$. This is quite important in view of the following observation, already known to Libri (1832):

**(1A)** *Let $p > 2$ be a prime. If there exist infinitely many primes $q$ such that (1.1) has only the trivial solution, i.e., $N(p,q) = 0$, then Fermat's theorem is true for the exponent $p$.*

PROOF. Assume that there exist nonzero integers $x$, $y$, $z$ such that

$$x^p + y^p + z^p = 0.$$

Let $q_1, \ldots, q_n$ be the prime divisors of $xyz$. Let $q$ be any prime, $q > \max\{q_1, \ldots, q_n\}$. Then, for each such prime $q$, $(x,y,z)$ is a nontrivial solution of (1.1), that is $N(p,q) > 0$. Thus, if $N(p,q) = 0$ then $q \leq \max\{q_1, \ldots, q_n\}$. This contradicts the hypothesis.                                                            □

However, the assumption made in (1A) is actually false. For example, Libri proved that

$$N(3,q) \geq q - 8 - \sqrt{4q - 27} \quad \text{when } q \equiv 1 \pmod 3.$$

So, if $q \equiv 1 \pmod 3$, $q \geq 19$, then $N(3,q) > 0$.

Pépin computed in 1880 the exact value of $N(3,q)$ (see also Landau, 1913):

$$N(3,q) = q - 8 - l,$$

where $4q = l^2 + 27m^2$ ($l$, $m$ are integers) and $l \equiv 1 \pmod 3$. For example, $N(3,19) = 4$, $N(3,31) = 19$, $N(3,37) = 18$, $N(3,43) = 27$, $N(3,61) = 52$.

In 1887, Pellet showed that for every prime exponent $p$ there exists a number $q_0(p)$ such that if $q$ is a prime, $q \geq q_0(p)$, then $N(p,q) \neq 0$. However, his proof did not provide any indication of the value of $q_0(p)$, nor about the number $N(p,q)$. In a later paper of 1911, Pellet gave bounds for the number $N(p,q)$.

In 1909, Cornacchia gave an upper bound for $q_0(p)$. In the same year, Dickson indicated a more accurate upper bound, namely

$$q_0(p) \leq (p - 1)^2(p - 2)^2 + 6p - 2. \tag{1.2}$$

Dickson also computed a lower bound for the number $N(p,q)$.

Dickson's proof involves rather lengthy computations in the cyclotomic field. A shorter proof, yielding the less accurate upper bound

$$q_0(p) \leq (p!)e + 1 \tag{1.3}$$

is due to Schur (1917). It is reproduced in LeVeque's book (1956).

Concerning the number of solutions, already in 1837 and 1838 Lebesgue used the methods of Gauss and Libri to determine a general formula for the number of pairwise incongruent solutions of a congruence modulo a prime. He applied this formula to the congruences

$$a_1 X_1^m + a_2 X_2^m + \cdots + a_k X_k^m \equiv b \pmod q, \tag{1.4}$$

where $q \equiv 1 \pmod m$, $m \geq 2$.

Also in 1909, Hurwitz studied the congruence (1.4) for a prime exponent $m = p$, and $b = 0$, and thereby extended the results of Dickson.

In connection with Waring's problem, Hardy and Littlewood studied in 1922 the special case where $a_1 = \cdots = a_k = 1$. This congruence is the starting point in the theory which led to the Riemann hypothesis for function fields over finite fields, the beautiful work of Hasse and Weil. A very illuminating paper on this topic is that of Weil (1949). It will be clear, as I proceed, that many of the methods used in evaluating the number of solutions of congruences have evolved from original ideas of Gauss.

So much for the history of this question. I'll now explain Dickson's theorem, and give an outline of the proof. To begin with, I will isolate the trivial special case.

**(1B)** *Let $p > 2$ be a prime. Let $q$ be a prime such that $q = 6mp + 1$ ($m$ integer) or $p$ does not divide $q - 1$. Then*

$$X^p + Y^p + Z^p \equiv 0 \pmod{q}$$

*has a nontrivial solution.*

PROOF. If $p \nmid q - 1$, if $a, b$ are integers such that $ap + b(q - 1) = 1$, choosing integers $x, y, z$, not multiples of $q$, such that $x + y + z \equiv 0 \pmod q$, then $x^{ap} \equiv x \pmod q$, $y^{ap} \equiv y \pmod q$, $z^{ap} \equiv z \pmod q$. Hence $(x^a, y^a, z^a)$ is a nontrivial solution of the congruence.

Now, if $q = 6mp + 1$, if $h$ is a primitive root modulo $q$, then $(h^{2mp})^3 \equiv 1 \pmod q$. Let $\bar{h} = h \pmod q$. So $\bar{h}^{2mp}$ is a primitive cubic root of 1 in the field $\mathbb{F}_q$. Thus

$$1 + \bar{h}^{2mp} + \bar{h}^{4mp} = 0 \quad (\text{in } \mathbb{F}_q).$$

Hence

$$1 + h^{2mp} + h^{4mp} \equiv 0 \pmod q. \qquad \square$$

From now on, I may assume that $q = kp + 1$, where $k$ is even and moreover $3 \nmid k$.

The proof of Dickson's theorem involves the Jacobi cyclotomic sums, already used by Kummer (see Lecture VII). I recall the main definitions and properties. Let

$g = $ a primitive root modulo $p$,
$h = $ a primitive root modulo $q$,
$\zeta = $ a primitive $p$th root of 1,
$\rho = $ a primitive $q$th root of 1.

The $p$ periods of $k$ terms in the cyclotomic field $\mathbb{Q}(\rho)$ are:

$$\begin{aligned}
\eta_0 &= \rho + \rho^{h^p} + \rho^{h^{2p}} + \cdots + \rho^{h^{(k-1)p}}, \\
\eta_1 &= \rho^h + \rho^{h^{p+1}} + \rho^{h^{2p+1}} + \cdots + \rho^{h^{(k-1)p+1}}, \\
&\vdots \\
\eta_{p-1} &= \rho^{h^{p-1}} + \rho^{h^{2p-1}} + \rho^{h^{3p-1}} + \cdots + \rho^{h^{kp-1}}.
\end{aligned} \qquad (1.5)$$

Consider also the Gauss sums, for every $j = 0, 1, \ldots, p - 1$.

$$\tau_j = \langle \zeta^j, \rho \rangle = \sum_{t=1}^{q-1} \zeta^{j\,ind_h(t)} \rho^t, \tag{1.6}$$

where $ind_h(t) = s$, $0 \le s \le q - 2$, when $t \equiv h^s \pmod q$. In particular

$$\tau_0 = \langle 1, \rho \rangle = \sum_{t=1}^{q-1} \rho^t = -1. \tag{1.7}$$

I recall that

$$\tau_j \tau_{p-j} = q \tag{1.8}$$

$$\tau_j \bar\tau_j = q \tag{1.9}$$

($\bar\tau_j$ denotes the complex conjugate of $\tau_j$). Hence

$$\tau_{p-j} = \bar\tau_j. \tag{1.10}$$

The periods may also be easily expressed in terms of the Jacobi sums:

$$\eta_i = \frac{1}{p} \sum_{j=0}^{p-1} \zeta^{-ji} \tau_j \quad \text{(for } i = 0, 1, \ldots, p - 1\text{)}. \tag{1.11}$$

With all these concepts, I now indicate an expression for $N(p,q)$ in terms of the periods:

**Lemma 1.1.**

$$N(p,q) = \frac{1}{q}\left[(q-1)^3 + (q-1)p^2 \sum_{i=0}^{p-1} \eta_i^3\right].$$

PROOF. If $x$, $y$, $z$ are integers, $1 \le x, y, z \le q - 1$, then

$$\sum_{t=0}^{q-1} \rho^{t(x^p + y^p + z^p)} = \begin{cases} 0 & \text{when } x^p + y^p + z^p \not\equiv 0 \pmod q, \\ q & \text{when } x^p + y^p + z^p \equiv 0 \pmod q. \end{cases}$$

So

$$qN(p,q) = \sum_{x,y,z=1}^{q-1} \left( \sum_{t=0}^{q-1} \rho^{t(x^p + y^p + z^p)} \right)$$

$$= \sum_{t=0}^{q-1} \left( \sum_{x,y,z=1}^{q-1} \rho^{tx^p} \rho^{ty^p} \rho^{tz^p} \right)$$

$$= \sum_{t=0}^{q-1} \left( \sum_{x=1}^{q-1} \rho^{tx^p} \right)^3$$

$$= (q-1)^3 + \sum_{t=1}^{q-1} \left( \sum_{x=1}^{q-1} \rho^{tx^p} \right)^3.$$

Now if $t = h^i$ and $x = h^j$, where $0 \le i, j \le q - 2$, then

$$\sum_{x=1}^{q-1} \rho^{txp} = \sum_{j=0}^{q-2} \rho^{h^{i+pj}} = p\eta_i.$$

Since $i \equiv i' \pmod p$ implies that $\eta_i = \eta_{i'}$ it follows that

$$qN(p,q) = (q-1)^3 + k\sum_{i=0}^{p-1} p^3 \eta_i^3,$$

where $q = kp + 1$. This concludes the proof. $\qquad\square$

Based on the above standard lemma, Klösgen in 1970 simplified the proof of Dickson's theorem. I'll sketch this, omitting computational details.

(1C) *The number $N(p,q)$ of pairwise incongruent nontrivial solutions of (1.1) satisfies*:

$$(q-1)[q+1-3p-(p-1)(p-2)\sqrt{q}]$$
$$< N(p,q) < (q-1)[q+1-3p+(p-1)(p-2)\sqrt{q}].$$

*Hence, if $q \ge (p-1)^2(p-2)^2 + 6p - 2$, then $N(p,q) > 0$.*

PROOF. By (1.11) and the lemma:

$$N(p,q) = \frac{1}{q}\left[(q-1)^3 + (q-1)p^2 \frac{1}{p^3}\sum_{i=0}^{p-1}\left\{\sum_{j=0}^{p-1}\zeta^{-ji}\tau_j\right\}^3\right].$$

This may be brought to the form:

$$N(p,q) = \frac{1}{q}\left[(q-1)^3 + \frac{q-1}{p}\sum_{j_1,j_2,j_3=0}^{p-1}\tau_{j_1}\tau_{j_2}\tau_{j_3}A_{j_1j_2j_3}\right],$$

where

$$A_{j_1j_2j_3} = \sum_{i=0}^{p-1}\zeta^{-i(j_1+j_2+j_3)} = \begin{cases} 0 & \text{when } j_1 + j_2 + j_3 \not\equiv 0 \pmod p, \\ p & \text{when } j_1 + j_2 + j_3 \equiv 0 \pmod p. \end{cases}$$

Since $\tau_0 = -1$,

$$N(p,q) = \frac{q-1}{q}[(q-1)^2 - 1 - 3q(p-1) + S],$$

where

$$S = \sum_{\substack{j_1,j_2,j_3=1 \\ j_1+j_2+j_3 \equiv 0 (\bmod\, p)}}^{p-1} \tau_{j_1}\tau_{j_2}\tau_{j_3}.$$

So

$$\left|\frac{N(p,q)}{q-1} - (q+1-3p)\right| = \frac{1}{q}|S|.$$

But $|\tau_j| = \sqrt{q}$. Also, for every $j_1$, $1 \le j_1 \le p - 1$, there are $p - 2$ pairs $(j_2, j_3)$ such that $1 \le j_2, j_3 \le p - 1$ and $j_1 + j_2 + j_3 = p$ or $2p$. Hence

$$|S| \le (p - 1)(p - 2)q^{3/2}.$$

Hence

$$\left| \frac{N(p,q)}{q - 1} - (q + 1 - 3p) \right| \le (p - 1)(p - 2)\sqrt{q}.$$

This leads at once to the first inequalities. The last assertion is now easily deduced. □

An application of the general Hasse and Weil theorem (the Riemann hypothesis for projective curves over finite fields) yields the same bound for $q_0(p)$. Indeed, let $N^*(p,q)$ be the number of points of the curve $X^p + Y^p + Z^p = 0$ in the projective three-dimensional space over the field with $q$ elements. Thus $N(p,q) = N^*(p,q) - 3p$, since there are $3p$ points with one coordinate equal to zero. The curve is nonsingular, with genus $g = (p-1)(p-2)/2$. The general formula is

$$|N^*(p,q) - (q + 1)| < 2g\sqrt{q}.$$

Hence if $q + 1 > (p - 1)(p - 2)\sqrt{q} + 3p$, then $N(p,q) > 0$. A simple computation leads to the same upper bound for $q_0(p)$.

It should be pointed out that the bound for $q_0(p)$ is not the best possible. For example, by actual computation:

$q_0(3) \le 20$, while $N(3,q) = 0$ only for $q = 7, 13$;
$q_0(5) \le 172$, while $N(5,q) = 0$ only for $q = 11, 41, 71, 101$;
$q_0(7) \le 940$, while $N(7,q) = 0$ only for $q = 29, 71, 113, 491$.

The generalization of Dickson's theorem due to Hurwitz is the following:

**(1D)** *Let $p$, $q$ be distinct odd primes, $q = kp + 1$. Let $a$, $b$, $c$ be integers and let $N$ denote the number of solutions $(x,y,z)$, $1 \le x, y, z \le q - 1$, of the congruence*

$$aX^p + bY^p + cZ^p \equiv 0 \pmod{q}. \tag{1.12}$$

*Then*

$$(q - 1)[(q + 1) - (p - 1)(p - 2)\sqrt{q} - pv]$$
$$< N < (q - 1)[(q + 1) + (p - 1)(p - 2)\sqrt{q} - pv],$$

*where $h$ is a primitive root modulo $q$,*

$$a \equiv h^r \pmod{q}, \qquad b \equiv h^s \pmod{q}, \qquad c \equiv h^t \pmod{q},$$

$0 \le r, s, t \le q - 2$, *and*

$$v = \begin{cases} 0 & \text{when } r, s, t \text{ are pairwise incongruent modulo } p, \\ 3 & \text{when } r, s, t \text{ are pairwise congruent modulo } p, \\ 1 & \text{otherwise.} \end{cases}$$

*If $q \ge (p - 1)^2(p - 2)^2 + 2(pv - 1)$, then $N > 0$.*

## 2. The Local Fermat's Theorem

One of the methods to study diophantine equations consists of searching for solutions in $q$-adic fields (for every prime $q$). In some cases, if the equation has a solution in every $q$-adic field $\hat{\mathbb{Q}}_q$ then it also has a solution in $\mathbb{Q}$. For example, this happens for quadratic equations. In such instances, it is said that the Hasse principle *or the local-global principle* holds.

But, for Fermat's equation, I'll show that the local global principle is not satisfied. The basic result to be used is the lemma of Hensel (1908), which I will state in its original stronger form.

Let $\hat{\mathbb{Q}}_q$ be the field of $q$-adic numbers, $\hat{\mathbb{Z}}_q$ the ring of $q$-adic integers, and $v_q$ the $q$-adic valuation of $\hat{\mathbb{Q}}_q$.

**Lemma 2.1** (Hensel). *Let $F(X)$ be a monic polynomial with coefficients in $\hat{\mathbb{Z}}_q$. Let $G_0(X)$, $H_0(X)$ be monic polynomials with coefficients in $\mathbb{Z}$, and let*

$$R = \mathrm{Res}(G_0(X), H_0(X)) \in \mathbb{Z}$$

*be the resultant of these polynomials.*

*If $r = v_q(R) \geq 0$ and if $F(X) \equiv G_0(X)H_0(X) \pmod{q^s}$, where $s > 2r$, then there exist polynomials $G(X)$, $H(X) \in \hat{\mathbb{Z}}_q[X]$ such that*

$$G(X) \equiv G_0(X) \pmod{q^{s-r}},$$
$$H(X) \equiv H_0(X) \pmod{q^{s-r}},$$

*and $F(X) = G(X)H(X)$.*

For example, if the residue classes of $G_0(X)$, $H_0(X)$ modulo $q$ are relatively prime polynomials in $\mathbb{F}_q[X]$, then $R = 1$, $r = 0$ and $s$ may be taken to be equal to 1.

The most useful special case is the following:

**Lemma 2.2** (Hensel). *If $F(X)$ is a monic polynomial with coefficients in $\hat{\mathbb{Z}}_q$ and if $a \in \mathbb{Z}$ is a simple root of the congruence*

$$F(X) \equiv 0 \pmod{q},$$

*then there exists a $q$-adic integer $\alpha \in \hat{\mathbb{Z}}_q$ such that $\alpha \equiv a \pmod{q}$ and $F(\alpha) = 0$.*

I owe the proof of the following result to Brettler (1974):

**(2A)** *For every prime $q$ and every prime $p$, Fermat's equation $X^p + Y^p + Z^p = 0$ has a nontrivial solution in the field of $q$-adic numbers.*

PROOF. If $p = 2$ this is trivial, since there are already nontrivial solutions in $\mathbb{Z}$. Henceforth I assume $p \neq 2$.

*First Case. $q \neq p$.*

Let $F(X) = X^p + q^p - 1$. Then

$$X^p + q^p - 1 \equiv (X - 1)(X^{p-1} + X^{p-2} + \cdots + X + 1) \pmod{q}.$$

Since $1 \bmod q$ is not a root of $X^{p-1} + X^{p-2} + \cdots + X + 1$ modulo $q$, by Lemma 2.2 there exists a $q$-adic integer $\alpha$ such that $\alpha \equiv 1 \pmod{q}$ and $\alpha^p + q^p + (-1)^p = 0$.

*Second Case. $q = p$.*

Let

$$F(X) = X^p + p^p + 1,$$
$$G_0(X) = X + 1,$$
$$H_0(X) = X^{p-1} - X^{p-2} + X^{p-3} - \cdots + 1.$$

The resultant $R = \mathrm{Res}(G_0(X), H_0(X)) = H_0(-1) = p$, so its $p$-adic value is $v_p(R) = 1$. Since $G_0(X)H_0(X) = X^p + 1$,

$$F(X) \equiv G_0(X)H_0(X) \pmod{p^p}.$$

Noting that $p \geq 3 > 2v_p(R)$, by Lemma 2.1 there exist monic polynomials $G(X), H(X) \in \hat{\mathbb{Z}}_p[X]$ such that

$$G(X) \equiv G_0(X) \pmod{p^{p-1}},$$
$$H(X) \equiv H_0(X) \pmod{p^{p-1}},$$

and $F(X) = G(X)H(X)$.

Then $G(X) = X + \alpha$, where $\alpha \in \hat{\mathbb{Z}}_p$, $\alpha \equiv 1 \pmod{p^{p-1}}$ and $-\alpha^p + p^p + 1 = 0$. $\qquad\square$

Another problem would be to determine when Fermat's equation has a solution in units of the field of $q$-adic numbers. It is immediate from the definitions that the following holds:

**(2B)** *If $p$, $q$ are odd primes (not necessarily distinct), the following conditions are equivalent:*

a. *There exist units $\alpha$, $\beta$, $\gamma \in \hat{\mathbb{Z}}_q$ such that $\alpha^p + \beta^p + \gamma^p = 0$.*
b. *There exist integers $x_0$, $y_0$, $z_0$, not multiples of $q$, such that*

$$x_0^p + y_0^p + z_0^p \equiv 0 \pmod{q^{1+e}}.$$

c. *For every $n \geq 0$ there exist integers $x_n$, $y_n$, $z_n$, not multiples of $q$, such that*

$$x_n^p + y_n^p + z_n^p \equiv 0 \pmod{q^{n+1+e}}$$

*and $x_{n+1} \equiv x_n \pmod{q^{n+1}}$, $y_{n+1} \equiv y_n \pmod{q^{n+1}}$, $z_{n+1} \equiv z_n \pmod{q^{n+1}}$,*

*where*

$$e = \begin{cases} 0 & \text{if } q \neq p \\ 1 & \text{if } q = p \end{cases}.$$

This leads to the study of Fermat's congruence modulo the powers of a prime.

# 3. The Problem Modulo a Prime-Power

I will consider in this section the congruence

$$X^{p^m} + Y^{p^m} + Z^{p^m} \equiv 0 \pmod{p^n}, \tag{3.1}$$

where $p$ is an odd prime and $n > m \geq 1$. When does it have a solution in integers not divisible by $p$?

It is possible to assume that $n = m + 1$. Indeed, a simple argument shows that if $x$, $y$, $z$ are integers, not multiples of $p$, and such that

$$x^{p^m} + y^{p^m} + z^{p^m} \equiv 0 \pmod{p^{m+1}},$$

then for every $r \geq 0$ there exist integers $x_r$, $y_r$, $z_r$, not multiples of $p$, such that

$$x_r^{p^m} + y_r^{p^m} + z_r^{p^m} \equiv 0 \pmod{p^{m+1+r}}$$

and $x_{r+1} \equiv x_r \pmod{p^{r+1}}$, $y_{r+1} \equiv y_r \pmod{p^{r+1}}$, $z_{r+1} \equiv z_r \pmod{p^{r+1}}$.

Thus, from now on, I shall take $n = m + 1$ in the congruence (3.1). With the same methods it is also possible, and in fact quite interesting, to study the congruence

$$X_1^{p^m} + X_2^{p^m} + \cdots + X_k^{p^m} \equiv 0 \pmod{p^{m+1}}. \tag{3.2}$$

As usual, a (nontrivial) solution $(x_1, \ldots, x_k)$ consists of integers satisfying the congruence and such that $1 \leq x_i \leq p^{m+1} - 1$, $p \nmid x_i$ (for all $i = 1, \ldots, k$).

Two solutions $(x_1, \ldots, x_k)$, $(y_1, \ldots, y_k)$ are equivalent if there exists some integer $a$, not a multiple of $p$, $1 \leq a \leq p^{m+1} - 1$, and a permutation $\pi$ of $\{1, 2, \ldots, k\}$ such that $y_i \equiv a x_{\pi(i)} \pmod{p^{m+1}}$ for $i = 1, \ldots, k$.

I denote by $\bar{a}$ the residue class of $a$ mod $p^{m+1}$ and by $(\mathbb{Z}/p^{m+1})^\times$ the multiplicative group of invertible residue classes modulo $p^{m+1}$.

Let $U$ be the subgroup of all $p^m$th powers $\bar{a}^{p^m}$ and let $V$ be the subgroup of all $\bar{b}$ such that $b \equiv 1 \pmod{p}$. It is an elementary fact that $(\mathbb{Z}/p^{m+1})^\times \cong U \times V$. Let $h \geq 1$ and

$$hU = \left\{ \sum_{i=1}^h \bar{a}_i \,\middle|\, \bar{a}_i \in U \quad \text{for } i = 1, \ldots, h \right\}.$$

Then (3.2) has a (nontrivial) solution exactly when $\bar{0} \in kU$.

Let $g$ be a primitive root modulo $p$, $1 < g < p$ and let $r \equiv g^{p^m} \pmod{p^{m+1}}$, $1 < r < p^{m+1}$. Then $U = \{\bar{r}, \bar{r}^2, \bar{r}^3, \ldots, \bar{r}^{p-1} = \bar{1}\}$. In other words, given $g$, every element $\bar{x}^{p^m} \in U$ is uniquely equal to some power $\bar{r}^i$ (with $0 \leq i \leq p - 2$). So every solution of (3.2) corresponds bijectively to a representation of $\bar{0}$ as sums of powers of $\bar{r}$ in $(\mathbb{Z}/p^{m+1})^\times$:

$$r^{i_1} + r^{i_2} + \cdots + r^{i_k} \equiv 0 \pmod{p^{m+1}}$$

with $0 \leq i_t \leq p - 2$ (for $t = 1, 2, \ldots, k$).

Two representations $(r^{i_1}, r^{i_2}, \ldots, r^{i_k})$ and $(r^{j_1}, r^{j_2}, \ldots, r^{j_k})$ are equivalent when they correspond to equivalent solutions of (3.2). In other words, there is a permutation $\pi$ of $\{1, 2, \ldots, k\}$ and an integer $h$, $0 \le h \le p - 2$, such that

$$i_t \equiv j_{\pi(t)} + h \pmod{p - 1}$$

for $t = 1, \ldots, k$.

A representation $(r^{i_1}, \ldots, r^{i_k})$ of 0 is *normalized* when

$$i_1 = 0 \le i_2 \le \cdots \le i_k \le p - 2.$$

It is clear that every representation is equivalent to one which is normalized. However, it may be shown that an equivalence class of representations may contain more than one which is normalized.

A *cyclic solution* of (3.2) is a solution $(x_1, x_2, \ldots, x_k)$ such that there exists an integer $a$, $p \nmid a$, for which

$$x_1 \equiv 1 \pmod{p^{m+1}},$$
$$x_2 \equiv a \pmod{p^{m+1}},$$
$$\vdots$$
$$x_j \equiv a^{j-1} \pmod{p^{m+1}},$$
$$\vdots$$
$$x_k \equiv a^{k-1} \pmod{p^{m+1}}.$$

The corresponding representation of 0 is

$$1 + r^i + r^{2i} + \cdots + r^{(k-1)i} \equiv 0 \pmod{p^{m+1}}, \tag{3.3}$$

where $a \equiv r^i \pmod{p^{m+1}}$, $0 \le i \le p - 2$.

It is easily seen that if $p \equiv 1 \pmod{k}$, then there is a cyclic representation, namely taking $i = (p - 1)/k$. In particular, if $k = 3$, $m = 1$, taking $i = (p - 1)/3$, then

$$1 + r^i + r^{2i} \equiv 0 \pmod{p^2}. \tag{3.4}$$

In this case $r^i$ is a cubic root of 1 $\pmod{p^2}$.

Klösgen showed (in 1970) the following criterion for Fermat's theorem, based on the existence of a noncyclic representation of 0:

**(3A)** *Let $m \ge 1$, let $p$ be a prime, $p \equiv 1 \pmod{3}$. If there exist integers $x$, $y$, $z$, not multiples of $p$, such that*

$$x^{p^m} + y^{p^m} + z^{p^m} = 0,$$

*then 0 has a noncyclic representation modulo $p^{3m+1}$.*

PROOF. To begin, Klösgen uses a generalization by Inkeri (1946) of Fleck's congruence [see Lecture IV, (3A)]. Namely, $x^p \equiv x \pmod{p^{2m+1}}$. Raising to the $p$th power repeatedly, this gives $x^{p^{3m}} \equiv x^{p^m} \pmod{p^{3m+1}}$. Similarly, $y^{p^{3m}} \equiv y^{p^m} \pmod{p^{3m+1}}$ and $z^{p^{3m}} \equiv z^{p^m} \pmod{p^{3m+1}}$.

By hypothesis, $x^{p^{3m}} + y^{p^{3m}} + z^{p^{3m}} \equiv 0 \pmod{p^{3m+1}}$. Let $w$ be an integer such that $wx \equiv y \pmod{p}$. From $x + y + z \equiv 0 \pmod{p}$ it follows that $z \equiv x(p - 1 - w) \pmod{p}$, so $x^{p^{3m}}(1 + w^{p^{3m}} + (p - 1 - w)^{p^{3m}}) \equiv 0 \pmod{p^{3m+1}}$. It suffices to show that the solution $(1, w, p - 1 - w)$ is not a cyclic solution.

Since $(x^{p^{m-1}})^p + (y^{p^{m-1}})^p + (z^{p^{m-1}})^p = 0$, it follows from Pollaczek's result (1917) that $x^{2p^{m-1}} + (xy)^{p^{m-1}} + y^{2p^{m-1}} \not\equiv 0 \pmod{p}$. Hence $x^2 + xy + y^2 \not\equiv 0$ $\pmod{p}$. This implies that the solution is not cyclic, because if

$$(p - 1 - w)^{p^{3m}} \equiv w^{2p^{3m}} \pmod{p^{3m+1}},$$

then $(1 + w)^{p^{3m}} + w^{2p^{3m}} \equiv 0 \pmod{p}$ hence $1 + w + w^2 \equiv 0 \pmod{p}$, which is a contradiction.                                                                       □

For example, let $p \equiv 1 \pmod 3$. If the only representation of 0 modulo $p^4$ is the cyclic representation, then the first case of Fermat's theorem holds for the exponent $p$.

Klösgen described all the normalized representations equivalent to a given one. In 1965, Peschl also showed how to obtain new normalized solutions from a given one (under certain hypotheses), but I'll refrain from entering into more details.

Instead, I now turn to the study of the number of solutions of (3.2). The method is similar to the one already indicated for the congruence modulo $p$, which gave Dickson's theorem.

Let $p > 2$ be a prime number, let $k \geq 3$ and $m \geq 0$. I denote by $F(p,m,k)$ the number of $(x_1, \ldots, x_k)$ such that $1 \leq x_i \leq p - 1$ (for $i = 1, \ldots, k$) and

$$x_1^{p^m} + x_2^{p^m} + \cdots + x_k^{p^m} \equiv 0 \pmod{p^{m+1}}.$$

Similarly, if $a$ is an integer, $1 \leq a \leq p - 1$, let $F(p,m,k;a)$ be the number of $k$-tuples as above, such that

$$x_1^{p^m} + x_2^{p^m} + \cdots + x_k^{p^m} \equiv ap^m \pmod{p^{m+1}}.$$

Let $N(p,m,k)$ be the number of $(x_2, \ldots, x_k)$ such that $1 \leq x_i \leq p - 1$ (for $i = 2, \ldots, k$) and

$$1 + x_2^{p^m} + \cdots + x_k^{p^m} \equiv 0 \pmod{p^{m+1}}.$$

For $k = 3$ and $m = 1$, I shall simply write: $F(p) = F(p,1,3)$, $F(p;a) = F(p,1,3;a)$, $N(p) = N(p,1,3)$.

For $m = 0$, it is not difficult to see:

$$F(p,0,k) = (p - 1)\frac{(p - 1)^{k-1} + (-1)^k}{p}, \tag{3.5}$$

$$N(p,0,k) = \frac{(p - 1)^{k-1} + (-1)^k}{p}. \tag{3.6}$$

In particular,

$$F(p,0,3) = (p - 1)(p - 2),$$
$$N(p,0,3) = p - 2.$$

More generally:

**(3B)** *With above notations:*

1. $F(p,m,k) = (p - 1)N(p,m,k)$.

2. $F(p,m,k;1) = F(p,m,k;2) = \cdots = F(p,m,k;p-1)$. (*This number shall be denoted by* $F^*(p,m,k)$.)
3. $N(p,m,k) = N(p, m-1, k) - F^*(p,m,k)$.

In particular:

$$N(p,1,k) = \frac{(p-1)^{k-1} + (-1)^k}{p} - F^*(p,1,k).$$

Thus, the determination of the number of solutions of (3.2) depends on the determination of $F^*(p,j,k)$, for $j = 1, 2, \ldots, m$. To this extent, the above relations play only an auxiliary role.

To arrive at an explicit formula for $N(p,m,k)$, Klösgen introduced the periods of the cyclotomic field $\mathbb{Q}(\zeta)$, where $\zeta$ is a primitive $p^m$th root of 1 ($p$ an odd prime, $m \geq 1$).

Let

$h = $ primitive root modulo $p^m$.
$\rho = $ primitive $p^{m+1}$th root of 1, such that $\rho^p = \zeta$.

The Gaussian periods of $p$ terms are $\eta_i = \eta_i(p,m,h)$ defined as follows:

$$\eta_i = \rho^{h^i} + \rho^{h^{p^m + i}} + \rho^{h^{2p^m + i}} + \cdots + \rho^{h^{(p-2)p^m + i}} \quad \text{(for } i = 0,1,\ldots,p^m - 1\text{)}.$$
$$(3.7)$$

Clearly

$$\sum_{i=0}^{p^m - 1} \eta_i = 0. \tag{3.8}$$

Extend the definition of $\eta_j$ for every index $j$, by letting $\eta_j = \eta_i$, where $j \equiv i \pmod{p^m}$, $0 \leq i \leq p^m - 1$.

Up to a change of numbering, the Gaussian periods are independent of the choice of the primitive root $h$.

Each $\eta_i$ turns out to be a real number and

$$|\eta_i| < p - 1. \tag{3.9}$$

If $p \nmid t$ let $\text{ind}_h(t) = s$, where $0 \leq s \leq (p-1)p^m - 1$, and $t \equiv h^s \pmod{p^{m+1}}$. The Jacobi sums are defined as in (1.6):

$$\tau_j = \langle \zeta^j, \rho \rangle = \sum_{\substack{t=1 \\ p \nmid t}}^{p^{m+1} - 1} \zeta^{j \, \text{ind}_h(t)} \rho^t \quad \text{(for } j = 0,1,\ldots,p^m - 1\text{)}.$$

In particular

$$\tau_0 = \langle 1, \rho \rangle = 0.$$

More generally,

$$\tau_j \neq 0 \quad \text{if and only if} \quad p \nmid j. \tag{3.10}$$

Also

$$\overline{\tau}_j = \tau_{p^m - j} \tag{3.11}$$

($\bar{\tau}_j$ denotes the complex conjugate of $\tau_j$) and

$$\tau_j \bar{\tau}_j = \begin{cases} p^{m+1} & \text{when } p \nmid j, \\ 0 & \text{when } p \mid j. \end{cases} \tag{3.12}$$

As a generalization of (1.11), the following connection between the periods and the Jacobi sums may be shown.

Consider the matrix $Z = (\zeta^{ij})_{i, j = 0, 1, \ldots, p^m - 1}$ and the vectors

$$\tau = \begin{pmatrix} \tau_0 \\ \tau_1 \\ \vdots \\ \tau_{p^m - 1} \end{pmatrix} \quad \text{and} \quad \eta = \begin{pmatrix} \eta_0 \\ \eta_1 \\ \vdots \\ \eta_{p^m - 1} \end{pmatrix}.$$

Then $Z\bar{Z} = p^m I$ ($I$ the identity matrix), that is,

$$\sum_{k=1}^{p^m} \zeta^{ik} \zeta^{-kj} = p^m \delta_{ij} \quad \text{(for } i, j = 0, 1, \ldots, p^m - 1\text{)}. \tag{3.13}$$

Also, $\bar{Z}\tau = p^m \eta$, that is,

$$\sum_{j=0}^{p^m - 1} \zeta^{-ij}\tau_j = p^m \eta_i \quad \text{(for } i = 0, 1, \ldots, p^m - 1\text{)} \tag{3.14}$$

and $\tau = Z\eta$, that is,

$$\sum_{j=0}^{p^m - 1} \zeta^{ij}\eta_j = \tau_i \quad \text{(for } i = 0, 1, \ldots, p^m - 1\text{)}. \tag{3.15}$$

As a corollary, the following special sums of periods vanish:

$$\sum_{j=0}^{p^k - 1} \eta_{i + jp^{m-k}} = 0 \tag{3.16}$$

for $1 \le k \le m$ and $i = 0, 1, \ldots, p^{m-k} - 1$.

Taking for example, $m = 2$, $k = 1$, then

$$\sum_{j=0}^{p-1} \eta_{i + jp} = 0. \tag{3.17}$$

It is also possible to evaluate certain sums of squares of periods:

$$\sum_{j=0}^{p^k - 1} \eta_{i + jp^{m-k}}^2 = p^k(p - 1) \tag{3.18}$$

for $1 \le k \le m$ and $i = 0, 1, \ldots, p^{m-k} - 1$.

In particular, taking $k = m$, then

$$\sum_{j=0}^{p^m - 1} \eta_j^2 = p^m(p - 1). \tag{3.19}$$

For $m = 1$, the Jacobi sums for $i = 0, 1, \ldots, p - 1$ may be evaluated as follows:

$$\tau_i = p\rho^{(ai)p}, \tag{3.20}$$

where $a(p - 1) = \text{ind}_h(1 + p)$.

In analogy with Lemma 1.1, Klösgen showed the following inductive expression for $F(p,m,k)$ and $N(p,m,k)$:

(3C) *With above notations*:

1. $F(p,m,k) = (1/p)F(p, m-1, k) + ((p-1)/p^{m+1})\sum_{i=0}^{p^m-1} \eta_i^k$.
2. $N(p,m,k) = (1/p)N(p, m-1, k) + (1/p^{m+1})\sum_{i=0}^{p^m-1} \eta_i^k$.

Taking $m = 1$, $k = 3$, this gives:

$$F(p) = \frac{(p-1)(p-2)}{p} + \frac{p-1}{p^2}\sum_{i=0}^{p-1} \eta_i^3, \tag{3.21}$$

$$N(p) = \frac{p-2}{p} + \frac{1}{p^2}\sum_{i=0}^{p-1} \eta_i^3. \tag{3.22}$$

Let $S(p^j,k) = \sum_{i=0}^{p^m-1} [\eta_i(p,j)]^k$. Then, the above recurrence relations become:

$$F(p,m,k) = \frac{p-1}{p^{m+1}}\left[(p-1)^{k-1} + (-1)^k + \sum_{j=1}^{m} S(p^j,k)\right], \tag{3.23}$$

$$N(p,m,k) = \frac{1}{p^{m+1}}\left[(p-1)^k + (-1)^k + \sum_{j=1}^{m} S(p^j,k)\right]. \tag{3.24}$$

The next task is to find an upper bound for the number of solutions $N(p,1,3) = N(p)$. In view of (3.22), what is required is to find an upper bound for the sum $\sum_{i=1}^{p-1} \eta_i^3$. Taking into account the relations (3.8) and (3.19), the question is answered by the following:

**Lemma 3.1.** *Let $n \geq 3$, let*

$$f(y_1, \ldots, y_n) = \sum_{i=1}^{n} y_i^3$$

*defined for all points $(y_1, \ldots, y_n)$ such that $\sum_{i=1}^{n} y_i = 0$ and $\sum_{i=1}^{n} y_i^2 = n(n-1)$.*

1. *If the function $f$ assumes an extreme value at the point $(y_1, \ldots, y_n)$, then there exists an integer $T$, $1 \leq T \leq n-1$, such that (up to a permutation of indices)*

$$y_1 = \cdots = y_T = (n-T)\sqrt{\frac{n-1}{T(n-T)}},$$

$$y_{T+1} = \cdots = y_n = -T\sqrt{\frac{n-1}{T(n-T)}}.$$

*If $y^T$ is the point with above coordinates, then:*

2. $f(y^T) = n(n-1)(n-2T)\sqrt{(n-1)/T(n-T)}$.
3. *If $T = 1$, then $f(y^T) = n(n-1)(n-2)$ is the absolute maximum of the function.*

4. *If $1 \leq T \leq n - 1$, then $f(y^T)$ is the absolute maximum of the function on the points $(y_1, \ldots, y_n)$ of the given domain, such that*

$$y_i \leq (n - T) \sqrt{\frac{n - 1}{T(n - T)}}.$$

With this lemma, it may be shown that

(3D) $N(p) < \max\{\eta_i | 0 \leq i \leq p - 1\}$.

Another question which may be answered is the asymptotic behavior of the number of solutions, as $k$ tends to $\infty$:

$$N(p,m,k) \sim \frac{(p - 1)^{k-1}}{p^{m+1}}.$$  (3.25)

Indeed, from the previous formulas:

$$\frac{N(p,m,k)}{(p - 1)^{k-1}/p^{m+1}} = 1 + \frac{(-1)^k}{(p - 1)^{k-1}} + (p - 1) \sum_{j=1}^{m} \frac{S(p^j,k)}{(p - 1)^k}$$

$$= 1 + \frac{(-1)^k}{(p - 1)^{k-1}} + (p - 1) \sum_{j=1}^{m} \sum_{i=1}^{p^j - 1} \left[ \frac{\eta_i(p,j)}{p - 1} \right]^k.$$

Since $|\eta_i(p,j)| < p - 1$ (by (3.9)) it follows that $\lim_{k \to \infty} [\eta_i(p,j)/(p - 1)]^k = 0$ proving the assertion.

Also, since $0 \leq N(p,m,k) \leq N(p, m - 1, k)$ [by (3B)], there exists an integer $m_0 = m_0(k,p) \geq 1$ such that $N(p,m,k) = N(p,m_0,k)$ for every $m \geq m_0$. This number may be interpreted as being equal to the number of solutions of the equation

$$1 + X_2 + \cdots + X_k = 0$$

by elements in $\hat{U}_p$ (the multiplicative group of $(p - 1)$th roots of 1 in the ring $\hat{Z}_p$ of $p$-adic integers).

Klösgen computed the values of $N(p,m,k)$ for various values of $m$, $k$. To illustrate these results, I give some examples below:

Table for $N(p,1,k)$

| $p$ \ $k$ | 3 | 4 | 5 | 6 | 7 |
|---|---|---|---|---|---|
| 5 | 0 | 9 | 0 | 100 | 35 |
| 7 | 2 | 15 | 60 | 340 | 1680 |
| 11 | 0 | 31 | 24 | 1600 | 5250 |
| 13 | 2 | 33 | 200 | 2260 | 21630 |
| 17 | 0 | 57 | 140 | 6220 | 50120 |
| 19 | 2 | 51 | 390 | 6880 | 101430 |

Table for $N(p,2,k)$ (with values underlined,
when not the same as in the preceding table)

| k  |     |     |     |      |       |
|----|-----|-----|-----|------|-------|
| p  | 3   | 4   | 5   | 6    | 7     |
| 5  | 0   | 9   | 0   | 100  | 0     |
| 7  | 2   | 15  | 60  | 340  | 1680  |
| 11 | 0   | 27  | 24  | 1090 | 2520  |
| 13 | 2   | 33  | 180 | 1930 | 15540 |
| 17 | 0   | 45  | 0   | 3160 | 945   |
| 19 | 2   | 51  | 300 | 4600 | 44520 |

The study of such congruences was also the object of numerous papers by Vandiver and by Hua, but their work is beyond the scope of my lectures. For the same reason, I shall also make only a passing reference to the work of Weil (1949), Igusa (1975) and others, on the number of solutions of diagonal forms $f(X) = X_1^d + \cdots + X_k^d$ over finite fields $\mathbb{F}_{p^s}$ and rings $\mathbb{Z}/\mathbb{Z}p^s$, where $p$ is a prime not dividing $d$, $s \geq 1$. Let $N_s$ (respectively $N_s^*$) denote the number of solutions of $f(X) = 0$ in $\mathbb{F}_{p^s}$ (respectively $\mathbb{Z}/\mathbb{Z}p^s$). Weil and Igusa proved respectively that $F(t) = 1 + \sum_{s=1}^{\infty} N_s t^s$, $F^*(t) = 1 + \sum_{s=1}^{\infty} N_s^* t^s$ are rational functions of the indeterminate $t$. Stevenson (1978) gave a new proof of Igusa's result, using the same method of Weil.

# Bibliography

1832  Libri, G.
    Mémoire sur la résolution de quelques équations indéterminées. *J. reine u. angew. Math.*, **9**, 1832, 277–294.

1837  Lebesgue, V. A.
    Recherches sur les nombres. *J. Math. Pures et Appl.* **2**, 1937, 253–292.

1838  Lebesgue, V. A.
    Recherches sur les nombres. *J. Math. Pures et Appl.*, **3**, 1838, 113–144.

1880  Pépin, T.
    Sur diverses tentatives de démonstration du théorème de Fermat. *C. R. Acad. Sci. Paris*, **91**, 1880, 366–368.

1887  Pellet, A. E.
    Mémoire sur la théorie algébrique des équations. *Bull. Soc. Math. France*, **15**, 1887, 61–102.

1908  Hensel, K.
    *Theorie der algebraischen Zahlen*, Teubner, Leipzig, 1908.

1909  Cornacchia, G.
    Sulla congruenza $x^n + y^n \equiv z^n \pmod{p}$. *Giornale di Mat.*, **47**, 1909, 219–268.

1909  Dickson, L. E.
    On the congruence $x^n + y^n + z^n \equiv 0 \pmod{p}$. *J. reine u. angew. Math.*, **135**, 1909, 134–141.

1909 Dickson, L. E.
Lower limit for the number of sets of solutions of $x^e + y^e + z^e \equiv 0 \pmod{p}$. *J. reine u. angew. Math.*, **135**, 1909, 181–188.

1909 Hurwitz, A.
Über die Kongruenz $ax^e + by^e + cz^e \equiv 0 \pmod{p}$. *J. reine u. angew. Math.*, **136**, 1909, 272–292. Reprinted in *Math. Werke*, II, Birkhäuser Verlag, Basel, 1933, 430–445.

1911 Pellet, A. E.
Réponse à une question de E. Dubouis. *L'Interm. des Math.*, **18**, 1911, 81–82.

1913 Landau, E.
Réponse à une question de C. Flye Sainte-Marie. *L'Interm. des Math.*, **20**, 1913, 154.

1917 Pollaczek, F.
Über den grossen Fermat'schen Satz. *Sitzungsber. Akad. d. Wien*, Abt. IIa, **126**, 1917, 45–59.

1917 Schur, I.
Über die Kongruenz $x^m + y^m \equiv z^m \pmod{p}$. *Jahresber. d. Deutschen Math. Verein.*, **25**, 1917, 114–117.

1922 Hardy, G. H. and Littlewood, J. E.
Some problems of "Partitio Numerorum": IV. The singular series in Waring's problem and the value of the number $G(k)$. *Math. Z.*, **12**, 1922, 161–188.

1946 Inkeri, K.
Untersuchungen über die Fermatsche Vermutung. *Annales Acad. Sci. Fennicae*, Ser. A, **I**, No. 33, 1946, 1–60.

1949 Weil, A.
Numbers of solutions of equations in finite fields. *Bull. Amer. Math. Soc.*, **55**, 1949, 497–508.

1956 LeVeque, W. J.
*Topics in Number Theory*, vol. II. Addison-Wesley, Reading, Mass., 1956.

1965 Peschl, E.
Remarques sur la résolubilité de la congruence $x^p + y^p + z^p \equiv 0 \pmod{p^2}$, $xyz \not\equiv 0 \pmod{p}$ pour un nombre premier impair. *Mém. Acad. Sci. Inscriptions et Belles Lettres de Toulouse*, **127**, $14^e$ série, 6, 1965, 121–127.

1970 Klösgen, W.
Untersuchungen über Fermatsche Kongruenzen. Gesellsch. f. Math. und Datenverarbeitung No. 36, 124 pages, 1970, Bonn.

1974 Brettler, E.
Towards a local Fermat problem. Unpublished manuscript, Kingston, 1974.

1975 Igusa, H.
Complex powers and asymptotic expansions, II. *J. reine u. angew. Math.*, **278/279**, 1975, 307–321.

1978 Stevenson, E.
The rationality of the Poincaré series of a diagonal form. Thesis, Princeton University, 1978.

# LECTURE XIII

# Variations and Fugue on a Theme

As composers sometimes do, when they strike a rich theme, mathematicians also like to consider variations of an interesting problem.

The theme is Fermat's problem, in all its force and splendor. The variations are analogous problems, sometimes reaching new depths, often only a caricature of the original problem. And the fugue, which usually ends the whole composition, keeps some of the original atmosphere. It begins with the well-known theme which, one might believe, has been completely exploited. Yet, with a happy use of technique, it gains in substance and soon transcends the original idea.

## 1. Variation I (In the Tone of Polynomial Functions)

Consider the problem of finding the solutions of the equation

$$X^n + Y^n = Z^n \tag{1.1}$$

(where $n \geq 2$) in the field of rational functions $K(t)$ (where $K$ is a field and $t$ is transcendental).

Since the equation is homogeneous, it is equivalent to find solutions in polynomials that is, elements of $K[t]$. But this ring is a unique factorization domain, so it suffices to look for the nontrivial solutions $(f,g,h)$, where $f, g, h \in K[t]$ and $\gcd(f,g,h) = 1$.

In 1879, Liouville studied the problem when $K = \mathbb{C}$, the field of complex numbers; he used analytical methods. In 1880, Korkine gave another proof, which is algebraic and valid for any field $K$ of characteristic 0. Another analytic proof with comments is in Shanks' book of 1962 (page 145), as well as in a paper of Greenleaf (1969). This latter proof actually is valid over any field whose characteristic does not divide the exponent in Fermat's equation.

First, I consider the case $n = 2$. If $K$ has characteristic 2, then any sum of squares is a square. So this case is trivial and may be left aside.

**(1A)** *Let $K$ be a field of characteristic distinct from 2, and assume that there exist elements $a$, $b$, $c \in K$ such that $a^2 + b^2 = c^2$. Then the equation $X^2 + Y^2 = Z^2$ has primitive nonconstant solutions $f$, $g$, $h$ in $K[t]$. If $f$ has degree 1, and $f^2 + g^2 = h^2$, then*

$$f = a + \frac{c \mp b}{a} dt,$$

$$g = b \pm dt \pm \left(\frac{c \mp b}{2a^2}\right) d^2 t, \qquad (1.2)$$

$$h = c + dt + \left(\frac{c \mp b}{2a^2}\right) d^2 t,$$

*where $d \in K$, $d \neq 0$.*

The proof of this result contains no secrets.

Now, I consider the case $n > 2$ and I present the proof of Greenleaf:

**(1B)** *Let $n > 2$ and let $K$ be a field of characteristic not dividing $n$. Any primitive solution of (1.1) in $K[t]$ consists of constant polynomials.*

PROOF. Without loss of generality, I may assume $K$ algebraically closed.

Suppose that (1.1) admits a primitive solution $(f,g,h)$, such that the maximum $m$ of the degrees of $f$, $g$, $h$ is positive. Among all such solutions, I consider one for which $m$ is minimum.

$K$ contains a primitive $n$th root $\zeta$ of 1, in view of the hypothesis. Let

$$g^n = h^n - f^n = \prod_{j=0}^{n-1} (h - \zeta^j f).$$

Since the polynomials $h - \zeta^j f$ are pairwise relatively prime and $K[t]$ is a unique factorization domain, each one is an $n$th power:

$$h - \zeta^j f = g_j^n, \quad \text{where } g_j \in K[t].$$

But $h - f$, $h - \zeta f$, $h - \zeta^2 f$ are in the vector space generated over $K$ by $h, f$. So there exist $a_0, a_1, a_2 \in K$, not all equal to 0, such that

$$a_0(h - f) + a_1(h - \zeta f) + a_2(h - \zeta^2 f) = 0.$$

Actually, these coefficients have to be all distinct from zero, because $f$, $h$ are relatively prime.

Put $k_j = \sqrt[n]{a_j g_j} \in K[t]$. Then

$$k_0^n + k_1^n + k_2^n = 0$$

with $\deg(k_j) < m$ $(j = 0,1,2)$, and $\max\{\deg(k_0), \deg(k_1), \deg(k_2)\} > 0$. This contradicts the choice of $m$. $\qquad\square$

Pierre Samuel showed me this proof, which is more sophisticated. If $f^n + g^n = h^n$, where $f$, $g$, $h \in K[t]$ are relatively prime nonconstant polynomials, then $(f/h, g/h)$ is a generic point of Fermat's curve $X^n + Y^n = 1$. Hence $K(f/h, g/h) \subset K(t)$. By Lüroth's theorem, Fermat's curve is rational, that is of genus 0, and this implies that $n = 1$ or 2.

## 2. Variation II (In the Tone of Entire Functions)

This time, the search is for solutions in entire functions. But I'll not consider the question in its entirety (never mind the pun). I'll just pick a few striking results and proofs.
    Consider the equation

$$X^2 + Y^2 = 1. \tag{2.1}$$

An obvious nontrivial solution in entire functions is $f(z) = \cos z$, $h(z) = \sin(z)$. Similarly, if $h(z)$ is any entire function, then $\cos(h(z))$, $\sin(h(z))$ is a solution in entire functions. But conversely, Iyer proved in 1939:

**(2A)** *If $f(z)$, $g(z)$ are entire functions satisfying (2.1), then there exists an entire function $h(z)$ such that $f(z) = \cos(h(z))$, $g(z) = \sin(h(z))$.*

PROOF. From $f(z)^2 + g(z)^2 = 1$ it follows that $[f(z) + ig(z)][f(z) - ig(z)] = 1$. Thus the two factors are entire functions without any zero. Then there exists $h(z)$, an entire function such that

$$f(z) + ig(z) = e^{ih(z)} \qquad f(z) - ig(z) = e^{-ih(z)}.$$

Therefore $f(z) = (e^{ih(z)} + e^{-ih(z)})/2 = \cos(h(z))$ and $g(z) = (e^{ih(z)} - e^{-ih(z)})/2i = \sin(h(z))$. $\qquad\square$

Now I consider the case where $n \geq 3$ and instead of (2.1), the equation is

$$X^n + Y^n = p(z), \tag{2.2}$$

where $p(z)$ is a nonzero polynomial function of degree at most $n - 2$. The following result was obtained by Gross in 1966. The special case where $p(z) = 1$ was first done by Iyer in 1939, and then by Jategaonkar in 1965:

**(2B)** *If $n \geq 3$, if $p(z)$ is a nonzero polynomial of degree at most $n - 2$, if $f(z)$ and $g(z)$ are entire functions such that $f(z)^n + g(z)^n = p(z)$, then $f(z)$, $g(z)$, and $p(z)$ are constants.*

PROOF. I'll give the proof under the stronger hypothesis that $p(z)$ has degree at most $n - 3$. For the case where $p(z)$ has degree $n - 2$ the proof is somewhat more technical.

Let $\zeta$ be a primitive $2n$th root of 1. Then

$$\frac{p(z)}{g(z)^n} = \prod_{j=1}^{n} \left( \frac{f(z)}{g(z)} + \zeta^{2j-1} \right).$$

The meromorphic function $p(z)/g(z)^n$ has at most $n - 3$ zeroes. So, at least three factors of the right-hand side never vanish. Thus, the meromorphic function $f(z)/g(z)$ misses three values. By the well-known theorem of Picard, $f(z)/g(z)$ is a constant.

Now, it is anticlimatic to conclude the proof of the statement.                    □

Obvious corollaries are the following:

**(2C)** *If $n \geq 3$ and $f(z)$, $g(z)$ are entire functions such that*

$$f(z)^n + g(z)^n = 1,$$

*then $f(z)$, $g(z)$ are constants.*

**(2D)** *If $n \geq 3$, if $f(z)$, $g(z)$, $h(z)$ are nonzero entire functions such that $h(z)$ never vanishes, and if*

$$f(z)^n + g(z)^n = h(z)^n,$$

*then there exist nonzero complex numbers $a$, $b$ such that $f(z) = ah(z)$, $g(z) = bh(z)$, $a^n + b^n = 1$.*

## 3. Variation III (In the Theta Tone)

In the preceding variation, it was quite exciting seeing that the (little) Picard theorem provided the key for the proof.

This time, the search is for solutions which are analytic in the interior of the unit disk. And the exponent is just equal to 4.

In 1829, Jacobi discovered a solution for

$$X^4 + Y^4 = Z^4$$

in theta functions. Since these functions play quite an important role in various branches of arithmetic, analysis, and geometry, I'll take the time to introduce them and indicate the main steps towards Jacobi's result.

It is convenient to introduce the theta functions as solutions of certain functional equations.

Let $t$ belong to the upper half-plane and let $q = e^{\pi i t}$, so $|q| < 1$, that is, $q$ is in the interior of the unit disk.

Let $k \geq 1$ be an integer, let $t$ be given in the upper half-plane, let $b = b(t) \in \mathbb{C}$, $b \neq 0$ and consider the vector space $V$ (depending on $k,t,b$) of all entire functions $f(z)$ satisfying

$$f(z + \pi) = f(z) \qquad f(z + \pi t) = be^{-2kiz}f(z). \qquad (3.1)$$

**(3A)** *V has dimension k.*

SKETCH OF THE PROOF. Let $f(z) \in V$. For every integer $n \neq 0$,

$$c_n = \frac{1}{\pi} \int_{-\pi/2}^{\pi/2} f(z)e^{-2\pi iz}\,dz$$

is the coefficient of the Fourier expansion $\sum_{n=-\infty}^{\infty} c_n e^{2niz}$ of the periodic function $f(z)$ in the interval $(-\pi/2, \pi/2)$.

For every integer $h \neq 0$, consider the rectangle $C_h$, with vertices $-\pi/2$, $\pi/2$, $(\pi/2) + ih$, $-(\pi/2) + ih$. By Cauchy's theorem and the hypothesis (3.1)

$$c_n = e^{2nh} \int_{-\pi/2}^{\pi/2} f(z - ih)e^{2niz}\,dz.$$

Thus

$$|c_n| \leq e^{2nh}\pi M_h,$$

where

$$M_h = \max_{-\pi/2 \leq z \leq \pi/2} \{|f(z - ih)|\}.$$

Writing $t = u + iv$ (where $u, v$ are real, $v > 0$), writing $h/\pi = mv + r$, with $0 \leq r < v$, and $m$ an integer, and writing $mu = s + u'$, with $-\frac{1}{2} < u' < \frac{1}{2}$, and $s$ an integer, it is not difficult to show that

$$M_h = b^{-m}e^{2km\pi v}A,$$

where

$$A = \max_{-\pi/2 \leq z \leq \pi/2} \{e^{2kmz}|f(z + \pi(u' - ir))|\}.$$

From these majorations, it is possible in a more or less standard way, to deduce that the Fourier series is absolutely and uniformly convergent in any bounded region. Thus

$$f(z) = \sum_{n=-\infty}^{\infty} c_n e^{2niz}.$$

The second functional equation of $f(z)$ gives relations between the coefficients of the Fourier expansion,

$$c_{n+k} = b^{-1}q^{2n}c_n.$$

Hence, if $n = sk + r$, with integers $r, s$, $s > 0$, $0 \leq r < k$, by iteration

$$c_n = c_{r+sk} = b^{-s}q^{2(sr + \frac{1}{2}s(s-1)k)}c_r.$$

This relation holds also for $s < 0$.
Let

$$f_r(z) = e^{2riz} \sum_{s=-\infty}^{\infty} b^{-s}q^{2(sr + \frac{1}{2}s(s-1)k)}e^{2skiz}.$$

From the uniqueness of the Fourier expansion, it follows that $f(z)$ is, in a unique way, a linear combination of the functions $f_0(z), f_1(z), \ldots, f_{k-1}(z)$, which belong to the vector space $V$, and therefore constitute a basis.     □

The special case where $k = 1$ and $b = q^{-1}$ gives the function denoted by

$$\Theta_3(z,q) = \sum_{s=-\infty}^{\infty} q^{s^2} e^{2siz}$$

$$= 1 + 2q \cos 2z + 2q^4 \cos 4z + 2q^9 \cos 6z + \cdots . \qquad (3.2)$$

Taking $k = 1$ and $b = -q^{-1}$ gives the function

$$\Theta_4(z,q) = \sum_{s=-\infty}^{\infty} (-1)^s q^{s^2} e^{2siz}$$

$$= 1 - 2q \cos 2z + 2q^4 \cos 4z - 2q^9 \cos 6z + \cdots . \qquad (3.3)$$

Similarly, considering the functional equations

$$f(z + \pi) = -f(z) \qquad f(z + \pi t) = be^{-2iz} f(z) \qquad (3.4)$$

it is seen that the space of such functions has dimension 1, and a generator is, for example, the function

$$f_0^*(z) = \sum_{s=-\infty}^{\infty} b^{-s} q^{s^2} e^{(2s+1)iz}.$$

Taking $b = -q^{-1}$, respectively $b = q^{-1}$, gives the functions

$$\Theta_1(z,q) = -iq^{1/4} \sum_{s=-\infty}^{\infty} (-1)^s q^{s^2+s} e^{(2s+1)iz}$$

$$= 2q^{1/4}(\sin z - q^2 \sin 3z + q^6 \sin 5z - q^{12} \sin 7z + \cdots) \qquad (3.5)$$

$$\Theta_2(z,q) = q^{1/4} \sum_{s=-\infty}^{\infty} q^{s^2+s} e^{(2s+1)iz}$$

$$= 2q^{1/4}(\cos z + q^2 \cos 3z + q^6 \cos 5z + q^{12} \cos 7z + \cdots). \qquad (3.6)$$

The functions $\Theta_i(z,q)$ ($i = 1,2,3,4$) are Jacobi *theta functions*. I denote also $\theta_i(q) = \Theta_i(0,q)$, so these are functions defined in the interior of the unit disk.

These functions are of course mutually related. For example, for every $q$ in the interior of the unit disk:

$$\Theta_1\left(z + \frac{\pi}{2}, q\right) = \Theta_2(z,q), \qquad \Theta_2\left(z + \frac{\pi}{2}, q\right) = -\Theta_1(z,q), \qquad (3.7)$$

$$\Theta_3\left(z + \frac{\pi}{2}, q\right) = \Theta_4(z,q), \qquad \Theta_4\left(z + \frac{\pi}{2}, q\right) = \Theta_3(z,q), \qquad (3.8)$$

$$\Theta_1\left(z + t\frac{\pi}{2}, q\right) = iq^{-1/4}e^{-iz}\Theta_4(z,q), \qquad \Theta_4\left(z + t\frac{\pi}{2}, q\right) = iq^{-1/4}e^{-iz}\Theta_1(z,q)$$

(3.9)

$$\Theta_2\left(z + t\frac{\pi}{2}, q\right) = q^{-1/4}e^{-iz}\Theta_3(z,q), \qquad \Theta_3\left(z + t\frac{\pi}{2}, q\right) = q^{-1/4}e^{-iz}\Theta_2(z,q)$$

(3.10)

and also

$$\Theta_1(0,q) = 0. \tag{3.11}$$

Between the squares of any three theta functions, there are also linear relations:

**(3B)** *The following four linear relations are satisfied:*

$$\Theta_4^2(0,q)\Theta_1^2(z,q) + \Theta_3^2(0,q)\Theta_2^2(z,q) = \Theta_2^2(0,q)\Theta_3^2(z,q), \tag{3.12}$$

$$\Theta_3^2(0,q)\Theta_1^2(z,q) + \Theta_4^2(0,q)\Theta_2^2(z,q) = \Theta_2^2(0,q)\Theta_4^2(z,q), \tag{3.13}$$

$$\Theta_3^2(0,q)\Theta_1^2(z,q) + \Theta_4^2(0,q)\Theta_3^2(z,q) = \Theta_2^2(0,q)\Theta_4^2(z,q), \tag{3.14}$$

$$-\Theta_2^2(0,q)\Theta_2^2(z,q) + \Theta_3^2(0,q)\Theta_3^2(z,q) = \Theta_4^2(0,q)\Theta_4^2(z,q). \tag{3.15}$$

PROOF. I'll prove only the first relation.

The squares of the theta functions satisfy the functional equations

$$f(z + \pi) = f(z) \qquad f(z + \pi t) = q^{-2}e^{-4iz}f(z). \tag{3.16}$$

By (3A) the space of entire functions satisfying these equations has dimension 2. So, the squares of any three theta functions satisfy a nontrivial linear relation.

If $a$, $b$, $c$ are complex numbers such that

$$a\Theta_1^2(z,q) + b\Theta_2^2(z,q) = c\Theta_3^2(z,q)$$

computing these functions at $z = \pi/2$, $t(\pi/2)$, $(\pi/2) + t(\pi/2)$ and taking into account the relations indicated above yields

$$a\Theta_2^2(0,q) + b\Theta_1^2(0,q) = c\Theta_4^2(0,q),$$
$$-aq^{-1/2}\Theta_4^2(0,q) + bq^{-1/2}\Theta_3^2(0,q) = cq^{-1/2}\Theta_2^2(0,q),$$
$$aq^{-1/2}\Theta_3^2(0,q) - bq^{-1/2}\Theta_4^2(0,q) = -cq^{-1/2}\Theta_1^2(0,q).$$

But $\Theta_1(0,q) = 0$, $\Theta_2(0,q) \neq 0$, hence

$$a = \frac{\Theta_4^2(0,q)}{\Theta_2^2(0,q)}c, \qquad b = \frac{\Theta_3^2(0,q)}{\Theta_2^2(0,q)}c, \quad \text{with } c \neq 0.$$

This gives the first relation. $\qquad\qquad\qquad\qquad\qquad\qquad\qquad\qquad\square$

It follows from relation (3.15) for $z = 0$ that

$$\Theta_2^4(0,q) + \Theta_4^4(0,q) = \Theta_3^4(0,q) \tag{3.17}$$

and this may be rewritten as:

(3C) *The theta functions* $\theta_2(q)$, $\theta_4(q)$, $\theta_3(q)$ *satisfy the relation*

$$\theta_2^4(q) + \theta_4^4(q) = \theta_3^4(q)$$

*in the interior of the unit disk.*

The proof whose main points I sketched may be found in the book of Bellman (1961). Another completely elementary proof has been published by van der Pol in 1955.

# 4. Variation IV (In the Tone of Differential Equations)

In 1962, Rodrigues-Salinas looked at the following differential equation, in analogy to Fermat's equation:

$$\frac{\partial^n u}{\partial x^n} + \frac{\partial^n u}{\partial y^n} = \frac{\partial^n u}{\partial z^n} \tag{4.1}$$

There is, surprisingly, a relationship between the solution of Fermat's equation

$$X^n + Y^n = Z^n \tag{4.2}$$

and the solution of the differential equation in functions $u(x,y,z)$, satisfying certain periodicity and boundary conditions.

Let $Q$ be the unit cube in the space $\mathbb{R}^3$, that is, the set of triples $(x,y,z)$ such that $0 \leq x \leq 1$, $0 \leq y \leq 1$, $0 \leq z \leq 1$. Suppose that $u(x,y,z)$ is a real-valued function of three real variables, with continuous partial derivatives up to the order $n$. Assume also

a. $u$ has period 1 in each variable.
b. There exists $M > 0$ such that, for any $x$, $y$, $z$:

$$\left| \frac{\partial^n u}{\partial x^n}(x,y,z) \right| \leq M, \qquad \left| \frac{\partial^n u}{\partial y^n}(x,y,z) \right| \leq M, \qquad \left| \frac{\partial^n u}{\partial z^n}(x,y,z) \right| \leq M.$$

c. $u$ is a solution of (4.1).

(4A) *With the above hypotheses, the Fourier expansion of* $u(x,y,z)$ *is convergent:*

$$u(x,y,z) = \sum_{p,q,r=-\infty}^{\infty} A_{p,q,r} e^{2\pi i(px+qy+rz)}$$

*and if* $p^n + q^n \neq r^n$, *then* $A_{p,q,r} = 0$.

The next results distinguish the cases where $n$ is odd or even. To state these we consider two further conditions on a solution $u$:

d. $u(x,y,z)$ vanishes on any of the faces of unit cube $Q$.
d'. $u(x,y,z)$ vanishes on two faces of $Q$.

(4B) *If $n$ is odd and if $u(x,y,z)$ satisfies the conditions* (a), (b), (c), *and* (d), *then $u(x,y,z)$ is identically zero.*

(4C) *If $n \geq 2$ is even, the following are equivalent*:

1. *There exist nonzero integers $a$, $b$, $c$ such that $a^n + b^n = c^n$.*
2. *There exists a function $u(x,y,z)$, not identically zero, satisfying* (a), (b), (c), *and* (d').

I'll not try to sketch the proof of these results, since they may be found in the paper of Rodrigues-Salinas. Let me just say that if $a^2 + b^2 = c^2$, then $u(x,y,z) = (\sin 2\pi ax)(\sin 2\pi by)(\sin 2\pi cz)$.


# 5. Variation V (Giocoso)

Great tragedies often have some scenes of humor, tense musical symphonies their moments of gaiety. Mathematicians of the most serious kind also like to be amused.

While collecting information about Fermat's theorem, I gathered various items which I have decided not to hide.

To begin, I wish to quote a paper of Orts Aracil (1961). He said:

As Fermat tried to find the solutions of $X^n + Y^n = Z^n$ in positive integers, he must have considered first the easier cases $n = 3$, $n = 4$, obtaining the result that no such solutions exist. In view of this, he decided to tackle directly the general equation and begun searching whether the equation admits solutions of the form

$$x = a^p, \qquad y = a^q, \qquad z = a^r,$$

where $a$, $p$, $q$, $r$ are positive integers.
He required the equality

$$(a^p)^n + (a^q)^n = (a^r)^n$$

that is

$$a^{np} + a^{nq} = a^{nr}$$

which may be written also:

$$(a^n)^p + (a^n)^q = (a^n)^r,$$

and putting $a^n = m$,

$$m^p + m^q = m^r.$$

Previously, in the paper, Orts Arcil considered the dual of Fermat's equation

$$n^X + n^Y = n^Z$$

and he proved:

**(5A)** *If $n \geq 3$, there are no positive integers $x$, $y$, $z$ such that $n^x + n^y = n^z$. However, for $n = 2$, if $2^x + 2^y = 2^z$, then $x = y$, $z = x + 1$.*

He says:

> Coming to the theme proper of this article, which constitutes the key of our conjecture, let us observe that if in Fermat's equation
>
> $$X^n + Y^n = Z^n$$
>
> the bases and exponents are permuted, it becomes
>
> $$n^X + n^Y = n^Z,$$
>
> an equation to which, ¡curiosa coincidencia!, exactly and literally the same statement may be applied: this equation has no positive integer solutions if $n \geq 3$.
>    And even more curious, is that the proof in this case is immediate, at the level of any student of the first courses of mathematics at the university.

He adds that he proved this point experimentally. And he concludes:

> This result [(5A)], no doubt known to Fermat, might have induced him, "sin más" to formulate his famous theorem.

"Caramba," it might have been indeed a "curiosa coincidencia". One which spurred a great progress to mathematics.


## 6. Variation VI (In the Negative Tone)

In 1967 and 1968, Thérond considered the equation $X^n + Y^n = Z^n$ where $n$ is a strictly negative integer. And he proved:

**(6A)** *If Fermat's theorem is true and if $n < -2$, then the equation $X^n + Y^n = Z^n$ has no solution in nonzero integers.*

**(6B)** *The equation $X^{-1} + Y^{-1} = Z^{-1}$ has primitive solutions $(x,y,z)$ which are of the form*

$$x = a(a + b), \qquad y = b(a + b), \qquad z = ab,$$

*where $a, b$ are nonzero integers, $\gcd(a,b) = 1$.*

**(6C)** *The equation $X^{-2} + Y^{-2} = Z^{-2}$ has primitive solutions $(x,y,z)$ which are of the form*

$$x = 2ab(a^2 + b^2), \qquad y = (a^2 - b^2)(a^2 + b^2), \qquad z = 2ab(a^2 - b^2),$$

*where $a$, $b$ are nonzero distinct integers such that $\gcd(a,b) = 1$.*

# 7. Variation VII (In the Ordinal Tone)

In 1950, Sierpiński proved that the last theorem of Fermat is false for ordinal numbers:

For every ordinal number $\mu$, there exist distinct ordinal numbers $\alpha$, $\beta$, $\gamma$, each larger than $\mu$ and such that

$$\alpha^n + \beta^n = \gamma^n \quad \text{for } n = 1, 2, 3, \ldots .$$

This is the first of various similar results showing that Fermat's equation

$$X^\lambda + Y^\lambda = Z^\lambda,$$

where $\lambda$ is any ordinal number, also admits solutions which are not trivial.

Once more, it is brought to light how strange is the arithmetic of ordinal numbers.

# 8. Variation VIII (In a Nonassociative Tone)

Nonassociative arithmetics? Yes, they have been invented, and perhaps soon they may enter into mathematics with some unexpected applications. While this day has not yet arrived, it is nevertheless fortunate that mathematicians know whether their Fermat's last theorem is true or false in a nonassociative arithmetic.

I suspect that not everyone is familiar with this new theory. So, I will expose its general ideas.

I trace the first paper to Etherington in 1939 (and again in 1949). Under another name, they were considered also by A. Robinson (1949) and Evans (1957). It was Evans who proved Fermat's last theorem in such nonassociative arithmetics; see also Minc (1959).

It is best to follow the method of Peano, used to define the natural numbers. Let $S$ be a set of elements which will be called *nanumbers* (nonassociative numbers), satisfying the following conditions:

a. There is a nanumber, denoted by 1.
b. There is a binary operation $+$ between nanumbers.
c. There are no nanumbers $a$, $b$ such that $a + b = 1$.

d. If $a, b, c, d$ are nanumbers and $a + b = c + d$, then $a = c, b = d$.
e. (Principle of nonassociative induction): If $S'$ is a subset of $S$, such that $1 \in S'$ and $S'$ is closed under addition, then $S' = S$.

The nanumbers are therefore represented as $1$, $1 + 1$, $1 + (1 + 1)$, $(1 + 1) + 1$, $1 + (1 + (1 + 1))$, $1 + ((1 + 1) + 1)$, $(1 + 1) + (1 + 1)$, $((1 + 1) + 1) + 1$, $(1 + (1 + 1)) + 1$, and so on.

The natural nanumbers are $1, 1 + 1$ denoted $2$, and similarly $1 + (1 + 1) = 1 + 2 = 3$, $1 + 3 = 4$, etc. . . . .

Each nanumber has a length, defined as follows: $|1| = 1$; if $a = b + c$, then $|a| = |b| + |c|$; this is well defined in view of the axioms (c) and (d).

I will now state various properties:

1. If $a + b = b + a$, then $a = b$.
2. For all $a, b$: $a \neq a + b$.
3. For all $a, b, c$: $a + (b + c) \neq (a + b) + c$.

The multiplication of nanumbers is defined as follows:

$$a1 = a \quad \text{and} \quad a(b + c) = ab + ac.$$

Clearly $|ab| = |a| \cdot |b|$. Among the properties, I note:

4. If $ab = 1$, then $a = 1$ and $b = 1$.
5. $1a = a$ for all $a$.
6. $a(bc) = (ab)c$ for all $a, b, c$.
7. If $ba = ca$, then $b = c$.
8. If $ab = ac$, then $b = c$.

From the associative property of multiplication, it is possible to define unambiguously exponentiation, where $n$ is a natural number: $a^1 = a$, $a^n = aa^{n-1} = a^{n-1}a$.

If $a = bc$, then $b$ is called a left factor of $a$ and $c$ is a right factor of $a$. If, moreover, $b, c$ are different from $1, a$, they are called proper factors. If $a \neq 1$ has no proper left factor, then it is called a prime nanumber. It follows that it has no proper right factor.

9. If $c$ is a proper left factor of $a + b$, then $c$ is a left factor of $a$ and of $b$.

In particular, $1 + b, a + 1$ are primes.

10. If the prime $c$ is a left factor of $ab$, with $a \neq 1$, then $c$ is a left factor of $a$.

And now, the fundamental theorem of the arithmetic of nanumbers.

11. Every nanumber is, in unique way, a product of prime nanumbers.
12. If a prime nanumber $p$ is a right factor of $ab$, and $b \neq 1$, then $p$ is a right factor of $b$.

With all this preparation, I give the proof of Fermat's last theorem for nanumbers:

**(8A)** *If $n$ is a natural number, $n \geq 2$, then the equation $X^n + Y^n = Z^n$ has no solution in nanumbers.*

PROOF. Assume that $x$, $y$, $z$ are nanumbers such that $x^n + y^n = z^n$.

Note that $x \neq 1$, otherwise $z^n = 1 + y^n$ would be prime [by (9)], which is not true, because $n \geq 2$. Similarly, $y \neq 1$. I note also that $|x|^n + |y|^n = |z|^n$.

Since $z$ is a proper left factor of $z^n = x^n + y^n$, by (9) $z$ is a left factor of $x^n$ and of $y^n$. By the fundamental theorem (11),

$$x = p_1 p_2 \cdots \cdot p_s \qquad (s \geq 1),$$
$$z = q_1 q_2 \cdots \cdot q_t \qquad (t \geq 1).$$

Since $zu = x^n$ (for some nanumber $u$), $q_1 \cdots q_t u = p_1 p_2 \cdots p_s x^{n-1}$.

If $t \leq s$, by the uniqueness of decomposition into a product of prime nanumbers, $q_1 = p_1, \ldots, q_t = p_t$, hence $x = z p_{t+1} \cdots p_s$ so $|x| = |z| \, |p_{s+1}| \cdots |p_s| \geq |z|$. On the other hand, $|z|^n = |x|^n + |y|^n > |x|^n$ so $|z| > |x|$, which is a contradiction.

Therefore $t > s$ and $q_1 = p_1, \ldots, q_s = p_s$ so $z = xa$, where $a = q_{s+1} \cdots q_t \neq 1$.

A similar argument gives $z = yb$ where $b \neq 1$. Therefore $|z|^n = |x|^n + |y|^n = (|z|^n/|a|^n) + (|z|^n/|b|^n)$, hence $1 = (1/|a|^n) + (1/|b|^n)$, where $|a|, |b| > 1$. This is clearly impossible.                                                       □

# 9. Variation IX (In the Matrix Tone)

Not much is known concerning the solution of Fermat's equation in square matrices. The question is trivial if singular matrix solutions are allowed. Indeed, if $k \geq 2$, if

$$A = \begin{pmatrix} 1 & 0 \\ 0 & 0 \end{pmatrix}, \qquad B = \begin{pmatrix} 0 & 0 \\ 0 & 1 \end{pmatrix}, \qquad C = \begin{pmatrix} 1 & 0 \\ 0 & 1 \end{pmatrix}$$

and if

$$\tilde{A} = \begin{pmatrix} A & 0 \\ 0 & 0 \end{pmatrix}, \qquad \tilde{B} = \begin{pmatrix} B & 0 \\ 0 & 0 \end{pmatrix}, \qquad \tilde{C} = \begin{pmatrix} C & 0 \\ 0 & 0 \end{pmatrix}$$

are $k \times k$ matrices, then they are idempotent, $\tilde{A} + \tilde{B} = \tilde{C}$, so $\tilde{A}^n + \tilde{B}^n = \tilde{C}^n$ for every $n \geq 1$. This was explicitly noted by Bolker in 1968 and, in special cases, by Domiaty in 1966.

Hence, I shall only be interested in solutions in nonsingular matrices. Bolker proved the first result of any interest:

**(9A)** *Let $R$ be a commutative ring with unit element. Let $m \geq 1$ and assume that there exist nonzero elements $x$, $y$, $z \in R$ such that $x^m + y^m = z^m$. Then there exist nonsingular $n \times n$ matrices $A$, $B$, $C$ over $R$ such that*

$$A^{mn} + B^{mn} = C^{mn}.$$

PROOF. If $\pi$ is any permutation on $n$ letters let $P = \rho(\pi)$ be the $n \times n$ matrix over $R$:

$$P = (\delta_{\pi(i),j})_{i,j},$$

that is, the entry of $P$ position $(i,j)$ is $\delta_{\pi(i),j}$ (the Kronecker $\delta$-symbol):
     If $\pi$ is any circular permutation, $\rho(\pi) = P$, let

$$A = \text{diag}(x,1,\ldots,1)P,$$
$$B = \text{diag}(y,1,\ldots,1)P,$$
$$C = \text{diag}(z,1,\ldots,1)P.$$

Then the matrices $A$, $B$, $C$ are nonsingular and satisfy Fermat's equation for the exponent $nm$.          $\square$

Taking $m = 1$ and nonzero elements $x$, $y$, $z$ such that $x + y = z$, then the above construction gives $n \times n$ nonsingular matrices $A$, $B$, $C$ such that $A^n + B^n = C^n$.

In 1972, Brenner and de Pillis determined other solutions in nonsingular $n \times n$ matrices for the equation in (9A). These authors have also proved:

**(9B)** *If $A$, $B$, $C$ are nonsingular $n \times n$ matrices with entries in $\mathbb{Z}$, if $AB = BA$ and if $A^m + B^m = C^m$, where $m \geq 1$, then there exists a nontrivial triple of algebraic integers $a$, $b$, $c$, each of degree at most $n$ and such that $a^m + b^m = c^m$.*

This follows at once from the existence of a common eigenvector for the commuting matrices $A$, $B$.
     A partial converse, proved by Brenner and de Pillis is the following:

**(9C)** *If $a$, $b$, $c$ are integers in a quadratic number field, such that $a^m + b^m = c^m$, where $m \geq 2$, then there exist nonsingular $2 \times 2$ matrices $A$, $B$, $C$, with entries in $\mathbb{Z}$, such that $A^m + B^m = C^m$.*

In 1961, Barnett and Weitkamp studied in more detail the solution of Fermat's equation in $2 \times 2$ nonsingular matrices with rational coefficients. This is their main result.

**(9D)** *Let $n > 2$, $n \neq 4$. If Fermat's theorem holds for the exponent $n$ (when $n$ is odd) or for the exponent $n/2$ (when $n$ is even) if $A$, $B$ are nonsingular $2 \times 2$ matrices over $\mathbb{Q}$, which are not scalar matrices but such that $A^n$, $B^n$ are scalar matrices, then there exists no $2 \times 2$ nonsingular matrix $C$, with entries in $\mathbb{Q}$, such that*

$$A^n + B^n = C^n.$$

Barnett and Weitkamp gave also many examples of matrices $A$, $B$, $C$ such that $A^n + B^n = C^n$:

1. Take

$$A = \begin{pmatrix} -1 & -1 \\ 0 & 1 \end{pmatrix}, \qquad B = \begin{pmatrix} 1 & 2 \\ -1 & -1 \end{pmatrix},$$

$$C = \begin{pmatrix} -2 & -3 \\ 1 & 2 \end{pmatrix}, \qquad n \geq 3, n \equiv 3 \pmod 4.$$

2. Take

$$A = \begin{pmatrix} -3 & -5 \\ 2 & 3 \end{pmatrix}, \qquad B = \begin{pmatrix} 1 & 2 \\ -1 & -1 \end{pmatrix},$$

$$C = \begin{pmatrix} -2 & -3 \\ 1 & 2 \end{pmatrix}, \qquad n \geq 1, n \equiv 1 \pmod 4.$$

In the above examples the matrices have fourth power equal to the identity matrix.

3. If $r$ is any rational number, and if

$$A = \begin{pmatrix} r + \dfrac{2}{3} & 1 \\ -r^2 - \dfrac{r}{3} - \dfrac{34}{9} & \dfrac{1}{3} - r \end{pmatrix},$$

$$B = \begin{pmatrix} r & 1 \\ 2r - r^2 - 1 & 2 - r \end{pmatrix},$$

$$C = \begin{pmatrix} -2 & 0 \\ r + \frac{25}{21} & -1 \end{pmatrix},$$

then $A^3 + B^3 = C^3$.

And there are more examples and methods, but I'll not discuss this matter any further.

# 10. Fugue (In the Quadratic Tone)

After all the variations, the long awaited fugue comes and it is written in the quadratic tone, which most suits it. What I really mean, is that I'll now consider the solution of Fermat's equation in algebraic number fields. And here again, the trivial cases have to be excluded. For example, $1^n + 1^n + (\sqrt[n]{-2})^n = 0$ tells us that in $\mathbb{Q}(\sqrt[n]{-2})$ the equation

$$X^n + Y^n + Z^n = 0 \tag{10.1}$$

has a nontrivial solution. So, the problem is interesting only when the algebraic number field has degree less than the exponent.

How about the simplest situation: To find out whether (10.1) has a solution in quadratic number fields? This study has been done only for a few exponents: $n = 2, 3, 4, 6, 9$.

If $n = 2$ it makes a difference to consider (10.1) with exponent 2, or the equation $X^2 + Y^2 = Z^2$. This equation has nontrivial solutions in ordinary integers, but it may be asked whether there are solutions of the type $(a + b\sqrt{m}, a - b\sqrt{m}, c)$, where $a, b, c \in \mathbb{Z}$, $m$ is a square-free integer, $m \neq 0, 1$. Aigner answered this question in 1934:

**(10A)** *There exist nonzero integers* $a, b, c \in \mathbb{Z}$ *such that*

$$(a + b\sqrt{m})^2 + (a - b\sqrt{m})^2 = c^2$$

*if and only if* $m$ *has no prime factor* $p$ *such that* $p \equiv \pm 3 \pmod 8$.

On the other hand, for the equation $X^2 + Y^2 + Z^2 = 0$, it is only interesting to investigate the solutions in an imaginary quadratic field. The following result was first proved by Nagell in 1972; a simpler proof was given by Szymiczek (1974):

**(10B)** *If* $m > 0$ *is square-free, then the equation* $X^2 + Y^2 + Z^2 = 0$ *has a nontrivial solution in the imaginary quadratic field* $\mathbb{Q}(\sqrt{-m})$ *if and only if* $m \not\equiv -1 \pmod 8$.

Another way of looking at this result is the following. The level of an imaginary field is the smallest number $s$ of squares such that $-1$ is the sum of $s$ squares. Since the field is imaginary, such number $s$ exists, as was shown by Artin and Schreier (1927). Thus, the above theorem says that the level of $\mathbb{Q}(\sqrt{-m})$ is at most 2 if and only if $m \not\equiv -1 \pmod 8$.

Aigner settled the case of exponent 4 in 1934; a new proof was given by Fogels in 1938:

**(10C)** *The equation* $X^4 + Y^4 = Z^4$ *has a nontrivial solution in the field* $\mathbb{Q}(\sqrt{m})$ *(where* $m$ *is a square-free integer,* $m \neq 0, 1$*) if and only if* $m = -7$.

In this situation, every nontrivial solution is equivalent (that is proportional with a nonzero multiplier) to one of the following solutions (with arbitrary signs):

$$(\pm(1 + \sqrt{-7}), \pm(1 - \sqrt{-7}), \pm 2).$$

The proof is somewhat elaborate, but involves no real difficulty.

Since I'm not giving the proofs for the exponents 6, 9 (which depend on the case of the exponent 3), I may as well state the results right now. Once more, they were obtained by Aigner in 1957:

**(10D)** *If* $m \neq 0, 1$ *is any square-free integer, then* $X^6 + Y^6 = Z^6$ *has only the trivial solution in* $\mathbb{Q}(\sqrt{m})$.

**(10E)** *If $m \neq 0, 1$ is any square-free integer, then $X^9 + Y^9 = Z^9$ has only the trivial solution in $\mathbb{Q}(\sqrt{m})$.*

The proof of (10D) is also based on a result due to Nagell (1924):

$$X^3 - 1 = 3Y^2 \tag{10.2}$$

has only the trivial solution $(1,0)$ over the field $\mathbb{Q}$ of rational numbers. I stress that there are no other rational solutions. It is already not totally easy to show that there are no other integral solutions of (10.2) than $(1,0)$. See a proof given by Obláth (1952).

Concerning the rational solutions, the analogous equation $X^3 + 1 = Y^2$ had been treated already by Euler. Nagell has in fact considered the equation $X^3 + 1 = DY^2$, where $D \neq 0$ is an arbitrary integer. Taking $D = -3$, the equation is equivalent to (10.2).

These considerations illustrate the fact—to be expected—that to solve diophantine equations over quadratic fields, leads to the solutions over $\mathbb{Q}$, or $\mathbb{Z}$, of some associated diophantine equations.

Let me also remark that (10D) and (10E) would be automatically true, if the Fermat cubic

$$X^3 + Y^3 + Z^3 = 0 \tag{10.3}$$

had only the trivial solution. But, as I shall indicate, this is far from being the case. The questions become: for which quadratic fields does (10.3) have nontrivial solutions, and in this situation, to describe all possible solutions.

To fix the terminology, two solutions $(x,y,z)$ and $(x',y',z')$ in a field $K$ are $K$-equivalent if there exists $a \in K$, $a \neq 0$, such that $x' = ax$, $y' = ay$, $z' = az$. A solution $(x,y,z)$ is called a *quadratic solution* if $x$, $y$, $z$ belong to some quadratic extension of $\mathbb{Q}$.

In 1915, Burnside described all the nontrivial quadratic solutions of Fermat's cubic; they may be parametrized by the rational numbers different from $0, -1$. This was rediscovered by Duarte in 1944, who gave a simpler proof.

**(10F)**

1. *If $k \in \mathbb{Q}$, $k \neq 0, -1$, and*

$$x_k = 3 + \sqrt{-3(1 + 4k^3)},$$
$$y_k = 3 - \sqrt{-3(1 + 4k^3)},$$
$$z_k = 6k,$$

*then $x_k^3 + y_k^3 + z_k^3 = 0$.*
2. *If $a \in \mathbb{Q}$, $a \neq 0$, if $k' = ak$ (with $k \neq 0$), then the solutions corresponding to $k$ and $k'$ by the above method are equivalent if and only if $a = 1$.*
3. *If $(x,y,z)$ is a nontrivial solution in the quadratic field $\mathbb{Q}(\sqrt{m})$, there exists $k \in \mathbb{Q}$, $k \neq 0, -1$, such that*

$$m = -3(1 + 4k^3)u^2, \tag{10.4}$$

*where u is a rational number and $(x,y,z)$ is $\mathbb{Q}(\sqrt{m})$-equivalent to the solution $(x_k, y_k, z_k)$.*

This parametrization falls short of the expectations, because $k$ is a rational parameter. It gives of course many quadratic solutions, for example:

$k = 2$ gives $(3 + 3\sqrt{-11}, 3 - 3\sqrt{-11}, 12)$,
$k = 1$ gives $(3 + \sqrt{-15}, 3 - \sqrt{-15}, 6)$,
$k = -2$ gives $(3 + \sqrt{93}, 3 - \sqrt{93}, -12)$.

On the other hand, it is not obvious at once that there is a solution in $\mathbb{Q}(\sqrt{-2})$, since it is not obvious that

$$-2 = -3(1 + 4k^3)u^2$$

has a rational solution $k$. But, in fact taking $k = \frac{1}{2}$, $u = \frac{2}{3}$, this gives the solution $(2 + \sqrt{-2}, 2 - \sqrt{-2}, 2)$.

In order to study the quadratic solutions of the Fermat cubic, it is helpful to remark that any solution is equivalent (in the field in question) to a conjugate solution, which is one of the form

$$x = a + b\sqrt{m},$$
$$y = a - b\sqrt{m},   \tag{10.5}$$
$$z = c.$$

This was established by Fueter in 1930 and again by Aigner in 1952.
And now comes a very crucial fact shown by Fueter in 1913:

**(10G)** *The Fermat cubic has a conjugate solution in $\mathbb{Q}(\sqrt{m})$ if and only if it has a conjugate solution in $\mathbb{Q}(\sqrt{-3m})$.*

PROOF. I give this proof, since I shall require the explicit formulas to pass from the solution in one field to the other.

Let (10.5) be a conjugate solution in $\mathbb{Q}(\sqrt{m})$. Then

$$-c^3 = 2a(a^2 + 3mb^2).   \tag{10.6}$$

Writing $(a + b\sqrt{m})^3 = \alpha + \beta\sqrt{m}$ (with $\alpha$, $\beta \in \mathbb{Z}$), then $(a - b\sqrt{m})^3 = \alpha - \beta\sqrt{m}$ so

$$(a^2 - b^2m)^3 = \alpha^2 - \beta^2 m.$$

From $x^3 + y^3 + z^3 = 0$ it follows that

$$\alpha = a(a^2 + 3b^2m) = -\tfrac{1}{2}c^3,$$
$$\beta = b(3a^2 + b^2m).   \tag{10.7}$$

Then

$$\xi = 3\alpha + \beta\sqrt{-3m}$$
$$\eta = 3\alpha - \beta\sqrt{-3m}   \tag{10.8}$$
$$\zeta = 3c(a^2 - b^2m)$$

is a conjugate solution in $\mathbb{Q}(\sqrt{-3m})$. It may be verified that this solution is not trivial.

To show the converse, I just note that $\mathbb{Q}(\sqrt{m}) = \mathbb{Q}(\sqrt{-3(-3m)})$. $\quad\square$

From a more sophisticated point of view, the above method of deducing solutions in $\mathbb{Q}(\sqrt{-3m})$ from solutions in $\mathbb{Q}(\sqrt{m})$ corresponds to the complex multiplication by $\sqrt{-3}$ on the elliptic curve $X^3 + Y^3 + Z^3 = 0$. More information on this point may be obtained in Cassels's survey article (1966), §§ 24 and 26.

Repeating the procedure used in the proof of (10G) yields the following fact:

**(10H)** *If Fermat's cubic has a nontrivial solution in the field $\mathbb{Q}(\sqrt{m})$, then it has infinitely many (pairwise nonequivalent) solutions in this field.*

I will not prove this, but illustrate with a typical numerical example. Beginning with the solution $(2 + \sqrt{-2}, 2 - \sqrt{-2}, 2)$ the proof of (10G) gives the solution

$$(-6 + 5\sqrt{6}, -6 - 5\sqrt{6}, 18).$$

Repeating the procedure, I obtain the new solution

$$(-4374 + 1935\sqrt{-2}, -4374 - 1935\sqrt{-2}, -3078).$$

It may be shown that the new solution is never equivalent to the given one.

From these considerations, the main question is to decide whether in a given field $\mathbb{Q}(\sqrt{m})$ there exists a nontrivial solution. As I said, this amounts to solving (10.4) in rational numbers $m, u$. This task cannot be handled directly as such.

From Fueter's result, I may assume that $3 \nmid m$. Four cases are possible, according to the sign of $m$, and $m \equiv \pm 1 \pmod{3}$. For some unexplained reason, the results seem very much to differ according to the case. In some of the cases the class number of the quadratic field plays a role, in others it doesn't. I shall denote by $H(m)$ the class number of the quadratic field $\mathbb{Q}(\sqrt{m})$, where $m$ is square-free, $m \neq 0$. In 1913, Fueter proved:

**(10I)** *If $m$ is a square-free integer, $m < 0$, $m \equiv -1 \pmod{3}$, and if there exists a nontrivial solution for Fermat's cubic equation, then 3 divides the class number $H(m)$.*

Fueter's proof involved a long analysis of behavior of primes in appropriate extensions. He gave also actual examples of solutions in various fields. For example, in $\mathbb{Q}(\sqrt{-31})$, which has class number 3, he computed the following solutions:

$$\left(\frac{9 + \sqrt{-31}}{2}\right)^3 + \left(\frac{9 - \sqrt{-31}}{2}\right)^3 + 3^3 = 0$$

and

$$(36 + 5\sqrt{-31})^3 + (36 - 5\sqrt{-31})^3 + 42^3 = 0.$$

Incidentally, applying the method of (10H) to the first of the above solutions does not lead to the second one. So, it is not true that the method indicated will provide all solutions from a given one. And one wonders how many "independent" solutions will suffice to describe all, after repeated application of (10H). This is still unknown.

Fueter also considered the case where $m < 0$, $m \equiv 1 \pmod 3$ and he gave examples of solutions both when $H(m)$ is a multiple, or not a multiple of 3:

In $\mathbb{Q}(\sqrt{-2})$, of class number 1, there are the solutions

$$(2 + \sqrt{-2})^3 + (2 - \sqrt{-2})^3 + 2^3 = 0$$

and

$$(16 + \sqrt{-2})^3 + (16 - \sqrt{-2})^3 + (-20)^3 = 0$$

(once more the second solution is not derived from the first one by applying the method of (2H)).

In $\mathbb{Q}(\sqrt{-23})$, of class number 3, Fueter found the solution

$$\left(\frac{9 + \sqrt{-23}}{2}\right)^3 + \left(\frac{9 - \sqrt{-23}}{2}\right)^3 + (-3)^3 = 0.$$

In 1952, Aigner took up the case where $m > 0$. His first result makes essential use of an important theorem of Scholz (1932), which connects the 3-rank of the class groups of $\mathbb{Q}(\sqrt{m})$ and $\mathbb{Q}(\sqrt{-3m})$.

Let $\mathscr{C}$ be the class group of $\mathbb{Q}(\sqrt{m})$ (where $m > 0$) and let $\mathscr{C}'$ be the class group of $\mathbb{Q}(\sqrt{-3m})$. Let $\mathscr{C}_3$ (respectively $\mathscr{C}'_3$) be the subgroup of elements of order dividing 3 in $\mathscr{C}$ (respectively $\mathscr{C}'$). So $\mathscr{C}_3$, $\mathscr{C}'_3$ are finite-dimensional vector spaces over the field $\mathbb{F}_3$ with 3 elements. Let $s = \dim_{\mathbb{F}_3}(\mathscr{C}_3)$, $s' = \dim_{\mathbb{F}_3}(\mathscr{C}'_3)$. Then Scholz proved:

**(10J)** *With the above notations*

$$s \leq s' \leq s + 1.$$

In particular, if 3 does not divide $H(-3m)$, then 3 does not divide $H(m)$.

The above theorem is not an isolated fact, but a special case of Leopoldt's reflection theorem (1958), explained in Lecture IX.

Back to Aigner—he proved (1952):

**(10K)** *If $m$ is a square-free integer, $m > 0$ and $m \equiv 1 \pmod 3$, if there exists a nontrivial solution of Fermat's cubic in $\mathbb{Q}(\sqrt{m})$, then 3 divides $H(-3m)$.*

As an example of this situation, Aigner showed that $(486 + 11\sqrt{58})^3 + (486 - 11\sqrt{58})^3 - 630^3 = 0$; in this case $H(-174) = 12$.

Next, he observed that if $m > 0$, $m \equiv -1 \pmod 3$, then the existence of a nontrivial solution is independent of the divisibility of the class number by

3, as the following examples illustrate:
In $\mathbb{Q}(\sqrt{2})$, with $H(-6) = 2$:

$$(18 + 17\sqrt{2})^3 + (18 - 17\sqrt{2})^3 - 42^3 = 0.$$

In $\mathbb{Q}(\sqrt{82})$, with $H(-246) = 12$:

$$(2 + \sqrt{82})^3 + (2 - \sqrt{82})^3 - 10^3 = 0.$$

The situation being fairly mysterious, Aigner proceeded to investigate more deeply the problem in a series of very interesting papers (1952, 1955, 1956). His first result is the following:

**(10L)** *Let $m$ be a square-free integer, $3 \nmid m$, $m > 0$. Assume the following condition: $m$ is a product of primes $q_i$ such that $q_i \equiv 1$ (mod 3) and 2 is not a cubic residue modulo $q_i$. Then Fermat's cubic has only the trivial solution in $\mathbb{Q}(\sqrt{m})$ and in $\mathbb{Q}(\sqrt{-m})$.*

For example, since 2 is not a cubic residue modulo 61 then Fermat's cubic has only the trivial solution in $\mathbb{Q}(\sqrt{-61})$; yet $H(-61) = 6$. This shows that the converse of Fueter's result (10I) is false.

Similarly, 2 is not a cubic residue modulo 67, so Fermat's cubic has only the trivial solution in $\mathbb{Q}(\sqrt{67})$; yet $H(-201) = 12$. This shows that the converse of Aigner's theorem (10K) is false.

Next Aigner improved his last result as follows:

**(10M)** *Let $p$ be a prime such that $p \equiv 5$ (mod 6) and $\varepsilon = \sqrt[3]{2} - 1$ is not a square modulo $p$. Let $m = 2pq_1 \cdots q_r$, with $r \geq 0$, where $q_i$ are distinct primes satisfying the conditions: $q_i \equiv 1$ (mod 3), 2 is not a cubic residue modulo $q_i$. Then Fermat's cubic has only the trivial solution in $\mathbb{Q}(\sqrt{m})$ and in $\mathbb{Q}(\sqrt{-m})$.*

These results are, however, only the most special cases of a theory developed by Aigner in 1956. I cannot enter into any of the details, but I hope to explain clearly his results.

The object is to find integers $k$ which will guarantee that the only solution of Fermat's cubic in certain fields will be trivial. Accordingly, an integer $k > 0$ (not a multiple of 3) is an *obstructing* integer ($= Unmöglichkeitskernzahl$) for Fermat's cubic when Fermat's cubic has only the trivial solution in $\mathbb{Q}(\sqrt{m})$ and in $\mathbb{Q}(\sqrt{-m})$ for every $m = kq_1 \cdots q_r$, with $r \geq 0$, where $q_i$ are distinct primes satisfying the conditions:

1. $q_i \equiv 1$ (mod 3) for $i = 1, \ldots, r$,
2. 2 is not a cubic residue modulo $q_i$.

Thus, the two preceding theorems say that 1 is an obstructing integer and if $p$ is a prime, $p \equiv 5$ (mod 6) and $\varepsilon = \sqrt[3]{2} - 1$ is not a square modulo $p$, then $2p$ is an obstructing integer.

To obtain results about obstructing integers, Aigner required considerations in the cubic field $K = \mathbb{Q}(\sqrt[3]{2})$. This field has an integral basis

$\{1, \sqrt[3]{2}, \sqrt[3]{4}\}$. Every integer $\alpha \in K$ is therefore uniquely written in the form $\alpha = a + b\sqrt[3]{2} + c\sqrt[3]{4}$, with $a, b, c \in \mathbb{Z}$. This is abbreviated as $\alpha = (a,b,c)$.

The class number of $K$ is 1. It has a fundamental unit $\varepsilon = \sqrt[3]{2} - 1$. The norm of $\alpha = (a,b,c)$ is $N(\alpha) = a^3 + 2b^3 + 4c^3 - 6abc$. Let $\bar{\alpha}$ be defined by $\alpha\bar{\alpha} = N(\alpha)$. Then $\bar{\alpha} = (a^2 - 2bc, 2c^2 - ab, b^2 - ac)$.

An algebraic integer $\alpha = (a,b,c)$ is even when $2\,|\,a$, that is, $\sqrt[3]{2}$ divides $\alpha$. It is odd when $2 \nmid a$, that is, $\sqrt[3]{2} \nmid \alpha$. If $\alpha$ is odd, then either $\alpha \equiv \varepsilon \pmod{\sqrt[3]{4}}$ or $\alpha \equiv 1 \pmod{\sqrt[3]{4}}$. This latter case happens if and only if $2 \nmid b$ and then $\alpha$ is called a primary odd integer. If $\alpha, \beta$ are primary, so are $\alpha\beta$ and $\bar{\alpha}$.

The following rational integers play a role: $d_1(\alpha) = a^2 - 2bc$, $d_2(\alpha) = b^2 - ac$, and $d_3(\alpha) = 4c^2 - 2ab$.

A first result is the following: If $\alpha = (a,b,c)$ is a primary integer, then the Jacobi symbol

$$\left(\frac{d_2(\alpha)}{N(\alpha)}\right) = +1.$$

Next, Aigner showed that if $\alpha = (a,b,c)$, $\alpha' = (a',b',c')$ are square-free relatively prime primary integers such that $N(\alpha)$, $N(\alpha')$ are relatively prime square-free integers, then

$$\left(\frac{\bar{\alpha}}{\alpha'}\right) = \left(\frac{\bar{\alpha'}}{\alpha}\right).$$

This is a reciprocity law, which allowed Aigner to define a symbol $T(p/q)$.

Namely, first let $p, q$ be distinc  primes, $p \equiv 5 \pmod 6$, $q \equiv 5 \pmod 6$. Then there exist primary numbers $\alpha, \alpha'$, unique up to square factors, such that $\alpha$ divides $p$, $\alpha'$ divides $q$. Since

$$\left(\frac{\bar{\alpha}}{\alpha'}\right) = \left(\frac{\bar{\alpha'}}{\alpha}\right),$$

Aigner defined unambiguously

$$T\left(\frac{p}{q}\right) = \left(\frac{\bar{\alpha'}}{\alpha}\right).$$

So the symbol $T(p/q)$ has value 1 or $-1$, and

$$T\left(\frac{p}{q}\right) = T\left(\frac{q}{p}\right).$$

This symbol was extended by Aigner for all pairs of primes, irrespectively of their residue modulo 6, but I'll not enter into this definition.

Then Aigner proved (1956):

(10N) *If $p, q$ are distinct primes, such that $p \equiv 5 \pmod 6$, $q \equiv 5 \pmod 6$ and*

$$T\left(\frac{p}{q}\right) = -1,$$

*then, k = pq is an obstructing number for Fermat's cubic.*

For example $1135 = 5 \times 227$ and $1927 = 41 \times 47$ are obstructing numbers, so Fermat's cubic has only the trivial solution in $\mathbb{Q}(\sqrt{\pm 1135})$ and in $\mathbb{Q}(\sqrt{\pm 1927})$, irrespective of the values of the class number.

On the other hand, $T(\frac{5}{17}) = +1$ and there is the solution in $\mathbb{Q}(\sqrt{85})$:

$$(1 + \sqrt{85})^3 + (1 - \sqrt{85})^3 - 8^3 = 0.$$

For several primes $p_i$, $p_i \equiv 5 \pmod 6$, when is $k = p_1 \cdots p_r$ an obstructing number? Aigner found a sufficient condition in terms of a graph $\Gamma$ associated with $k$. The vertices of $\Gamma$ are points corresponding to the primes $p_1, \ldots, p_r$ and labelled by these primes. There is an edge between $p_i$, $p_j$ exactly when

$$T\left(\frac{p_i}{p_j}\right) = -1.$$

A graph $\Gamma$ is called an *obstructing graph* when the following condition is satisfied: If $V \neq \varnothing$ is any set of vertices of $\Gamma$, there exists a vertex $v_0$ such that the total number of edges from $v_0$ to the vertices in $V$ is odd. (I note that $v_0$ may be chosen in $V$.)

Examples of obstructing graphs are the following:

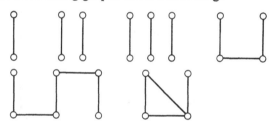

Examples of nonobstructing graphs are:

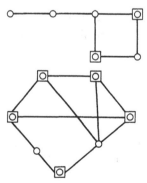

(the boxed vertices constitute the subset $V$ of the definition)

In this context, the main theorem is the following:

**(10O)** *If the graph of k is an obstructing graph, then k is an obstructing integer for Fermat's cubic.*

So the problem becomes one of determining conditions for a graph to be obstructing. Aigner found the following graph-theoretical theorem:

**(10P)** *If T is an obstructing graph, then the number of vertices must be even.*

As a corollary:

**(10Q)** *If k is the product of an odd number of distinct primes $q_i$ where $q_i = 2$ or $q_i \equiv 5 \pmod 6$, then k is not an obstructing number.*

A more precise analysis allowed Aigner to show:

**(10R)** *If $k \equiv 2 \pmod 3$, then k is not an obstructing integer.*

The following examples of obstructing integers are among the innumerous possibilities which illustrate Aigner's results:

$64042 = 2 \times 11 \times 41 \times 71$ has the graph

$$
\begin{array}{cc}
2 & 41 \\
| & | \\
11 & 71
\end{array}
$$

$21505 = 5 \times 11 \times 17 \times 23$ has the graph

$$
\begin{array}{cc}
5 & 17 \\
| & | \\
11 - & 23
\end{array}
$$

$3910 = 2 \times 5 \times 17 \times 23$ has the graph

$$
\begin{array}{cc}
2 & 17 \\
|\backslash & | \\
5 - & 23
\end{array}
$$

I stop here, refraining from giving still other results from Aigner. At any rate, this should be enough to incite the reader's curiosity about this problem.

# Bibliography

1829   Jacobi, C. G. J.
     Fundamenta Nova Theoriae Functionum Ellipticarum. *Gesammelte Werke*, I, Königsberg, 1829, 49–239. Königl. Preussischen Akad. d. Wiss., Reimer, Berlin, 1881.

1879   Liouville, R.
     Sur l'impossibilité de la relation algébrique $X^n + Y^n + Z^n = 0$. *C.R. Acad. Sci. Paris*, **87**, 1879, 1108–1110.

1880   Korkine, A.
     Sur l'impossibilité de la relation algébrique $X^n + Y^n + Z^n = 0$. *C.R. Acad. Sci. Paris*, **90**, 1880, 303–304.

1913 Fueter, R.
Die Diophantische Gleichung $\xi^3 + \eta^3 + \zeta^3 = 0$. Sitzungsberichte Heidelberg Akad. d. Wiss., 1913, 25. Abh., 25 pp.

1915 Burnside, W.
On the rational solutions $X^3 + Y^3 + Z^3 = 0$ in quadratic fields *Proc. London Math. Soc.*, **14**, 1915, 1–4.

1924 Nagell, T.
Über die rationale Punkte auf einigen kubischen Kurven. *Tôhoku Math. J.*, **24**, 1924, 48–53.

1927 Artin, E. and Schreier, O.
Eine Kennzeichnung der reell abgeschlossenen Körper. *Hamburg Abhandl.*, **5**, 1927, 225–231. Reprinted in Artin's *Collected Papers* Addison-Wesley, Reading, Mass., 1965, 289–295.

1930 Fueter, R.
Über kubische diophantische Gleichungen. *Comm. Math. Helv.*, **2**, 1930, 69–89.

1932 Scholz, A.
Über die Beziehung der Klassenzahlen quadratischer Zahlkörper zueinander. *J. reine u. angew. Math.* **166**, 1932, 201–203.

1934 Aigner, A.
Über die Möglichkeit von $x^4 + y^4 = z^4$ in quadratischen Körpern. *Jahresber. d. Deutschen Math. Verein.* **43**, 1934, 226–229.

1938 Fogels, E.
Über die Möglichkeit einiger diophantischer Gleichung 3. und 4. Grades in quadratischen Körpern. *Comm. Math. Helv.* **10**, 1938, 263–269.

1939 Etherington, I. M. H.
On non-associative combinations. *Proc. R. Soc. Edinburgh*, Section A, **59**, 1939, 153–162.

1939 Iyer, G.
On certain functional equations. *J. Indian Math. Soc.*, **3**, 1939, 312–315.

1944 Duarte, F. J.
On the equation $x^3 + y^3 + z^3 = 0$. *Bul. Acad. Ci. Fis. Mat. Nat. (de Venezuela)* **8**, 1944, 971–979.

1949 Etherington, I. M. H.
Non-associative arithmetics. *Proc. R. Soc. Edinburgh*, Section A, **62**, 1949, 442–453.

1949 Robinson, A.
On non-associative systems. *Proc. Edinburgh Math. Soc.*, **8**, 1949, 111–118.

1950 Sierpiński, W.
Le dernier théorème de Fermat pour les ordinaux. *Fund. Math.* **37**, 1950, 201–205.

1952 Aigner, A.
Weitere Ergebnisse über $x^3 + y^3 = z^3$ in quadratischen Körpern. *Monatsh. f. Math.*, **56**, 1952, 240–252.

1952 Aigner, A.
Ein zweiter Fall der Unmöglichkeit von $x^3 + y^3 = z^3$ in quadratischen Körpern mit durch 3 teilbaren Klassenzahl. *Monatsh. f. Math.*, **56**, 1952, 335–338.

1952 Obláth, R.
Über einige unmögliche diophantische Gleichungen. *Matem. Tidsskrift*, ser. A, 1952, 53–62.

1955 van der Pol, B.
Démonstration élémentaire de la relation $\theta_3^4 = \theta_0^4 + \theta_2^4$. *Enseignment Math.*, (2), 1, 1955, 258–261.

1956   Aigner, A.
   Die kubische Fermatgleichung in quadratischen Körpern. *J. reine u. angew. Math.*
   **195**, 1956, 3–17.

1956   Aigner, A.
   Unmöglichkeitskernzahlen der kubischen Fermatgleichung mit Primfaktoren der
   Art $3n + 1$. *J. reine u. angew. Math.* **195**, 1956, 175–179.

1957   Aigner, A.
   Die Unmöglichkeit von $x^6 + y^6 = z^6$ und $x^9 + y^9 = z^9$ in quadratischen Körpern.
   *Monatsh. f. Math.*, **61**, 1957, 147–150.

1957   Evans, T.
   Non-associative number theory. *Amer. Math. Monthly*, **64**, 1957, 299–309.

1958   Leopoldt, H. W.
   Zur Struktur der $l$-Klassengruppe Galoisscher Zahlkörper. *J. reine u. angew.
   Math.*, **199**, 1958, 165–174.

1959   Minc, H.
   Theorems on non-associative number theory. *Amer. Math. Monthly*, **66**, 1959,
   486–488.

1961   Barnett, I. A. and Weitkamp, H. M.
   The equation $X^n + Y^n + Z^n = 0$ in rational binary matrices. *An. Stir. Univ. "Al. I.
   Cuza", Iasi*, Seçt 1, (NS) **7**, 1961, 1–64.

1961   Bellman, R.
   *A Brief Introduction to Theta Functions*, Holt, Rinehart and Winston, New York,
   1961.

1961   Orts Aracil, J. M.
   A conjecture concerning Fermat's last theorem (in Spanish). *Mem. Real Acad. Ci.
   Art. Barcelona*, **34**, 1961, 17–25.

1962   Rodrigues-Salinas, B.
   On Fermat's last theorem and the equation $(\partial^n u / \partial x^n) + (\partial^n u / \partial y^n) = \partial^n u / \partial z^n$ (in
   Spanish). *Univ. Lisbôa, Rev. Fac. Ci*, A, (2), **9**, 1962, 35–43.

1962   Shanks, D.
   *Solved and Unsolved Problems in Number Theory*, I, Spartan, Washington, 1962.
   Reprinted by Chelsea Publ. Co., New York, 1979.

1965   Jategaonkar, A. V.
   Elementary proof of a theorem of P. Montel on entire functions. *J. London Math.
   Soc.* **40**, 1965, 166–170.

1966   Cassels, J. W. S.
   Diophantine equations with special reference to elliptic curves. *J. London Math.
   Soc.*, **41**, 1966, 193–291.

1966   Domiaty, R. Z.
   Solutions of $x^4 + y^4 = z^4$ in $2 \times 2$ integral matrices. *Amer. Math. Monthly*, **73**,
   1966, 631.

1966   Domiaty, R. Z.
   Lösungen der Gleichung $x^n + y^n = z^n$ mit $n = 2^m$ im Ring gewisser ganzzahliger
   Matrizen. *Elem. Math.*, **21**, 1966, 5–7.

1966   Gross, F.
   On the functional equation $f^n + g^n = h^n$. *Amer. Math. Monthly*, **73**, 1966, 1093–
   1096.

1966   Gross, F.
   On the equation $f^n + g^n = 1$. *Bull. Amer. Math. Soc.*, **72**, 1966, 86–88.

1967   Thérond, J. D.
L'hypothèse de Fermat pour les exposants négatifs. *Enseignement Math.* (2), **13**, 1967, 247–252.

1968   Bolker, E. D.
Solutions of $A^k + B^k = C^k$ in $n \times n$ integral matrices. *Amer. Math. Monthly*, **75**, 1968, 759–760.

1968   Thérond, J. D.
L'hypothèse de Fermat pour les exposants négatifs. *Enseignement Math.*, (2), **14**, 1968, 195–196.

1969   Greenleaf, N.
On Fermat's equation in $\mathbb{C}$ $(t)$. *Amer. Math. Monthly*, **76**, 1969, 808–809.

1972   Brenner, J. L. and de Pillis, J.
Fermat's equation $A^p + B^p = C^p$ for matrices of integers. *Math. Mag.*, **45**, 1972, 12–15.

1972   Nagell, T.
Sur la résolubilité de l'équation $x^2 + y^2 + z^2 = 0$ dans un corps quadratique. *Acta Arithm.*, **21**, 1972, 35–43.

1974   Szymiczek, K.
Note on a paper by T. Nagell. *Acta Arithm.* **25**, 1974, 313–314.

# Epilogue

1. In June 1993, Wiles announced the proof of Fermat's last theorem (FLT). His manuscript, scrutinized by various experts, had flaws, which needed corrections. Undeterred, and with the help of Taylor, Wiles found a way out of the difficulties and has now made public two manuscripts, one coauthored by Taylor. They contain, so the author claims, a complete proof of FLT. This is now being examined carefully but there are reasons to hope that the proposed proof is indeed correct. This should be known with certitude very soon and for most mathematicians, it will represent the end of the saga.

Wiles deserves the admiration of all mathematicians for this achievement, which comes as the final step in a new strategy that I shall evoke shortly.

For some mathematicians, who are not satisfied with the battleground of elliptic curves and modular forms, judged extraneous to the problem —perhaps justly—the task is to find another proof, who knows, using the methods that were previously described in this book. There is of course no harm in trying it.

Only one aspect could be viewed as negative in the solution of the problem: the disappearance of a towering problem, which stimulated and channeled much research in number theory. It is a good thing that mathematicians can invent more problems than they are able to solve.

I shall now indicate, in a telegraphic style, some of the more remarkable developments since 1979, date of the first printing of this book.

1. In 1983, Faltings proved Mordell's conjecture. As a consequence, for every $n > 3$ there exists at most finitely many triples of relatively prime integers $(x, y, z)$, such that $x^n + y^n = z^n$. Some consequences were given by Filaseta and by Tzermias. Another consequence was obtained by Granville and Heath-Brown, independently: the set of exponents $n$ for which $X^n + Y^n = Z^n$ has only trivial solution, has natural density 1.

# Bibliography

1983  Faltings, G.
      Endlichkeitssätze für Abelsche Varietäten über Zahlkörpern. *Invent. Math.*, **73**,
      1983, 349–366.

1984  Filaseta, M.
      An application of Faltings' results to Fermat's last theorem. *C.R. Math. Rep.*
      *Acad. Sci. Canada*, **6**, 1984, 31–32.

1985  Granville, A.
      The set of exponents for which Fermat's last theorem is true, has density one.
      *C.R. Math. Rep. Acad. Sci. Canada*, **7**, 1985, 55–60.

1985  Heath-Brown, D. R.
      Fermat's last theorem is true for almost all exponents. *Bull London Math. Soc.*,
      **17**, 1985, 15–16.

1989  Tzermias, P.
      A short note on Fermat's last theorem. *C.R. Math. Rep. Acad. Sci. Canada*, **11**,
      1989, 259–260.

1990  Brown, T. C., and Friedman, A. R.
      The uniform density of sets of integers and Fermat's last theorem. *C.R. Math.*
      *Rep. Acad. Sci. Canada*, **12**, 1990, 1–6.

2. In 1988, Monagan and Granville showed that if the first case of Fermat's
last theorem is false for the exponent $p$, then $l^{p-1} = 1 \ (mvd \ p^2)$ for every prime
$l \leq 89$. As a consequence, the first case holds for every exponent $p < 6.93 \times$
$10^{17}$, according to later improvements of the older method of Gunderson.

# Bibliography

1988  Granville, A., and Monagan, M. B.
      The first case of Fermat's last theorem is true for all prime exponents up to 714,
      591, 116, 091, 389. *Trans. Amer. Math. Soc.*, **306**, 1988, 329–359.

1989  Tanner, J. W., and Wagstaff, S. S., Jr.
      New bound for the first case of Fermat's last theorem. *Math. Comp.*, **53**, 1989,
      743–750.

1990  Coppersmith, D.
      Fermat's last theorem (Case 1) and the Wieferich criterion. *Math. Comp.*, **54**,
      1990, 895–902.

3. Adleman and Heath-Brown, complemented by a paper of Fouvry, proved
that there exists an infinite set of prime exponents $p$ for which the first case of
Fermat's last theorem is true.

# Bibliography

1985  Adleman, L. M., and Heath-Brown, D. R.
      The first case of Fermat's last theorem. *Invent. Math.*, **79**, 1985, 409–416.

1985  Fouvry, E.
Théorème de Brun-Titchmarsh. Application au théorème de Fermat. *Invent Math.*, **79**, 1985, 383–407.
1986  Granville, A.
Powerful numbers and Fermat's last theorem. *C.R. Math. Rep. Acad. Sci. Canada*, **8**, 1986, 215–218.

4. The deeper study of the cyclotomic fields in its various aspects—units, class group, class number, irregular primes, etc.—continued with substantial progress. In particular, extensive calculations allowed to establish that Fermat's last theorem is true for every odd prime up to $4 \times 10^6$.

# Bibliography

1985  Thaine, F.
On the first case of Fermat's last theorem. *J. Nb. Th.*, **20**, 1985, 128–142.
1988  Thaine, F.
On the class groups of real abelian number fields. *Annals Math.*, **128**, 1988, 1–18.
1989  Terjanian, G.
Sur la loi de réciprocité des puissances k-ièmes. *Acta Arith.*, **54**, 1989, 87–125.
1990  Agoh, T.
On the Kummer-Mirimanoff congruences. *Acta Arith.*, **55**, 1990, 141–156.
1990  Skula, L.
Some consequences of the Kummer system of congruences. *Comm. Math. Univ. Sancti Pauli*, **39**, 1990, 19–40.
1991  Thaine, F.
On the relations between units and Jacobi sums in prime cyclotomic fields. *Manu. Math.*, **73**, 1991, 127–151.
1992  Agoh, T.
Some variations and consequences of Kummer-Mirimanoff congruences. *Acta Arith*, **62**, 1992, 73–96.
1992  Buhler, J., Crandall, R. E., and Sompolski, R. W.
Irregular primes to one million. *Math. Comp.*, **59**, 1992, 717–722.
1992  Skula, L.
Fermat's last theorem and the Fermat quotient. *Comm. Math. Univ. Sancti Pauli.*, **41**, 1992, 35–54.
1992  Sun, Z. H., and Sun, Z. W.
Fibonacci numbers and Fermat's last theorem. *Acta Arith.*, **60**, 1992, 371–388.
1993  Buhler, J., Crandall, R. E., Ernvall, R., and Metsänkylä, T.
Irregular primes and cyclotomic invariants to four million. *Math. Comp.*, **61**, 1993, 151–153.
1993  Granville, A.
On the Kummer-Wreferich-Skula criteria for the first case of Fermat's last theorem. In: *Advances in Number Theory* (eds. F. Q. Gouvêa and N. Yui), 479–498. Oxford University Press, New York, 1993.

1993   Jha, V.
The Stickelberger Ideal in the Spirit of Kummer with Application to the First Case of Fermat's Last Theorem. Queen's Papers in Pure and Applied Math. (1993), 181 pages. Queen's University, Kingston (Ontario).

1994   Agoh, T.
On the Kummer system of congruences and the Fermat quotient. *Expo. Math.*, **12**, 1994, 243–253.

1994   Skula, L.
The orders of solutions of the Kummer system of congruences. *Trans. Amer. Math. Soc.*, **343**, 1994, 587–607.

1994   Thaine, F.
On the $p$-part of the ideal class group of $\mathbb{Q}(\zeta + \zeta^{-1})$ and Vandiver's conjecture. Preprint (1994).

1995   Dilcher, K., and Skula, L
A new criterion for the first case of Fermat's last theorem. *Math. Comp.*, **64**, 1995, 363–392.

5. The "abc" conjecture of Masser and Oesterlé—which has attracted much attention for its numerous consequences—implies almost trivially that Fermat's last theorem is true for every sufficiently large exponent.

# Bibliography

1984   Mason, R. C.
Diophantine Equations over Function Fields. *London Math. Soc. Lecture Notes Ser.*, **96**, Cambridge University Press, 1984.

1985   Masser, D. W.
Some open problems. *Symp. Anal. Nb. Th.*, Imperial College, London, 1985 (unpublished).

1988   Oesterlé, J.
Nouvelles approches au "théorème" de Fermat. *Sém. Bourbaki 40ième année*, **694**, Février 1988, 1987–1988.

1990   Masser, D. W.
More on a conjecture of Szpiro. *Astérisque*, **183**, 1990, 19–24.

1991   Elkies, N. D.
ABC implies Mordell. *Duke Math J., Intern. Math. Res. Notes*, **7**, 1991, 99–109.

6. The new research avenue began when Hellegouarch and later, independently, Frey associated to every prime $p > 3$ and pair of non-zero coprime integers $A$, $B$, the elliptic curve of equation $y^2 = x(x - A^p)(x - B^p)$, now known as Frey's curve. Ribet showed that there is no integer $C \neq 0$ such that $A^p + B^p = C^p$, provided Frey's curve, which is a semi-stable elliptic curve, is a modular curve, as it should be, according to the conjecture of Taniyama and Shimura. Wiles, with the assistance of Taylor, established the validity

of the above conjecture and this, together with Ribet's work, proves Fermat's last theorem (final verification is still pending).

# Bibliography

For the convenience of the reader, the references are organized as follows:

## A. Elliptic curves, Modular Forms: Basic Texts

1962 Gunning, R. C.
*Lectures on Modular Forms.* Princeton University Press, Princeton, NJ, 1962.

1972 Ogg, A.
Survey of modular functions of one variable. In: *Modular Functions of One Variable* (ed. W. Kuyk). *Springer Lect. Notes in Math.,* **320**, 1972, 1–35.

1974 Tate, J.
The arithmetic of elliptic curves. *Invent. Math.,* **23**, 1974, 179–206.

1976 Lang, S.
*Introduction to Modular Forms.* Springer-Verlag, New York, 1976.

1984 Koblitz, N.
*Introduction to Elliptic Curves and Modular Forms.* Springer-Verlag, New York, 1984.

1986 Silverman, J. H.
*The Arithmetic of Elliptic Curves.* Springer-Verlag, New York, 1986.

1986 Cornell, G., and Silverman, J. H. (eds.)
*Arithmetic Geometry.* Springer-Verlag, Berlin, 1986.

1989 Miyake, T.
*Modular Forms,* Springer-Verlag, New York, 1989.

1989 Hida, H.
*Theory of p-adic Heike algebras and Galois representations.* Sugaku Expositions, 1989, 75–102.

1991 Cassels, J. W. S.
*Lectures on Elliptic Curves.* Cambridge University Press, Cambridge, 1991.

1992 Tate, J., and Silverman, J. H.
*Rational Points on Elliptic Curves.* Springer-Verlag, New York, 1992.

## B. Expository

1986 Wagon, S.
The evidence: Fermat's last theorem. *Math. Intell.,* **8**, No. 1, 1986, 59–61.

1993 Ram Murty, M.
Fermat's last theorem, an outline. *Gaz. Soc. Math. Québec,* **16**, No. 1, 1993, 4–13.

1993 Ram Murty, M.
*Topics in Number Theory. Mehta Res. Inst. Lect. Notes,* No. 1, Allahabad, 1993.

1994 Gouvêa, F. Q.
"A marvelous proof." *Amer. Math. Monthly,* **101**, 1994, 203–222.

1994 Cox, D. A.
Introduction to Fermat's last theorem. *Amer. Math. Monthly,* **101**, 1994, 3–14.

1994   Granville, A., and Monagan, M. P.
The status of Fermat's last theorem—mid 1994. Preprint (1994).

1994   Rubin, K., and Silverberg, A.
Wiles' Cambridge lecture. *Bull. Amer. Math. Soc.*, **11**, 1994, 15–38.

1995   Ribenboim, P.
Fermat's last theorem before June 23, 1993. To appear in Proc. Fourth Conference Can. Nb. Th. Assoc. (Halifax, July 1994) (ed. K. Dilcher). Amer. Math. Soc., Providence RI (1995).


C. Research

1975   Hellegouarch, Y.
Points d'ondre 2p sur les courbes elliptiques. *Acba Arith.*, **26**, 1975, 253–263.

1982   Frey, G.
Rationale Punkte auf Fermatkurven und getwisteten Modulkurven. *J. reine u. angew. Math.*, **331**, 1982, 185–191.

1986   Frey, G.
Elliptic curves and solutions of $A - B = C$. In *Sém. Th. Noubres, Paris, 1985–1986* (ed. C. Goldstein). *Progress in Math*. Birkhäuser, Boston, 1986, 39–51.

1986   Frey, G.
Links between stable elliptic curves and certain diophantine equations. *Ann. Univ. Saravrensis* 1 No. 1 (1986), 1–40.

1987   Frey, G.
Links between elliptic curves and solutions of $A - B = C$. *J. Indian Math. Soc.*, **51**, 1987, 117–145.

1987   Frey, G.
Links between solutions of $A - B = C$ and elliptic curves In Number Theory Ulm (ed. H.-P. Schlikewei and E. Wirsing). *Springer Lect. Notes in Math.* **180**. Springer-Verlag, Heidelberg, 1987.

1987–1990   Ribet, K. A.
On modular representations of Gal $(\overline{\mathbb{Q}}|\mathbb{Q})$ arising from modular forms. Preprint 1987. *Invent Math.*, **100**, 1990, 115–139.

1987   Serre, J. P.
Sur les représentations modulaires de degreé 2 de Gal $(\overline{\mathbb{Q}}|\mathbb{Q})$. *Duke Math. J.*, **54**, 1987, 179–230.

1989   Mazur, B.
Deforming Galois representations. In: *Galois Groups over* $\mathbb{Q}$ (eds Y. Ihara, K. A. Ribet, and J. P. Serre). *Math. Sci. Res. Inst. Publ.*, Vol. 16. Springer-Verlag, New York, 1989.

1990   Ribet, K. A.
From the Taniyama–Shimura conjecture to Fermat's last theorem. *Ann. Sci. Univ. Toulouse*, (5) **11**, 1990, 115–139.

1991   Kolyvagin, V.
Euler systems. In: *The Grothendieck Festschrift*, Vol. 2, 435–483. Birkhäuser, Boston, 1991.

1992   Flach, M.
A finiteness theorem for the symmetric square of an elliptic curve. *Invent. Math.*, **109**, 1992, 307–327.

1993 Lenstra, H.W. Jr.
Complete intersections and Gorenstein rings. Preprint (September 27, 1993).

1993 Ramakrishna, R.
On a variation of Mazur's deformation functor. *Compos. Math.*, **87**, 1993, 269–286.

1994 Wiles, A.
Modular elliptic curves and Fermat's last theorem. Preprint (October 7, 1994).

1994 Taylor, R., and Wiles, A.
Ring theoretic properties of certain Heike algebras. Preprint (October 7, 1994).

In the above preprints, there are other relevant references.

# Index of Names

# Subject Index